History, Philosophy and Theory of the Life Sciences

Raphael Falk

Zionism and the Biology of Jews

Springer

History, Philosophy and Theory of the Life Sciences

Volume 19

Editors
Charles T. Wolfe, Ghent University, Belgium
Philippe Huneman, IHPST (CNRS/Université Paris I Panthéon-Sorbonne), France
Thomas A.C. Reydon, Leibniz Universität Hannover, Germany

Editorial Board
Marshall Abrams (University of Alabama at Birmingham)
Andre Ariew (Missouri)
Minus van Baalen (UPMC, Paris)
Domenico Bertoloni Meli (Indiana)
Richard Burian (Virginia Tech)
Pietro Corsi (EHESS, Paris)
François Duchesneau (Université de Montréal)
John Dupré (Exeter)
Paul Farber (Oregon State)
Lisa Gannett (Saint Mary's University, Halifax)
Andy Gardner (Oxford)
Paul Griffiths (Sydney)
Jean Gayon (IHPST, Paris)
Guido Giglioni (Warburg Institute, London)
Thomas Heams (INRA, AgroParisTech, Paris)
James Lennox (Pittsburgh)
Annick Lesne (CNRS, UPMC, Paris)
Tim Lewens (Cambridge)
Edouard Machery (Pittsburgh)
Alexandre Métraux (Archives Poincaré, Nancy)
Hans Metz (Leiden)
Roberta Millstein (Davis)
Staffan Müller-Wille (Exeter)
Dominic Murphy (Sydney)
François Munoz (Université Montpellier 2)
Stuart Newman (New York Medical College)
Frederik Nijhout (Duke)
Samir Okasha (Bristol)
Susan Oyama (CUNY)
Kevin Padian (Berkeley)
David Queller (Washington University, St Louis)
Stéphane Schmitt (SPHERE, CNRS, Paris)
Phillip Sloan (Notre Dame)
Jacqueline Sullivan (Western University, London, ON)
Giuseppe Testa (IFOM-IEA, Milano)
J. Scott Turner (Syracuse)
Denis Walsh (Toronto)
Marcel Weber (Geneva)

More information about this series at http://www.springer.com/series/8916

Raphael Falk

Zionism and the Biology of Jews

 Springer

Raphael Falk
Department of Genetics & Program
  of History and Philosophy of Science
The Hebrew University of Jerusalem
Jerusalem, Israel

Edited, updated translation of Hebrew version הציונות והביולוגיה של היהודים RESLING
Publishing House, 2006. Danacode 585-235 (no isbn).

ISSN 2211-1948   ISSN 2211-1956   (electronic)
History, Philosophy and Theory of the Life Sciences
ISBN 978-3-319-57344-1   ISBN 978-3-319-57345-8   (eBook)
DOI 10.1007/978-3-319-57345-8

Library of Congress Control Number: 2017937542

© Springer International Publishing AG 2017

This work is subject to copyright. All rights are reserved by the Publisher, whether the whole or part of the material is concerned, specifically the rights of translation, reprinting, reuse of illustrations, recitation, broadcasting, reproduction on microfilms or in any other physical way, and transmission or information storage and retrieval, electronic adaptation, computer software, or by similar or dissimilar methodology now known or hereafter developed.

The use of general descriptive names, registered names, trademarks, service marks, etc. in this publication does not imply, even in the absence of a specific statement, that such names are exempt from the relevant protective laws and regulations and therefore free for general use.

The publisher, the authors and the editors are safe to assume that the advice and information in this book are believed to be true and accurate at the date of publication. Neither the publisher nor the authors or the editors give a warranty, express or implied, with respect to the material contained herein or for any errors or omissions that may have been made. The publisher remains neutral with regard to jurisdictional claims in published maps and institutional affiliations.

Printed on acid-free paper

This Springer imprint is published by Springer Nature
The registered company is Springer International Publishing AG
The registered company address is: Gewerbestrasse 11, 6330 Cham, Switzerland

In memoriam:
Lisa and Walter Falk,
Rivka and Benjamin Auren (Aharonovich),
Zionists.

Contents

Preface xi

Acknowledgments xv

1 Introduction 1
    1.1 Humans Vary; Are Jews Distinct? 7
    1.2 Who Is a Jew? 9
    1.3 From Anthropology and Eugenics to Population Genetics 12

2 From Emancipation to "Scientific Racism" 17
    2.1 Jews as a Distinct Entity 22
    2.2 The Biologization of Race 24
    2.3 Anti-Semitism 31
    2.4 Judaism as a Historic Entity 34

3 Heredity or Environment? 37
    3.1 Heredity or Society? 43
    3.2 Racism 44
    3.3 Eugenics 45

4 The Response: Zionism 49
    4.1 Theodor Herzl 55
    4.2 Max Nordau 57
    4.3 Zeev Jabotinsky 61
    4.4 Martin Buber 63
    4.5 Arthur Ruppin 65

5 A Jewish Race Notwithstanding? 71
    5.1 The Zionist Claim 75
    5.2 People of the Middle-East? 79
    5.3 A Political-Social Perspective 86

6	***Eidoth***		93
	6.1	The Middle Eastern Jew: The Jewish Prototype?	97
	6.2	On Khazars and Ashkenazim	100
	6.3	The Merger of *Eidoth*: Assimilation or Amalgamation?	104
	6.4	Jewish Diseases	108
	6.5	Immigrants and Natives	113
7	**Pioneers as Eugenic Agents**		119
	7.1	Hebrew Work – An Insurmountable Challenge	124
	7.2	Education and Racial Hygiene	126
	7.3	Jewish Intelligence (and Disease)	133
	7.4	The "Demographic Issue"	137
8	**The Inagathering of Exiles**		143
	8.1	Medical Anthropology and Population Genetics	145
	8.2	Common Relatives *versus* Common Genes	151
	8.3	The Genetics of the Israeli Melting Pot	162
	8.4	From Single-Genes to Systems Polymorphisms	171
9	**From DNA to Politics**		175
	9.1	Similar but Different	177
	9.2	The Trail of Y-Chromosome Haplotypes	182
	9.3	Towards Genome-Wide Association Studies	188
	9.4	DNA Sequence Analyses	193
	9.5	Politics *versus* Science	195
	9.6	Common Origins or Common Network?	197
10	**Coda: Zionism and the Biology of the Jews Tomorrow**		203
	10.1	A Jewish State or a State for the Jews?	207
Bibliography			211
Index			227

List of Figures

Fig. 2.1	Lavater's characterization of nationals by physiognomy (Lavater 1984)	19
Fig. 2.2	Characterization of Jewish facial patterns using Galton's technique of superimposing film negatives (Efron 1994)	28
Fig. 3.1	The title page of Theodor Herzl's *Altneuland* (1902)	47
Fig. 6.1	Classification of human types by three criteria (Sheldon 1954)	96
Fig. 6.2	From Ruppin's (1930a) gallery of Jewish types: Albert Einstein ("Mediterranean impact"); Lord Reading and Sir Herbert Samuel (England); and a Jewess, the daughter of a Sephardic father and an Ashkenazi mother, born in Jerusalem	97
Fig. 7.1	Jewish National Fund poster, 1947	120
Fig. 8.1	"Jewish facial form" in a Bakaïri – Papuan native (Stratz 1903)	151
Fig. 8.2	Haeckel's tree life in *The Evolution of Man* (1879)	161
Fig. 8.3	Graphic representation of frequencies of blood type in Jewish (full circles) and non-Jewish (open circles) communities. Full lines connect Jews and non-Jews of similar sites, broken lines connect Jews of different background (modified from Fig. 4 of Mourant et al. 1978)	164
Fig. 8.4	Muhsam's (1964) attempt to identify the frequencies of the ABO blood type of the Jewish forefathers: vectors from "Gentile environments" to the corresponding "genuine" Jewish *eidoth*. *Lower right*: expected; *upper left*: observed	166

Fig. 9.1 Multivariate analysis of genetic variants of various populations based on Y-chromosome hapolotype data (after Hammer et al. 2000). *Solid triangles* represent Jewish populations; *solid squares* represent Middle Eastern populations.. 190

Fig. 9.2 Model of the reconstruction of the Ashkenazi Jewish (AJ) and European (FL) demographic history. The *wide arrow* represents an admixture event (Carmi et al. 2014).......................... 194

Preface

> Biased scientists are inevitable, biased results are not (M. Weisberg and D. B. Paul, 2016, quoting J. E. Lewis et al., 2011, confronting S. J. Gould, 1978, discussing S. G. Morton's, 1839-1844 studies of human cranial capacity).

Zionism, a national sociocultural doctrine, and biology, an empirical natural science, appear to be two alien conceptual domains. Yet, toward the end of the nineteenth century, biology became increasingly a science that provided empirical foundations even to the philosophy of human social relations. Thus, it is natural to examine Zionism, the late-nineteenth-century political movement dedicated to the return of the Jewish people to their homeland, not only from the perspective of historians but also from that of biologists. Looking at the claims that Jews constitute a people with common biological roots may provide further justification to their political aspirations.

The nineteenth century witnessed the dawn of European nationalism. In the spirit of *Liberté, Egalité, Fraternité*, various political-cultural associations of peoples demanded the political right to their own homeland, from Greece and Italy in the 1820s up to Czechoslovakia and Poland in the 1920s. The longing of the Jews for the return to their homeland in the Land of Israel has been a predominant theme of their prayers for generations. Thus, the Zionist dream of a Jewish state, which was realized in 1948, should be considered a late event of the European nationalist sentiments.

Although Jews have suffered persecutions for almost two millennia, Jewish identity became "biological" only in the last decades of the nineteenth century. *AntiSemitism*, a term coined in the 1870s, also engendered the assertion that the physical features of Jews and their sociocultural traits are functions of their biological distinctiveness. With the biological rationalization of sociocultural discrimination, the perennial hatred of Jews became *racism*. Although this hatred achieved its climax manifestation in modern times in Nazi Germany, even today anti-Semitic racism is persistent in various Western as well as in numerous Middle Eastern societies.

The period from 1945 to 1950 witnessed massive waves of migrations within formerly Nazi-occupied Europe by refugees who repeatedly brought up notions of nationalism. The Jewish survivors of the Holocaust, many of whom did not have a home to return to, formed a very particular group among these refugees. Equally, discrimination and at times persecutions of the Jewish communities in North Africa and in the Middle East made life there as hard and unbearable as ever. The Zionist State of Israel offered citizenship in the historic homeland of the Jews, regardless of one's country of origin. But subsequent developments made many to consider Zionism as leftover of European colonialism rather than a liberating movement of people returning to its homeland. Such claims were further stressed by Israel's Law of Return of 1950 that gives the right to live in Israel and gain Israeli citizenship to any Jew.

But who is a Jew? Are the Jews of Morocco, Lithuania, Yemen, Ethiopia, Greece, or Iran of common descent? Are not indigenous Palestinians, who vehemently resisted the Zionist project and for many of whom the founding of Israel turned out to be the traumatic event of losing their homes and lands, the legal residents of the land? Is it possible that these Palestinians are the descendants of the ancient Jews?

Israel of the first decades of the twenty-first century in many respects played a central role in the development of genomics and the applications of biological technological innovations, including DNA sequencing and analyses. The history related in this book may help to make sense why Israel has wholeheartedly embraced a wide range of reproductive genetic techniques. Indeed, the uptake of such technologies, which is encouraged and supported by the government, is probably higher than in any other country. Indeed, Germany stands at the opposite extreme, and its caution regarding practices that involve the selection of fetuses and embryos is usually explained as a residual reaction to its coercive Nazi past. The shadow of the Third Reich certainly hovers over every discussion of the appropriate use of these practices. One might expect Israelis' reservations regarding what amounts to eugenic policies to be at least as strong as those expressed by Germans, and one might expect the discourse around the use of genetic services in Israel to be strongly influenced by revulsion at assumptions about which lives were – and which were not – worth living. That this is in fact very far from being the case is made intelligible by the history detailed in the present text, which shows that Zionism was actually closely intertwined with eugenics for a long time.

This volume is a revised and edited English version of my book *Tzionut Vehabiologia shel Hayehudim* (*Zionism and the Biology of Jews*) that was published in 2006. As noted, I am not a professional historian but rather a geneticist by training. I spent half of my career experimenting with *Drosophila* flies before I eventually became involved in the study of the history and philosophy of evolutionary biology, genetics, and eugenics.

In our world of scientific-technocratic reasoning, biological research – more specifically, molecular genomic research – has become a major tool that enables us to examine and perhaps validate the link between communities of present-day Jews and between them and the ancient residents of the Land of Israel. Establishing this linkage would help rationalize the ardent hopes of Jews throughout the centuries of returning to Zion. Detailed analyses of the genomes of individuals have identified specific DNA sequences that may indicate the common lineage of many Jewish communities and of other overlapping Mediterranean populations. A *vertical* depiction of the phylogenies often appears to sustain the traditional Jewish historical lore of the lineal descent of contemporary Jews from the historic residents of the Land of Israel. The same genetic relationships, however, also hint at secondary *horizontal* associations through intermarriages between Jewish and non-Jewish communities that by virtue of domicile came in close contact.

In this book, I attempt to present developments by taking the reader on an abbreviated tour of my own extended study of the hereditary relations of political Zionism. I narrate the story how it began with the promise to provide a haven for persecuted Jews, how the hard-gained realization of this promise allowed a national homeland to be established, how the original version became transformed with the establishment of the state, and how the extreme elements of the original vision became policies of implementing cleric-nationalistic claims of *inherent* rights of the Jewish people. However, I am also aware and acknowledge the plight of the Palestinian inhabitants in this land. It partly reflects my own history as a son of committed Zionist parents who emigrated from Germany to Palestine in 1933.

Experimental scientists pride themselves of being followers of Francis Bacon's (1561–1626) inductive method of investigating nature, presumably without prejudice. But clearly, this is impossible; we view the world through a lens that is polarized by our dispositions, inclinations, and preconceived notions. As a student of the evolution of scientific concepts, I based much of my narration on secondary readings of the sources. I have been continually surprised to discover how difficult it is to admit the extent to which many of us – in the natural sciences and in the sciences of man – are influenced by our preconceived ideas.

Jerusalem, Israel					Raphael Falk

Acknowledgments

My interest in the genetics of Jewish communities in Israel stems from the days when I was engaged in the dynamics of population genetics. My Ph.D. instructor, Elisabeth Goldschmidt, gave me the opportunity to study the dynamics of change in migrant populations by following the genetics and observing the sociocultural integration of the different, well-delineated immigrant isolates who settled in Israel. In particular, I was fortunate to be able to assist Goldschmidt in 1961, when I was secretary of the conference that she headed on *The Genetics of Migrant and Isolate Populations*.

My colleague of many years, Jacob Wahrman, helped me to develop a critical perspective on ideas and observations and to appreciate the advances in biological science, in general, and genetics, in particular. Of special value was the help of my wife Ruma, whose analytic and unrelenting attention to details helped me to recognize and overcome many of the misconceptions that I had adopted. My son Oren gave me critical and constructive comments. My family, our daughter Raya and her late husband Avisar, their daughter Lili and her husband Elisha, Raya's son Yoav, as well as our twin great-grandchildren Neta and Gefen, and Oren's family, Yael, Naomi, and Daniella, all have been constant sources of support. Many colleagues, students, and especially my assistants – Shula Baker, Miriam Broit, and Ana Rahat – helped me do the research, gave me advice and information, read sections of my drafts, and corrected some of my mistakes and errors. Judith Atidia, Eitan Bloom, Michael Brandeis, Snait Gissis, Oren Harman, Daphna Hirsch, Shaul Katz, Bat-Sheva Kerem, Alexandre Métraux, Ariella Oppenheim, Diane B. Paul, Sam S. Schweber, Dudy Tzfati, and Issachar Unnah are only some of those who were involved in the Hebrew version, advising, reading, and commenting on selected sections of the present text. Amit Baskin patiently helped me with the proof reading. Finally, Miriam Greenfield and Dania Valdez meticulously edited my English text.

I am also grateful to the readers who thoroughly reviewed the text, not only correcting mistakes but also suggesting important improvements and updates.

All remaining errors and mistakes are of course my own responsibility.

Chapter 1
Introduction

The longing for Zion has been a manifest attribute of Jews for generations. As a part of daily prayers, this longing of the heart has been expressed primarily as a wish for relief from persecutions and the other hardships of Jewish life in the Diaspora.[1] Yet, there were always individuals and groups who took this longing literally: some immigrated to Zion to spend their last days there and to be buried in the Holy Land; others were content with a bag of soil from Zion under their heads in their final resting place in the Diaspora. Still others, who could not tolerate the hardships of daily existence and persecution in the Diaspora, stimulated by religious sentiments, made the pilgrimage and actually settled in the Holy Land. Yehudah Halevy (1075–1141), the scholar and poet, travelled from Spain to the Holy Land in 1140, but died shortly after his arrival. A few individual Jews from Yemen came as pilgrims as early as the fifteenth century, and a larger contingency immigrated in the winter of 1881–1882. Jews from North Africa, under the leadership of Rabbi David Ben-Shimon, established the "Maghreb Community" in Jerusalem In 1860. Pious Jews from Eastern Europe settled individually in the holy cities of Tiberias, Safad, Jerusalem and Hebron. These are only a few instances of the realization of the eternal sentiment of "If I forget thee, O Jerusalem, let my right hand wither" (Psalms, 137:5). There was, however, a fundamental change in conception, though not in motivation, when in July 1882 a group of previous students from Russia, settled on a barren hill south of Jaffa and founded the colony of Rishon-LeZion, literally, the First to Zion.

Towards the end of the nineteenth century, economic and social pressures and repeated pogroms against Russian and Rumanian Jewish communities led more groups of Jewish students to conclude that the problem of the Jews could not be resolved by maintaining isolated communities of pious persons in the Holy Land, completely dependent on the Jewry of the Diaspora. Israel Belkind (1861–1929), the leader of the immigrant students, referred to this as "Holy Cities Zionism." Reflecting the nationalist mood and conceptions popular at the time, these young-

---

[1] The very term Diaspora, spelled with a capital D, indicates this feeling of longing for the homeland, Zion.

sters strove to form a center of gravity for Jews as an autonomous nation, a place where they immanently belonged and where they would not be considered strangers. They founded the Bilu movement – the Hebrew acronym for 'House of Jacob, Let's Go' (Isaiah 2:5) – and took an oath to live by three principles: national resurrection, settlement in Eretz-Israel,[2] and cultivation of the land. This idea of becoming a nation in the homeland was an outgrowth of the Age of Romanticism and Nationalism that flourished throughout Europe in the second half of the nineteenth century. In 1882 Leo Pinsker (1821–1891) wrote *Autoemancipation*, and in 1896 Theodor Herzl (1860–1904) wrote *Der Judenstaat*; these books reflected and radiated the romantic spirit of Jewish national resurrection.

In August 1897, the return-to-Zion notion of a Jewish settlement in the ancient homeland was formally inaugurated at the First Zionist Congress in Basel. At that Congress, the aspirations of the emancipated, semi-assimilated Jews, who considered themselves part of the Western European socio-political web, conjoined with tribulations of East European Jews, who were as a rule more integrated in traditional Jewish culture than were their Western brethren. This encounter at Basel provided the leverage to stimulate Jews to settle in Eretz-Israel during the first decades of the twentieth century. As events unfolded in Europe, Eretz-Israel turned out to be one of the few places that provided a haven for some of the Jews in Nazi Europe. On May 15, 1948, the State of Israel was founded in accordance with the United Nations' decision of November 29, 1947, and on July 5, 1950, the Law of Return, was legislated, thereby securing the right of every Jew to become a citizen of the State. Thus, the Jewish populations of Israel provided a unique opportunity to conduct a study: The Genetics of Migrant and Isolate Populations (see e.g., Goldschmidt 1963).[3] However, Israel, as a national home of the Jewish People that was founded on the

---

[2] David Vital in a prefatory note on nomenclature to his book (1987, p. *xiii*) writes: "There remains the question of the name of the Land. […] 'Erez-Israel' (literally: 'Land of Israel') to denote the country that a great many people do indeed think of as 'Palestine'. The difficulty is that for centuries, until the British took it over at the end of the Great War, it was an exceedingly loose geographical expression at best; and no political or administrative unit of that name, or covering that territory even approximately, existed." The formal name of the country, which from the time of the capitulation by the Romans was Palestine, became later, a district in the Ottoman Empire, called Phalastin in the Arabic version. I shall follow Vital who thought "it right to use the term 'Palestine' when the context or the documents required it and [rather] 'Eretz-Israel' when it seemed the more appropriate."

[3] In September, 1961, a group of geneticists from fifteen countries met in Jerusalem "to discuss the genetic differentiation among the Jewish groups in Israel." The conference was organized by Prof. Elizabeth Goldschmidt of the Hebrew University in Jerusalem, and Prof. Chaim Sheba of the Tel-Hashomer Hospital, and the present author as its secretary. The conference, *The Genetics of Migrant and Isolate Populations*, may be considered the founding event of human (and medical) genetics in Israel, because it was centered on an exhibition that invited not only academia, but also medical doctors all over the country, to examine their empirical data and present data of their own that might have genetic relevance (see Goldschmidt 1963, pp. 251–355).

In June 2011 a conference celebrating fifty years of the late Sheba's contribution to human and medical genetics in Israel was convened under a similar title. It concentrated on efforts to utilize the genetic relationships between Jews of different origins to trace their history since leaving their fatherland two thousand years ago.

humanistic ideas of *fin de siècle* Europe, was never acknowledged as such by the native population of Palestine, who viewed it essentially as a rudiment act of European Imperialism, while its neighbor states considered its citizens heathen foreign intruders. The constant threat to its existence and the continuous, ever increasing need to maintain a balance of power with its neighbors caused many Jews in Israel to turn away from the humanistic, often utopian ideologies of political Zionism and to become hard-core nationalists, increasingly driven by religious zeal.

Correspondingly, as the conflict of the Zionist State with the Arab world intensified, so did the wish to prove "scientifically," by biological-genetic means, the immanent physical, historical connection of the Jewish people to Zion. Genetics, it was hoped, would uphold not only the historical evidence, but would also provide biological evidence that the dispersed Jewish ethnic groups (*eidoth*)[4] of today are indeed one people whose roots trace back to Eretz-Israel.

Over the ages Jews, in various contexts and at different times were recognized as a people of a distinct religion, or as a highly knit people with unique socio-cultural bonds. Yet, what ultimately maintained the Jews' identity were the claims of their genealogical relationships: Jews were perceived as the descendants of Abraham, Isaac, and Jacob – the three Patriarchs – not only spiritually but primarily biologically. According to the official law in the State of Israel, Judaism is basically a blood-relationship: a Jew is one born to a Jewish mother (in acknowledgment of the difficulty of proving fatherhood). Nonetheless, over history there have always been individuals and communities who converted to Judaism. Consequently, even the formal religious law recognizes as Jews those who 'properly' converted to Judaism according to Rabbinic guidelines.

With the Age of Enlightenment in the second half of the eighteenth century, the claim for universal values seemed to open the way for the emancipation of the Jews. It was argued that what distinguished Jews from gentiles was their cultural history and their social conditions as communities, rather than their biological essence. Nevertheless, discrimination of Jews did not diminish: During the nineteenth century, at least in Western Europe, while the social and religious distinctive markers of the Jews gradually relaxed, while Jews formally enjoyed emancipation, much of the emphasis of discrimination shifted to biological markers, claiming that Jews were inherently different.

The nineteenth century saw major social shifts due to industrialization and commercialization that caused notable population movements both horizontally in terms of urbanization and vertically in terms of social categorization. After the publication of Charles Darwin's (1809–1882) *Origin of Species* in 1859 – and the discussions of its role in social relations in human societies by thinkers like Herbert Spencer (1820–1903) in Britain and Ernst Haeckel (1834–1919) in Germany – sociopolitical movements were increasingly interpreted biologically, in terms of hereditary inequalities among human beings. Thus, to the extent that religious and cultural arguments for segregation and persecution of Jews lost force, biological claims for persecution held sway: Jews were considered to be of a different "race" – a sociocultural invention of a presumed "biological entity" – and their specific traits were

---

[4] Ethnic-group is the term I shall use interchangeably with the Hebrew term *eidah*, in plural: *eidoth*.

part of their biological essence. Hatred of the Jews became hatred of the Semitic race, namely anti-Semitism. The term anti-Semitism was coined in the 1870s by the German publicist Wilhelm Marr (1819–1904). Marr's Darwinian philosophy was explicit: "Anyone who cannot hold his own has to go" (Zimmermann 1986, p. 67). The notion that the biological differences of people are responsible for their social differences gained credit and spread exponentially.

"Scientific racism," however, is older than Marr's concepts. The idea that human beings may be classified into five races that differ in origin had been raised towards the end of the eighteenth century; albeit that many thinkers of the Enlightenment and later maintained that differences between populations and societies were primarily differences in culture rather than in *material essence*. The prominent German philosopher Johann Gottfried Herder (1744–1803) created the notion of a nation, or *Volk,* as a meaningful entity. A *Volk* was conceived as a being, an entity distinguished by landscape, climate, language, tradition, foreign intercourse, and consequently, also by heredity. Thus, in the first half of the nineteenth century the idea of the *Volk* became increasingly emotionally charged with essentialist patriotic notions largely colored with biological nuances. Even if the *identifying* properties of the races were in the realm of culture, language or religion, it was sill argued that their *essences* were biological. The Nazis were those who eventually brought such racial theories to their full horrific application.

Herder argued that Jews living in Germany should enjoy the full rights and obligations of Germans, and that the non-Jews of the world owed a debt to the Jews for centuries of abuse that could only be discharged by actively assisting those Jews who wished to do so to regain political sovereignty in their ancient homeland of Israel. He refused to adhere to a rigid racial theory, writing that "notwithstanding the varieties of the human form, there is but one and the same species of man throughout the whole earth." In other words, Herder conceived Jewry as an example of a community of individuals of national character, maintained by a religious and traditional culture, rather than by race. He even conceived of the establishment of a political entity in Palestine, the site where their culture originated, as a guarantee for the persistence of Jewry (see, for example, Hess 2002, p. 55). Half a century later, Moses Hess (1812–1875), a close associate of Karl Marx and Friedrich Engels, became one of the first Jews who, disappointed by their emancipation, explicitly called for their national revival in Palestine. In his book *Rom und Jerusalem, die Letzte Nationalitätsfrage* [Rome and Jerusalem: The Last National Issue] (1862), he noted that "Jews are first of all a race," more in terms of Herder's *Volk*, than those of social-Darwinism, which was not yet conceived, and called for the Jews to reestablish their Jerusalem just as the Italians, under the leadership of Mazzini, established their Rome (Avineri 1986).

However, by the 1870s and 1880s, the claims that Jews belonged to a race that could be discerned in terms of the natural sciences, were repeatedly brought up, and hatred against them became more ethnic in character. Against this background, the plight of the Jews became increasingly a political issue by the end of the century. Contrary to many of the assimilated or integrated Jews of the Age of Enlightenment and the Age of Romanticism who focused on the cultural aspect of being Jewish, the

Zionists-to-be stressed that Jews were not merely members of a cultural or a religious entity, but an integral biological entity, even though they had been dispersed and had no country of their own. In other words, the Zionists adopted the concept of *Volk* in terms of a nation-race as molded by the notion of Blood and Soil (*Blut und Boden*) – current in central Europe of the time. Accordingly they demanded the materialization of their nationality rights in a country of their own. *Blut und Boden* became one of the popular inciting slogans of the Nazi Party. Undoubtedly many of those who expressed themselves at the turn of the century in terms of *Blut und Boden* were referring to the abstract, Hegelian term, rather than to the anthropological or biological notion, and surely not to the later National Socialist interpretation of the term. Yet, considering the positivist attempts to impose social and humanitarian principles upon the principles of the natural sciences, it is difficult to accept that persons who adopted this term did not see the real life consequences of such an expression.

Nevertheless, one should not forget that Zionism, contrary to the traditional ideological longing for Zion, was from its beginning a pragmatic political movement. The desire of Zionism to bring about the "normalization" of the Jews as an organic part of society through a change in lifestyle, was a very enlightened notion. Although the concept of race had not been well defined (as is also the case today[5]), racial identity – contrary to racial ideology – was a common, widely accepted biological truth among socialists, liberals, and humanists, and was considered a corollary of the inherent variability of natural species. Accordingly, in 1930, thinkers like Albert Einstein related to the Jews as a race that must also become a nation:

> Before we can effectively combat anti-Semitism, we must first of all educate ourselves out of it and out of the slave-mentality which it betokens. […] Only when we have the courage to regard ourselves as a nation, only when we respect ourselves, can we win the respect of others; or rather, the respect of others will then come of itself. Anti-Semitism as a psychological phenomenon will always be with us so long as Jews and non-Jews are thrown together. But where is the harm? It may be thanks to anti-Semitism that we are able to preserve our existence as a race; that at any rate is my belief. (Einstein 1930, p. 23)

A blunt, unfortunate example of the adherence of the Zionists to the nineteenth-century notion of Blood and Soil as ground for their territorial rights is the statement by the poet Chaim Nachman Bialik at a press conference at The Hebrew University of Jerusalem in the beginning of 1934: "I too, like Hitler, believe in the power of blood." In Bialik's opinion, the Jewish race's will-power and Jewish blood are what could successfully undermine "the remnants of paganism in the Christian world" (Bialik 1934).

Whereas the end of the eighteenth century was characterized by a pervasive belief that humanity was involved in a process of progress and enlightenment, the nineteenth century ended with the feeling that society was immersed in degenerative processes. The conception of the Zionists was that the often blemished "Jewish characteristics" were indeed signs of degeneration – the consequences of centuries of life in exile. Their task was, therefore, to lead the Jews back to the path of progress.

---

[5] See Roberts 2011. But see also Wade 2014.

As late as the establishment of the State of Israel, the declared aim of the Zionist movement was to change the image of the "Diaspora Jew" to that of a "New Jew," to that of a "Normal" citizen of a modern Western country. Immigration and settlement in Eretz-Israel, the Land of Israel, were presumed to bring about the expected conversion of both body and soul. However, after the Holocaust and the establishment of the State of Israel, the emphasis was eventually diverted to the ingathering of the Jews from the Diaspora and their absorption into the Israeli melting pot. Thus, the issue of the biological essence of Jewish existence that accompanied Zionist activity from its beginnings shifted emphasis: In the early decades of the twentieth century the emphasis was on the nature of the traits, regardless of whether these so-called Jewish characteristics were hereditary or merely acquired responses to living conditions in the Diaspora; in the second half of the century the emphasis shifted to the variegated composition of the hereditary pool of the various Jewish parishes and *eidoth*.[6] Scientists discovered early on that it was possible to trace the hereditary nature of traits, especially such diseases that appeared to be practically restricted to well-defined Jewish communities. The massive migration during the 1950s and onward provided the scientific community with a unique opportunity to study the dynamics of whole populations in nearly laboratory conditions (see Goldschmidt 1963, and juxtapose it with Reuter 2006). However, socio-political interests, as well as the dramatic developments in molecular genetics, increasingly directed research to use hereditary traits, like diseases, when discussing phylogenetic relationships between Jewish communities and their origins.

In this book, I wish primarily to discuss two issues: the claims that there exists *a biology of Jews* on the one hand, and the attempts to integrate it into a consistent *history of national-political Zionism*, on the other hand. Both issues unfolded on the background of a romantic national culture of Western Europe in the nineteenth century: Jews, primarily from Eastern Europe, were sucked into the world of these notions and soon they took the lead in the re-formulation of Jewish and Zionist existence.

We can delineate three overlapping questions that have emerged during the two centuries of Jewish emancipation and the one century of national-political Zionism in the West: What is special or unique to the Jews? Who were the genuine Jews? And how can one nowadays identify Jews? Whereas the first question was the focus of attention mainly at the end of the nineteenth century and the beginning of the twentieth, interest in the second question grew during the period between the two world wars. Much of the interest in this issue was diverted to the question of whether the Ashkenazi or the Sephardi ethnic-groups better represented the genuine "Jewish type." After the establishment of the State of Israel, discussions centered on the third

---

[6] In conflicts like those in the Balkans, in Africa, in India, in South-East Asia or in Northern Ireland, and to some extent even in the Israeli-Arab conflict, a starting point is the *existence* of distinct ethnic or religious entities that struggle for the same piece of land. On the other hand, except for Nazi efforts to *diagnose* the biological belonging of individuals to national-ethnic entities, there is no other example known to me like the Zionists' of an intensive effort to prove the immanent biological belonging or non-belonging of communities to what is considered to be the Jewish entity.

question of how one can identify the common genetic denominator of Jewish communities.

1.1 Humans Vary; Are Jews Distinct?

In Spain before the fourteenth century, the position of the Catholic Church was that anyone converting to Catholicism would be accepted as an equal, having the same rights as anyone born into a Catholic family (Yerushalmi 1982). Yet, converted Jews were discriminated against, because it was asserted that by virtue of "blood" they remained Jews. This fifteenth century viewpoint was apparently the first instance of racism based on biology.

The primacy of rational thinking in the Age of Enlightenment motivated the demand for equality for all human beings, irrespective of their religion or tradition, and the upsurge of interest in the Sciences of Man, both in the social-sciences and in the life-sciences. Centuries of explorations had revealed to the Europeans that their countries and cultures were just a few among many. This compelled them to reexamine nature and the place of humans in it. The philosopher of science, Stephen Toulmin, has suggested that the 13th of April, 1769 marked the beginning of a new era. On this day, HMS *Discovery* anchored in Tahiti: Captain Cook's crew was commissioned by the Royal Society to carry out some measurements needed to confirm Newton's theory. But, as Toulmin relates, what left a deep impression on the crew was their anthropological observations of the natives: they lived free, comfortable, and happy lives in spite of being heathens who were following life patterns completely alien to those of the Europeans (Toulmin 1972, pp. 41–42). Thus, instead of establishing the unshakable, God-given lawfulness of the eternal image of the Newtonian universe, the expedition discovered the existence of consciousness of multiplicity of habits, traditions, ideas, even principles of human morality!

Acquaintance with the huge and unexpected variability of human and also non-human nature, which was initiated of course long before Captain Cook's voyage, generated a search for new methods of classification of the accumulated knowledge that expressed the Divine Order (see also Gissis 2011). The most important among these attempts was that of Carlos Linnaeus (1707–1778), who in 1735 attempted to organize all animal and plant life into one hierarchical taxonomic system: species converge to genera, which converge to families, orders, and phyla. Man was included in this system in the species *Homo sapiens*, which in turn converged to the genus *Homo*. Furthermore, for the first time, Linnaeus formally divided the human species into four races: the red, the yellow, the white, and the black. Race, according to this conception was a material biological entity, a distinct and consistent entity in the hierarchy of Nature, just like species and genera. Thus, Linnaeus restored the Divine Order of Nature: variability could be harnessed as distinct, essential biological entities. Not all human beings were identical, but they were all humans, and their unique socio-cultural habits reflected their inherent differences as humans, without violating the divine order of things. Linnaeus's system was accepted throughout most of the scientific community towards the end of the eighteenth century. Although it was acknowledged that much of the variability in nature is a direct consequence of the specific living conditions of organisms, it became obvious that at least part of this variability among living creatures is immanent and hereditary. The variability that characterized the human races was accordingly accepted as fundamental.

Thus, although the discrimination of human beings on the basis of what we nowadays call "racism" was not new, from the end of the eighteenth century scientists began to adopt notions that would be regarded as "scientific racism." Put differently, with the increasing prestige of scientific analyses efforts were made to adopt and justify in public notions, which had been initially anchored in society and culture, as empirical-scientific terms. As the call for *liberté, égalité, fraternité* among humans became louder, more interest was directed at the non-identity of humans who presumably belong to the same entity. The further had scientific observation and experimentation had been advanced in the nineteenth century, the more social and political differences among humans were expressed in terms of biological notions of race. Rejection and discrimination of people on ethical and social grounds – such as class-differences in England, slavery of Blacks in America, European lateralization of "Natives" in Asian and African colonies, as well as hatred of Jews wherever they were – were now justified on the basis of biological arguments.

The publication in 1859 of Darwin's *Origin of Species*, and even earlier, in 1809, of the *Philosophie zoologique* by Jean-Baptiste Lamarck (1744–1829) provided not only important support for claims that differences among humans, individuals as well as populations, were biologically based, but also new dynamic dimensions to the differences within species. Furthermore, the notion of evolution by natural selection was soon translated into terms of competition for social and cultural advancement, as if the "survival of the fittest" was nothing but the survival of the most socially, culturally, or economically successful competitors among humans. Members of the upper socio-economic classes in Europe comprised – in their own opinion – the evidence by being survivors of such a "struggle." The traditional claims concerning the inferiority of certain communities and the superiority of other races were justified by claims concerning the inborn properties and the operation of the Laws of Nature.

Barely a few years after the publication of the *Origin of Species*, Darwin's cousin, Francis Galton (1822–1911), came up with the assertion that the liberal social and economic processes in Western societies, while promoting equality, contradicted the powers of natural selection that had been shaping human society for eons. Thus, according to Galton, specific conscious efforts were needed to prevent the ongoing biological deterioration of the human species of Western Societies. The achievements of human culture and science led to the containment of the effectiveness of natural selection (or at least diminished it), that was in the long run essential for preserving evolution's success. In 1865 Galton coined the term *eugenics*, namely, "good breeding," to refer to the socio-scientific effort intended to prevent the biological deterioration of the human species caused by the relaxation of the selective forces that had shaped it. The eugenics movement, which arose from Galton's teachings, was popular in the first half of the twentieth century: The very idea that human culture had a crucial role in shaping human evolution, both at the level of individuals and that of communities, led to an increasing desire to control and even navigate human biological future. Thus, men of power exerted increasingly brutal means to direct the evolution of their fellow-men. Although such distortions were probably first legally applied in the United-States, they reached their abhorrent climax in the context of the *Rassenhygiene* of the Nazi regime. It must be admitted, however, that from the end of the nineteenth century to the 1940s, arguments based on Darwinian Theory for means to antagonise the so-called flood of the deleterious effects of Western Culture on the social and scientific achievements, had been at the forefront of the humanitarian consciousness (Paul 1995). As heterogeneous as the German scientists community was, they were all concerned with questions of human diversity and human evolution. In all of those traditions and schools, researchers with Jewish background tried to pursue academic careers but with little success. Although many researches with Jewish background in the German-speaking countries contributed actively to the scientific debate about the so-called "Jewish race" even before 1933, they had little success (Lipphardt 2012). Thus, no wonder that many Zionists considered their movement to be a eugenic outpost for the rescue of the Jewish biological pool from the degeneration it suffered living in the Diaspora.

1.2 Who Is a Jew?

In an interview in 1998, journalist Michael Sheshar asked two retired Israeli Supreme Court judges "Who is a Jew?"[7] The secular Judge Haim Cohen responded: "the definition must be given by every single Jew for himself. If a person says of himself that he is a Jew, for me he is a Jew. This is his autonomy and nobody can decide for him or instead of him whether he is a Jew or not. There is no need in definitions."[8] The orthodox retired judge Menachem Elon emphatically contested this argument. He relied on the decision of the Knesset concerning the Law of Return: "The definition of the concept Jew, in this context, is: 'He who was born to a Jewish mother, or converted, and does not belong to another religion'. This is the lawful definition in the State of Israel. And in my view this law is most essential. Otherwise we have no Jewish nation!"

In 2005, Alain F. Corcos published *The Myth of the Jewish Race: A Biologist's Point of View*. The author and his family managed to escape from the jaws of the pro-Nazi anti-Semitic Vichy regime in southern France in 1944. Following a detailed analysis of the history of the Jews and Jew hatred, the author summarizes: "Many Jewish and non-Jewish writers find it difficult to accept the idea that Judaism is

---

[7] *Yedion Irgun Olei Merkas Europa*, 139 (August–September, 1998).

[8] In a later radio interview, Justice Cohen said: "Judaism is a matter of religion for one and culture for another. I completely ignore the genes and biology. I respect the spirit that I received from my parents, and from my parents' parents."

simply a religion and that Jews who abandon the faith, [...] are no longer Jews" (Corcos 2005, p. 18).

Other scholars disagree. Solomon Zeitlin, Professor of Post Biblical Literature at Dropsie College, noted: "The question – who is a Jew – first arose after the French Revolution when the Jews were politically emancipated" (Zeitlin 1959, p. 241). According to him,

> [A]nyone who is born of a Jewish mother or one who embraced Judaism, regardless of whether he observes or does not observe the precepts is a Jew. Judaism is a universal religion and no one can exclude himself. The Jews are also united by their history and to a great degree by Hebrew culture. Since Judaism represents the genius of one people there is also the ethnic element which unites them. [...] The land of Israel is not only the cradle of Judaism but Judaism as we know it today was moulded there. Throughout the ages the Jews of the Diaspora longed for establishment of a messianic kingdom in the land of Israel. [...] If Israel should become an ordinary, democratic industrial state it would be a great tragedy for Jewry and humanity as a whole. (Zeitlin 1959, pp. 269–270)

Rejecting such strict a definition in their book *Jews and Words*, the Israeli author Amos Oz and his daughter Fania Oz-Salzberger, insisted that it has been the Hebrew language that had formed the thread that has kept Jews together across the generations: "Jewish continuity has always hinged on uttered and written words, [...] Ours is not a bloodline but a textline" (Oz and Oz-Salzberger 2014, p. 1).

These polarized positions virtually confined the attempts of contemporary Jews to define Jewishness. David Vital opens his book *Zionism: The Formative Years* by noting that the Diaspora has been the most significant characteristic of the Jews for many generations (Vital 1982, e.g., pp. 5ff. & 349ff.). According to him, what characterizes and defines Jews is that in the past they had one country from which they were expelled. For Vital, this Exile has two aspects. One is the historic aspect – the physical reality of the Diaspora of the Jews, which had various and diverse consequences with respect to both their life styles and their surroundings. The other is the theological aspect – the spiritual reality of a People with no homeland, which has induced extreme changes in their religious conception of Judaism. Vital believes that although the ancient Israelites in the Near East were gradually dispersed to countries throughout Europe, they maintained the basic unity and the mutual relationships of the Jewish social and spiritual structures patterned in the first century C.E. up to the era of modern science and the industrial revolution. Although he does not insist on a biological criterion for the Jewishness of individuals or communities, it is apparent that he believes that during most of the years of exile Jews maintained not only spiritual and social unity, but also their unity of blood.

At the beginning of the nineteenth century an historic consciousness became acknowledged and with it the concern for "historic truths" rather than "eternal truths." In order to understand the social realities of being Jewish it was necessary to consider the relevant historic processes that produced these social realities. The ahistorical "laws of nature" were gradually replaced by historic definitions, which were conceived on the foundation of a "bourgeois society" and its expectation for emancipation as an ongoing process. The future was conceived and shaped by the actions of persons. For the philosopher Johann Gottfried Herder, who heralded early Romanticism, history was the scene where the phenomenon of nation formation was taking place and where a special spirit was being shaped. For the philosopher Friedrich Hegel (1770–1831), who was a bridge between the Age of Enlightenment and that of Romanticism, history had intention: the "spirit of the world" was revealed in the process of the realization of freedom. Accordingly, "Europe as we see it today is not the result of chance, but rather the necessary result of the efforts of the 'spirit of reason' revealed in the history of the world" (see Livneh-Freudenthal 2001). Hegel and his school conceived of phenomena of the real world, of nations, states and persons, as being the processes that determine its character. In this spirit, Hegel claimed, the modern state is a step in the process of the realization of freedom, because the regime is the guarantee for the realization of the freedom of its citizens.

According to Vital's analysis, the nature of Judaism started to change in the mid-eighteenth century in Western Europe, during the Age of Enlightenment, when a process of juridical emancipation of the Jews was initiated. As a consequence, by the nineteenth century, old patterns of Jewish life started to crumble. The old definition of Jews changed when their lifestyle, their religion, and their professions, or the kinds of trade they were allowed to engage in, all collapsed. However, the rate of this change and its impact differed from place to place. Least affected were the Jews in the lands of Islam, and most affected were those in Western Europe. In the latter, the more "enlightened" yet absolute rulers sought to eliminate the "Jewish problem" completely by undermining the uniqueness of the Jews and integrating them in a system in which privileges (in contrast to rights) were dispensed by the authority. On the other hand, many communities, especially in Eastern Europe, maintained their traditional ways of life, which were only slightly – if at all – affected by the spirit of the times.

It must be kept in mind that an important component of emancipation – the slow and intricate process by which Jews won redress from generations of discrimination – was that it was not restricted to Jews; rather it was a breakthrough from prohibitions and barriers for society at large. The intellectual enlightenment that brought with it the emancipation of the Jews, however, did not recognize social plurality, namely the privilege of equal rights for diverse communities, parishes, or nations. The French Revolution secured complete freedom for individuals, but rejected rights for any organized groups other than that of all citizens. Already in, 1790 Count Clermont-Tonnerre claimed:

> [I]t will be argued, the Jews have their own judges and particular laws. But, I answer, this is your fault and you should not permit it. Jews, as individuals, deserve everything: Jews as a nation nothing. [...] Within the state there can be neither a separate political body nor an order. There can be only the individual citizen. [...] it is inconceivable that there should be in the state a society of non-citizens, a nation within the nation. (Katz 1980, p. 109)

At the time, the Jewish community requested that the city of Colmar postpone the disbanding of the community in order to allow the Jews of Alsace (the district where the city is located) to get used to the new patterns and lifestyle of their neighbors. This request was emphatically rejected by the revolutionary assembly. Most of its members held to the notion that there existed a human capacity to mold one's own nature and they believed in the redeeming effect of the revolution, thus they did not see a need to defer the cancellation of the old life patterns.

At the beginning of emancipation, there was hardly any conflict between the traditional conception of Jews and non-Jews alike that the Israeli nation and its beliefs are one and thus, that the ethnic and the religious could not be separated. However, the further emancipation progressed, the more the religious aspects of Jewish alienation decreased – although they never disappeared – but concomitantly, the historic-social aspects, and eventually the historic-national aspects of the hostility to Jews increased. Overall, the anti-Jewish sentiments of the population at large did not decrease. With the growing importance of the economic and social aspects of life in the nineteenth century, anti-Jewish sentiments received new justification, and gradually became increasingly rational and of scientific pretentions. The more the Jews assimilated into the non-Jewish population in terms of their traditions, their views and their education, the greater was the urge of the Jew-haters to invent justifications for discrimination and persecution. The only "rational" way out appeared to be the argument that the Jews differed from their non-Jewish neighbors in their very essence, in their biology, which was manifested in their facial features, their life style, and even their cultural expressions. Accordingly, it became evident that matters would not change even if Jews tried as hard as possible to simulate non-Jews. This attitude brought about a shift from Jew-hatred to race-hatred, or anti-Semitism. Apparently it is the biological essence of the Jews that is the root of their religious, cultural and social segregation, which justifies unrelenting hatred against them. Contrary to other movements, such as revolutionary socialism, political Zionism had to face the burgeoning biological aspects of anti-Semitism and accept it as a fact of life.

1.3 From Anthropology and Eugenics to Population Genetics

The pioneers who settled in Palestine conceived of themselves as the delegates of the Zionist movement, the pioneers of the revival of the nation not only through the renewal of its cultural, social, and economic aspects, but also of its essential appearance and the physical health of the younger generation and that of future generations. It is no wonder that already in the 1920s the leaders of the immigrants, and primarily the physicians and the educators among them, emphasized the eugenic aspects of their responsibility to improve the hygiene of the race. Eugenic arguments, including the need to control the immigration of persons with hereditary and other diseases, were repeatedly declared in the 1930s, and the importance given to eugenics did not decrease even when the Nazi regime mobilized most of the German medical-scientific community. Thus, a Jewish physician in London praised the Jewish tradition of improvement of the race as late as 1939 (Feldman 1939), probably unconscious of dissertations such as those of the German Ottmar von Verschuer (1896–1969) who carried out genetic research on twins and published in the virulently racist journal *Forschungen zur Judenfrage*. Notably, Arthur Ruppin, the "father of agricultural settlement" in Palestine, was deeply impressed by the German racial researcher Hans F. K. Günther (1891–1968). In spite of warnings of the dangers of the Nazi regime, Ruppin met Günther in 1933 and noted in his diary that

1.3 From Anthropology and Eugenics to Population Genetics

Günther agreed that "Jews are not inferior in their value, though they were different in value" (Doron 1980, p. 421). This paradox, of the association of the eugenic movement with anti-Semitism (primarily in Germany) and restrictive immigration (primarily in the United States: see e.g., Reuter 2006) on the one hand, and Zionism, which encouraged Jews to immigrate, on the other hand, is fascinating. Ruppin's connections with Günther continued late into the, 1930s,[9] though he radically changed his attitude in the 1940s, when he became aware of the Holocaust. Eventually the term 'eugenics' became taboo in Palestine, as well as all over the world, not only because of the horrendous acts carried out in the name of eugenics in Nazi Germany, but also because of the insight gained from the misuse of eugenics in many democratic countries (see for example Kevles 1985, and Bashford and Levine 2010, especially pp. 3–24, 539–558. See also Broberg and Roll-Hansen 1996). As previously noted, circles of economists and politicians (and the Church) had already usurped the eugenic movement in the early years of the twentieth century, and utilized it to promote their own purposes, while making "free" use of scientific arguments (see Muller 1933).

Advances in genetics increasingly changed the studies of populations. Starting in the 1940s, the arguments related imcreasingly to common gene pools rather than to the traditional physical anthropological evidence (such as facial features, body structure, etc.), when discussing common origins of communities and populations. Such studies also excited researchers in Israel who realized the unique opportunity that the mass-immigration to Israel offered. The dynamics of migrant populations also presented the possibility for a unique and special contribution to the national effort of absorbing diverse immigrant groups, and this further stimulated the research effort.

A special challenge for research was the study of the specific diseases, some rather frequent among members of specific ethnic-groups. Israel became, in a sense, an international center for the research of population dynamics and changes in gene frequencies, especially those related to diseases. It appeared fruitful to take advantage of gene frequencies as tools in the attempt to reconstruct the history of Jewish communities, to trace their origins, and to track their migrations, by mapping appearances of genes related to certain diseases. It soon became apparent, however, that the forces affecting gene frequencies in populations are diverse and complex. With the emergence of genomics and the introduction of methods for directly following DNA molecule sequences, it often turned out that what looked similar at the clinical level was simply not so at the molecular level. Yet with improvement at the level of comparative DNA sequencing (such as genome-wide-association studies) more evidence of existing "blood relations" was accumulating – henceforth called DNA-sequence relationships between communities. A problem that remained was

---

[9] Neta Levit in an article in the journal *Kfiya of the Israeli Society against Psychiatric Aggression* (Hebrew, issue no. 8, January 2003, pp. 6–11) accused the "wave of immigrant doctors" who came from, or at least were educated in Central Europe, of having "imported into the country the principles of the German theory of race breeding", and the Hebrew Medical Association in Palestine of being "a propaganda echo for ideas and proposal that advanced the German race breeding theory."

the interpretation of the data. Do the common sequences of DNA indicate a common genealogical history of the communities, namely, both being the progeny of the ancient population of Eretz-Israel, or do they primarily indicate mating patterns along socio-cultural lines, that is both partners belonging to the same religion or cultural entity, or just living next to each other?

In recent years some historians have been challenging the traditional account of the massive exile of the Jews and the traditional stories of the formation of the Diaspora. These scholars suggest that numerically only a small proportion of the population was expelled after the Roman destruction of the Second Temple in Jerusalem and the Jewish revolts in the centuries that followed. Israel Yuval of The Hebrew University of Jerusalem wonders to what extent the story of the exile from our homeland is a myth, a fantasy, or history.[10] Shlomo Sand of Tel Aviv University maintains that although there were always Jewish communities with common religious and social connections, there was no distinct Jewish nation until recently. Jews lived together in socio-religious communities, composed of one people and were joined by other populations, many of which converted to Judaism. There were also Jews who deserted their religion and society (Sand 2009). Such notions are not new. Already in 1917, it was suggested by David Ben-Gurion that the Palestinian farmers (*fellahin*) were the progeny of the Jews who remained in the country when the rest of the population was sent into exile, and that the "agricultural settlements that the Arabs found in Palestine in the seventh century were the remains of the Hebrew population that had remained in the country" (Ben-Gurion 1917). The Bilu leader, Israel Belkind, expressed similar ideas in the 1920s (See Ornan 1969; and Broshi 2004).

Although at the beginning of the twentieth century a considerable proportion of the Palestinian Arabs were evidently immigrants from adjacent countries, some may be the descendents of the ancient inhabitants of the country. Thus, some Palestinian Arabs, like the Jews, may claim a genetic relationship to the ancient inhabitants of the country. These notions are contested by many historians and, in recent decades, by some researchers who study the relationships between DNA sequences of diverse Jewish *eidoth* and of non-Jewish Middle-Eastern populations and other communities. Many are still impressed by the objectivity ascribed to methods of empirical science, such as the *Structure* computer program that constructs genealogical trees based on an impressively large number of independent genetic markers. They apparently forget that such programs were constructed on inherent assumptions; for example, that the genealogical relations between members of a population or populations are historical rather than social, which produces branching trees, rather than intertwined trellises (see Bolnick 2008, pp. 74ff; Templeton 2008).

Is it conceivable that present-day Jews, more than comprising a biological entity of common origin, comprise a socio-cultural entity with an acquired common tradition? Or, put differently, to what extent do the biological connections between communities as well as the relative biological specificities of communities in the diverse

---

[10] Prof. Israel Yuval's presentation at The Hebrew University of Jerusalem, May, 15, 2011. See also Yuval (2005).

countries of the Diaspora (irrespective of whether that isolation was forced or voluntary) reflect a common *source* of that variability, and to what extent do they echo secondary socio-cultural *connections*? Could the answer to these questions provide, or at least uphold, the foundations for the construction of a pluralistic society that encompasses not only the communities but all its citizens?

***

It seems unjustified to claim that the people who devoted so much thought and so much research to the fate of the Jews were guided exclusively by objective, rational considerations. Jewish thinkers and investigators were occupied by diverse social and political paths – some to complete assimilation, some by the wish to integrate and become 'Germans of Mosaic belief', and still others by the Zionist notion of return to their homeland. Whatever path they followed, they were all guided by subjective sentiments, beliefs, and emotions. There can be little doubt that also today the inclinations of the thinkers, rather than merely the sequences of events, guide research paths. In other words, the research community's attitude to the issue of the biology of the Jews has been and remains to a large extent emotional. Should our genetic findings guide our reading of the history, or should our historical convictions guide our interpretation of the genetic findings?

The conception of the nation-state has changed radically in the twentieth century. Today it is clear that if there is any possibility of defining a population with a relatively closed breeding system, such delineations would rarely correspond to the national lines. Most differences between gene pools of breeding populations and communities are *relative* frequencies of genetic combinations, even though expressed by small, relatively closed populations it is easier to detect unique genetic traits that may serve as effective 'markers'. This is also the case with respect to the Jews: many Jewish communities have characteristics of small and relatively closed populations who lived under unique circumstances and migrated from place to place, as told in the history books. Even though there are no 'Jewish genes', certain Jewish communities may have genetic 'markers' that may allow one to follow the trail of their blood relations over long eras. Sometimes, there is also evidence that the composition of the gene pool of communities that define themselves as Jewish can be differentiated from the gene pools of the populations among whom they live. Furthermore, there are increasing indications that a Jewish community's gene pool conforms to the claims that it is related to gene pools of the Near East (see, e.g., "Similar but Different" in Chap. 9). At the same time, there exists increasing evidence that the difference in the characteristics of communities reflects specific environmental conditions (selection), or breeding with members of populations that have no historical relations with Jewish communities (assimilation). Therefore, there is much interest in the biologic structure of the populations of Israel and the forces that have been shaping them. Yet it must be kept in mind that such data may also be subject to different sociological and historical interpretations.

It is not in the hands of the biologists to decide the 'Jewishness' of one community or another, even in the face of the most sophisticated molecular devices:

Judaism and biology are two domains, different in kind. It is however a fact of life that embracing 'science' as an arbitrator in resolving all kinds of difficulties is still common. In every generation, there are still Zionists as well as non-Zionists who are not satisfied with the mental and social notions that bind Jews together, and who seek to find the link between the national and the biological aspects of being Jews.[11]

I do not intend to present in this book an historical view or a comprehensive picture of the biological literature of the origins of the Jews and the blood relations between them. As I have experienced in recent years, the subject is emotionally loaded. My perspective is that of a biologist who tries to examine Zionist history. Even though I tried to be objective, I am aware that my personal biases affected the writing. Chapters 2 and 3 mainly discuss the period preceding the formal Zionist organization, the changes in the attitude of Jews during the nineteenth century, and the changes in the life sciences and their relevance to interpersonal relations. The foundations of the Zionist movement and the attitudes toward it as an organization of the people of the Jewish race are discussed in Chap. 4. Chapter 5 deals primarily with anthropological studies that tried to unravel the racial essence of the Jews and their sources. Chapter 6 discusses the Zionists' attitude toward the problem of Jewish *eidoth* and their relationship to the Israelis of the ancient past. In the 1950s, anthropological research of the biology of the Jews was largely replaced by population genetics research. Assertions of "Jewish diseases," their origins and their impact on the population, were extensively discussed in Chap. 7, together with the climax of eugenic interpretations of Zionism. The problems of the "Ingathering of Exiles", which emerged after the Holocaust and with the foundation of the State of Israel, are discussed in Chap. 8. During the first years of the State of Israel emphasis was on the commonalities, the similarities, and the differences between communities, as revealed in the different frequencies of their genetic make-up. But with time the emphasis shifted to differentiating genetic markers. Genetic research gradually progressed from Jewish diseases and similar characteristic markers to polymorphisms of molecular markers, at the protein level and increasingly at the level of DNA sequences. This progress also augmented the political aspect of these studies, serving as means to follow the origins of the Jews and their relationships to other nations of the region, as discussed in Chap. 9. Finally, in Chap. 10, I try to draw a lesson from the journey along the paths of Zionism and the biology of the Jews.

---

[11] An interesting aspect is that of orthodox-religious circles that seek support of the "biological" argument for the Jewishness (or for membership in the Ten Lost Tribes) of tribes and congregations all over the world. Rabbi Eliyahu Avichail, the founder of the "*Amishav*" (Hebrew for "My People Return") organization and the author of the book *Israel's Tribes*, followed on his journeys "the footprints of forgotten Jewish communities, who lost their contact with the Jewish world […] at the same time he also located tribes that have no biological relationship to the people of Israel but who want very much to join them" (Yair Sheleg, "All want to be Jewish", *Haaretz*, September, 17, 1999, p. 27). In recent years, Rabbi Avichail "discovered" the tribe of Menasheh among the Koki, Mizo and Chin in the Manipur mountains at the border between India and Burma. In a TV program on "the search after the lost tribes," Hillel Halkin, a demographer of cultures, claimed that whereas the Jews of Ethiopia converted to Judaism during the Middle Ages and are not of ancient Jewish stock, the Koki, Mizo and Chin people are direct progeny of the Biblical tribe of Menasheh.

Chapter 2
From Emancipation to "Scientific Racism"

> Finally, allow me, Ladies and Gentlemen, to mention one other disease, not mentioned so far, which is yet, very important. This is the ancient suffering that the composer Heinrich Heine called "The Jews' Disease." In his famous verse at the occasion of the inauguration of the Jewish hospital in Hamburg he wrote:
>
> *Ein Hospital für arme kranke Juden*
> *Für Menchenkinder, welche dreifach elend*
> *Behaftet mit den bösen drei Gebresten*
> *Mit Armut, Körperschmertz und Judentume.*[1]
>
> And I ask: Is Judaism a disease? Judaism by itself is not a disease! The disease of the Jews is nothing but the reflex of the world's morality. – If, however, disease is suffering, then indeed there exists a Jews' disease, a very severe one. (Zondak 1940)

The establishment of political Zionism and the institution of its biological aspects in the heart of Europe toward the end of the nineteenth century cannot be understood without noting the role that the Jews and Judaism played in the socio-political and intellectual developments of the era. Intellectual and formal attempts to emancipate the Jews may be traced back to the Age of Enlightenment, at the end of the eighteenth century, and to the post-French Revolution and Napoleonic reign at the beginning of the nineteenth century. As noted retrospectively by Arthur Ruppin, the beginning of the change in attitude towards the Jews in Western Europe and their emancipation was primarily due to "the sudden change of outlook, social and economic, which characterized the whole of the eighteenth century": The establishment of political Zionism and its biologist aspects in the heart of Europe cannot be understood without noticing the role Jews and Judaism played in the socio-political

---

[1] A hospital for sick and needy Jews,
For human beings, who are triply wretched,
With three great maladies afflicted:
With poverty, corporal pain, and Judaism.

The corner-stone for the Jewish hospital in Hamburg was set by Salomon Heine, the nephew of the poet Heinrich Heine (1797–1856), on June, 10, 1841.

and intellectual developments of the era. Formal attempts to emancipate the Jews may be traced back to the Age of Enlightenment, at the end of the eighteenth century, and to the post-French Revolution Napoleonic reign at the beginning of the nineteenth century. But it was the development of commerce and industry, which finally shook off the fetters of the mediaeval corporations by taking on an individualist and capitalist character that brought Christians and Jews into contact with one another. Not only were Jews and Christians associated in greater business enterprises, but the alleged Jewish profession of money-lending suddenly lost its unpleasant savour. Whereas the Jew had formerly only been able to sustain the credit of the consumer, and was thus condemned to the calling of a usurer, he now found himself in a position, by assisting the credit of the *producer*, to become a valuable aid to businessmen, and a promoter of industries which required his capital (Ruppin 1913, pp. 6–7).

Johann Caspar Lavater (1741–1803) was a theologian and Christian preacher whose keen observations may be recognized as the realization of the unfolding notions of his time. From early on, he labored to correlate a person's appearance, especially one's facial features, with the person's character. Physiognomy, the prediction of character based on appearance, is an ancient occupation; it was, however, Lavater who turned it into a fashion, a science for the masses, to the extent that it was rumored that a person could not employ a house servant without prior physiognomic evaluation. According to the 1853–1860 edition of the *Encyclopedia Britannica,* the study of human character based on facial features became a plague in many places – people did not dare to go out on the street without a mask. Lavater managed not only to convince the masses of the authenticity of physiognomy, thanks to the scientific halo that he imputed to it; he also influenced many scholars and thinkers of the Age of Enlightenment. Among his converted acquaintances were the poet and scientist Johann Wolfgang von Goethe (1749–1832), the philosopher Johann Gottfried Herder, and the Jewish scholar Moses Mendelssohn (1729–1786). Although Lavater's convictions were fundamentally religious – "God created man in his image, hence the human face speaks to us in the Image of God in a unique language" – he adopted the most modern methods of scientific diagnosis at the time and developed his own techniques to present his teachings. Lavater repeatedly made the point that his system of belief was established on solid, objective scientific foundations. Even though many of Lavater's early followers abandoned him when they saw through the mystic-religious motivations behind his claims, he firmly established physiognomy not only as a means to characterize *individuals*, but also as an instrument to characterize *groups* in society. Thus, he discerned the characteristics of Jesus and those of Judas Iscariot (based on the pictures by the artist Holbein!), as well as those of nations and peoples. The following passage, taken from his *Physiognomical Fragments for the Advancement of Human Knowledge and Human Love,* reflects some of the biases that he presented as objective, scientific facts (Fig. 2.1):

> Here on the right stand members of the European nations and on the left those from other parts of the world. […] The Frenchman, who stands on the tip of his toes (20), explains in his lucid language matters that the Englishman (18), who God knows, is incapable of com-

Fig. 2.1 Lavater's characterization of nationals by physiognomy (Lavater 1984)
From the legend: Standing to the left are members of European nations and to the right members of other nations: A Frenchman talks to an Engelisheman (18). An Italian (22). The German (25) stands next to a Dutchman (24). A Hungarian (28)
The Chinese (4) looking to gain a grain of gold. A Turk (14). A Russian merchant (17). A worried Armenian (12). A Pole (10). The face of the Negero (6) indicating a mixture of a beastial spirit and lustuous might of body

> prehending. The Italian (22), with animated facial expressions, endeavors to find out where all this leads to. The image of the cold mannered person, immersed in thoughts, humble, not passing judgment, with a full healthy body, is none other than the German (25) [...]
>
> The Chinese (4) who looks downward tries to find out how he may succeed, by cunning or by cheating, to turn the grain of sand into gold. The Turk (14) has an open wide face, just like his garment. The Russian merchant (17) tries to gain something from him. [...] The face of the Negro (6), with the flat nose, the protruding lips and the flame in his eyes, indicates a special mixture of bestial spiritual numbness with forceful carnal voluptuousness. (Lavater 1984. TRF[2])

Such "scientific racism" was quite typical of the time: Lavater, the preacher concerned with the human soul, ostensibly adopted Linnaeus's ideas of scientific hierarchical classification of the plant and animal world, and extended them to what he conceived to be essential traits of human variability. Contrary to the customary notion that the Age of Enlightenment shattered the foundations of theological reasoning and established rational thinking with critical doubt, here we witness another face of the Enlightenment, one which bestowed a mantle of rationality on old theological argumentation.

Lavater and his fellow physiognomists could not ignore the Jews; nor could the philosophers with their ideas of the equality of all humans. Thus, in 1791 Johann Grohmann wrote in his booklet, *Ideen zu einer physiognomischen Anthropolgie* [Ideas on a Physiognomic Anthropology]:

---

[2] TRF: English translation by the author.

> I myself do not know how to describe this expression, and at which facial lines should I look for, that one can hardly ever view a Jewish face, without a certain feeling of lamentable pity [*jammerenden Mitleid*] and regret [*Bedauren*] growing in oneself, and to draw one's own facial muscles into creases of similar fashion. (Schmölders 1997, p. 61. TRF)[3]

As a matter of fact, as Ruppin noted, the doctrines of the Enlightenment that were originally meant to liberate humanity from the yoke of religion often added to the hardships of the Jews. Christians who became atheist ceased to be Christians; whereas Jews who renounced their religion remained Jews. Jews "in less than fifty years completely abandoned Yiddish (the so-called Jargon) in favour of the pure language of the country, and approached as nearly as possible to the Christians in dress and custom. […] From this to complete renunciation of Judaism was but a step" (Ruppin 1913, p. 7). Nevertheless, when the Jews gave up their faith and their customs, virulent prejudices against Jews did not disappear, not even among the knights of human civil rights. "French and German freethinkers" who glorified "reason, its demand for a rationalistic basis for everything, its ardour for science, and its antagonism to metaphysics and any positive religion, stood in direct opposition to the contemporary spirit of Judaism" (Ruppin 1913, pp. 7–8). Voltaire (1694–1778), the philosopher who excelled in his campaign against intolerance and bigotry, proved to be the most virulent anti-Semite. He wrote about the Jews: "We find in them only an ignorant and barbarous people, who have long united the most sordid avarice with the most detestable superstition and the most invincible hatred for every people by whom they are tolerated and enriched" (Corcos 2005, p. 41).

Moses Mendelssohn may arguably be considered the pioneer of Jewish emancipation (Elon 2002). Mendelssohn arrived in Berlin in 1743, and soon got involved in the circles of educated intellectuals, attempting to conciliate or to harmonize Judaism and the culture of the Enlightenment. He became friendly with the writer and philosopher Gotthold Ephraim Lessing (1729–1781), who eternalized Mendelssohn's person in his play *Nathan der Weise* [Nathan the Wise]. In it, Lessing attempted not only to show a Jew in a positive light, but also to understand his soul.

Lavater, who never tired of searching for scientific evidence for the act of Creation, translated from French to German the biologist Charles Bonnet's (1720–1793) book on the order of living beings as the embodiment of God's Will. Dedicating the translation to Mendelssohn, Lavater urged him to draw the consequences from Bonnet's convincing Christian arguments and to adopt Christianity. This dedication enraged many of Lavater's intellectual friends to the extent that Lavater eventually apologized to Mendelssohn (see Altmann 1985).

As far as the matter of belief was concerned, Mendelssohn rejected Lavater's offer without commenting on its racial aspect, namely that Lavater was ready to accept Mendelssohn as one of his people *in spite* of his Jewish origins! As a matter of fact, Lessing expressed a similar racial sentiment through the protagonist in his play: "O how worthy of esteem the Jews would be if they were all like you" (Poppel 1976, p. 8).

---

[3] The author makes a point that much of the science of anthropology, phrenology, pathognomy and ethno-physiognomy "hatched" from the principles of physiognomy and also includes juridical notions, such as the use of finger-prints for unique identification of persons.

Mendelssohn's response instigated the ongoing battle of those Jews who wished to become element of society at large and to join the intellectual community without giving up their Jewish identity. Upon encountering a non-Jewish society that vehemently opposed the integration of the Jews, Mendelssohn and the Jews of the Enlightenment emphasized the literary, scholarly, educational, and liberating aspects of the notion of enlightenment, referring to it by using the Hebrew term of *haskala* (scholarliness). This emphasis on the scholastic aspect and the prominence that Jews attained in various facets of the Enlightenment, determined for better or for worse, the fate of Jews in the centuries to come. Mendelssohn, who was the pioneer in integrating the *maskilim* (enlightened) into non-Jewish society in the eighteenth century, attempted to maintain the Jewish-religious element in this process. He emphasized the dogmatic content of Judaism that is founded on universal values and that may be examined and verified by logical parameters. In his opinion, the Laws of God that bind together the members of the Jewish community present the foundation for order in the community. However, as Ruppin pointed out, these claims were not accepted as valid by many, even in the Jewish communities. If Jews are content merely with the ethics and universal monotheism of Judaism, Ruppin asked, why should a distinct Jewry be maintained? (Ruppin 1913, p. 6). Indeed, many, including Mendelssohn's progeny, drew the same conclusion and abandoned Judaism. Most Jews of the Emancipation were content with some relaxation of the formal ties that bound them to the Jewish religion and tradition. As it turned out, such relaxation of Jewish links often did not bear the fruit that they hoped for. In the eyes of many, they still remained Jews – if not because of their religion and their life style, then at least because of their hereditary essence, namely their racial affiliation. Such was the fate even of those who apparently successfully integrated into the community at large, like the poet Heinrich Heine, or the statesman Benjamin Disraeli (1804–1881), the composer Gustav Mahler (1860–1911), or the chemist Fritz Haber (1868–1934). The *maskilic* advice proffered by the poet Yehudah Leib Gordon (1830–1892) – "Be a Jew at Home, and a Man in the Street" – did not help either; they all remained Jews in the public eye. Few succeeded in erasing their Jewish origins within one or two generations.

Mendelssohn had a somewhat contemptuous attitude to Jewish history: His "sacred temple" was not history, but philosophy.[4] Other Jews who did enjoy emancipation, yet refused to abandon Judaism, argued for its preservation because of Judaism's immanent religious and ethical insights, which contributed to humanity at large. The inclination to become "Germans or Frenchmen of Mosaic Faith" was attempted repeatedly during the nineteenth century, and although it had many supporters among Jews, it had very few supporters among non-Jews. Demands for equality were increasingly based on the inborn *civil* rights of each individual, rather than on principles of cultural or ethnic pluralism, which are stylish today.

---

[4] Livneh-Freudenthal (August, 2001). "The historic perception of the founders of the Science of Judaism" (in Hebrew). Talk at the World Congress for Jewish-Studies.

2.1 Jews as a Distinct Entity

Although many spoke about the racial characteristics of the Jews, they did not distinguish between hereditary and acquired traits. To be precise, the distinction between hereditary and acquired properties was not analyzed, even at the conceptual level, until the first decade of the twentieth century (see Falk 2009). Yet, the emancipation of the Jews was always accompanied by persistent claims of the inherent distinctiveness of the Jews. As early as 1791, Karl Wilhelm Friedrich Grattenauer expressed his opinion that baptism of the Jews would be useless, "just like washing the head of a negro to become white" (Gilman 1985). The German philosopher, Johann Gottlieb Fichte (1762–1814), who had a "penchant for rhetorical overkill," was even more explicit in his virulence, when declaring in 1793, "I see no way to give the Jews civil rights except to cut off their heads in one night and replace them with heads containing not a single Jewish idea" (Sweet 1993, p. 44[5]). The basic assumption of comparative anthropology of the nineteenth century was that Judaism is a property, an inherent characteristic that cannot be changed: When anthropologists wished to investigate the source of a physical or a cultural characteristic in a population whether inherited or acquired, the recommended method was to relate the variability in the population at large to that of the corresponding variability in the local Jewish population, thus providing a kind of experimental control. The assumption being that the two populations were identical in everything except for the hereditary Jewishness of one (the "scientification of a myth," in Lipphardt's 2012, p. S76 terminology). Any difference in the distribution of the phenomenon among the two must be due to the *biological* essence of the populations, whereas identity of distribution in the two communities would indicate that the variability was due to environmental factors. As a matter of fact, although Jews lived not far from non-Jews, nothing in their life circumstances – physical, cultural, or social – was the same in the two communities, perhaps with the exception of the climate. Although there was no scientific basis for such comparisons, interest in biological factors, namely racial differences, gained increasing attention, corresponding to the increasing interest in the life sciences during the nineteenth century (Katz 1980). As a rule, the importance of the biology of the Jews increased proportionally to the dialectics of the Emancipation that instigated anti-Jewish sentiments. Finally, all the achievements of the Emancipation were reduced, in the words of Monica Richarz, to "a transition from the status of Protection Jews to that of second-grade citizens" (Richarz 1982, p. 1).[6]

The first legal act by a European state to grant Jews the right to consider themselves permanent inhabitants of the land of their domicile was the issue of the Edict of Tolerance by Joseph II, the Austrian Kaiser, in 1782. Of special interest is Prussia,

---

[5] Sweet (1993) comments that "Kant, like Fichte, denied Jewish validity as a moral, religious system and made derogatory statements about Jews as a nation of cheats. As for Herder, [...] he considered Jewish national character intrinsically alien to Europe."

[6] The "emancipated" Jews in Germany in the nineteenth century were considered to be *Schutzjuden*, who had to act by lobbying and intercession to protect their civil rights, which involved paying the local Baron a fee, like a yearly license for a *Schutzbrief* [writ of protection].

2.1 Jews as a Distinct Entity

where discussions on the status of the Jews continued from the death of Friedrich II in 1786 until Jews were finally granted citizenship in 1812. A formal declaration of the emancipation of the Jews was pronounced only at the twilight of the Enlightenment in Westphalia during the Napoleonic rule in 1808. However, shortly thereafter, with the fall of Napoleon, there was increasing pressure to consider the rights that had been bestowed upon the Jews as obsolete. Manipulations to accept Jews into non-Jewish society, on the one hand, and the increasing demands to reject them, on the other hand, became significant components in the socio-political discourse of the nations of Central Europe, especially after the Restitution and the Vienna-Congress of 1814–1815.

Typical of this new status of non-emancipation of the post-Enlightenment era is the following "submissive request-letter to the royal office at Ehrenburg for the gracious dispense of the protection-money," written by the *Schutzjude* Matthias Weinberg, an inhabitant of the northwestern German township of Sulingen, on December 15, 1832:

> […] that I together with my wife, single and alone, must nourish us from rag-picking […] in this needy position it was until now very difficult for me to bring up the protection money and the taxes demanded from the Israelites for the needed protection letter. This is not surprising since both I and my wife are very weak, […] my physician explicitly ordered me to totally avoid the trips to the neighbouring villages for rag-picking. […] Thus I must concentrate on the local population and limit myself to rag-picking here. […] In this needy and hopeless position, I allow myself to appeal to His Excellence and to his kindness with this daring request to exempt me from the duty to pay protection money in the future and provide me, from now on free of charge, with the needed Protection Letter. (Hilmar and Hilmar 1986, p. 22. TRF)[7]

At the personal level, the improvement in the status of many Jews after leaving the ghetto was merely a formal achievement. Even when Jews were recognized as citizens, in fact they remained distinctly lower in status. Denouncing them and abusing their religion and beliefs were still common practices. Justifications for continuing this attitude were prevalent also at the intellectual level (Katz 1980). The historian, Friedrich Christian Rühs (1781–1820), for example, unequivocally rejected the Jew's demands for rights. He maintained that Jewry already constitutes a nation complete with laws and aristocracy (Rabbis) and, therefore, cannot be granted citizenship in a Christian state. Those Jews who were loyal to their political religion constituted "a state within a state" and, thus, were incapable of being loyal citizens of a German State. "A people cannot become a whole [*in ein Ganzes*] except through the internal coalescence [*inniges Zusammenwachsen*] of all the traits of its character, by a uniform manner of their manifestations by thought, language, faith, by devotion to its constitution." And according to Rühs, this did not happen with the Jews (Katz 1980, p. 77). In his book *Ueber die Ansprüche der Juden auf das deutsche Bürgerrecht* [About the Claim of the Jews for German Citizenship],

---

[7] Sulingen happens to be the home of my father's family. The Hilmars – teachers at the local school – initiated a program dealing with the history of the local Jewish community. It started with the physical cleaning and restoration of the Jewish cemetery and continued with a booklet on the history of the local Jewish community.

published in 1816, Rühs characterized Jews as "a particular nation, a political association, a religious party." These three ethno-religious categories were precisely what constituted the collective aspect of Jewish existence that emancipation was supposed to eliminate in favor of acceptance of Jews as individual human beings and as citizens (Katz 1979, 1980).

Jacob Katz believed that Rühs' explanation of how a nation may become a whole reflects a comprehensive, romantic definition of nationalism. Rühs purported to detect *collective* characteristics, discernible in the traditional culture, religion, and politics of a nation. He saw these characteristics as historical lore that persons belonging to a foreign nation, Jews included, must adopt if they wish to join a *nation*. A Jew wishing to join the German nation must accept the religion that characterizes the Germans – namely, Christianity. Actually, Rühs expressed the common notions of the nineteenth century concerning the inheritance of acquired characters (see Chap. 3): Even though acquired characters *may* eventually become inherited, it would take many generations until such a change could materialize. Rühs mobilized "facts" from Jewish history in Spain and in Poland that allegedly proved that Jews remained for many generations recognizable by their characteristic patterns. Even in places where they achieved the status of free citizens, they did not abandon their devotion to commerce, peddling, and finance. Jews are inherently different. There is no point in giving them citizenship privileges (Katz 1980, p. 78).

2.2 The Biologization of Race

The concept of race entered the science of biology in the nineteenth century.[8] Linnaeus, who organized the living world in a hierarchical system of phyla, orders, families, genera, and species, classified all humans as one species in this system: *Homo sapiens*. But the need to force human variation into discernable categories remained. The anatomist Samuel Thomas von Sommerring (1755–1830), for example, suggested at the end of the eighteenth century that humans would be classified by age, sex, nationality, diet, susceptibility to disease, life style, and clothing. Other scientists soon narrowed this list to age, sex, and nationality. There was no clear understanding of what properties were immanent or inherent and which were the result of circumstantial variability, but there was a belief that people may be classified into several innate major types or prototypes, whereas the remaining variability was nothing but "noise," caused by living conditions, that did not affect the concept of the essential prototype. The naturalist Johann Friedrich Blumenbach (1752–1840), who investigated the different essences of species, extended the organizing principles of Linnaeus to the classification of humans into races. As early as 1775, Blumenbach asked: "what is it that changes with the passing of generations, producing one time a degenerate progeny and another time a preferred one, but

---

[8] The modern use of the notion of race probably stems from, 1684. See, e.g., Schiebinger (1993), pp., 117ff.

always different from its ancestors?" Whereas Linnaeus discerned four human types, Blumenbach claimed that variability between human beings may be arranged into five distinctive "varieties": Caucasians, Mongols, Ethiopians, Americans, and Malayans. For many years, Blumenbach's argument was the accepted basis for the division of humans into races. According to Blumenbach, all races are a degenerate extraction of the Caucasian race – the Georgians – whom he considered to be the most beautiful of all humans; all physiological evidence converged on their region as the birthplace of humankind, the proof being "the unsullied whiteness of its inhabitants" and "the symmetry of the Georgian skull." Eventually, due to changes of climate – upon leaving their native "temperate zone" – and changes of nutrition as a result of the Caucasian migrations, all other races of man were born (Schiebinger 1993, pp. 126–131).[9] Blumenbach's ideas greatly influenced his friend, Emanuel Kant (1724–1804) (see Lenoir 1982).

For Blumenbach, humans belonged to different, essential races, distinct from each other in essential hereditary properties that had been acquired over many generations. Obviously, in principle, these characteristic traits could likewise change over time. More importantly, although possibly more difficult to demonstrate empirically, the racial characteristics comprised not only physical properties, but also mental and cultural ones. Thus, Blumenbach not only provided the physical anthropologists and the cultural anthropologists with the explanatory foundations of variability, but most importantly, he provided these notions also to linguistic anthropologists. Linguistic anthropology, an outgrowth of cultural anthropology, gained much popularity in France of the nineteenth century, when the common origins of the languages later called Hindu-Europeans suggested that deductions may be drawn from the familiar relationships between languages to common racial origins. This theory of uncovering racial relationships via linguistic relationships is mainly attributed to the historian Ernest Renan (1823–1892). According to Renan's central thesis of 1848, the origin of the Hindu-European speaking peoples is derived from a common Aryan race. Another group of peoples, whose languages also have common origins, were the Semites, who comprise a race *per se*. Renan, however, went further than drawing ethnological-biological conclusions from the ethnological-cultural data; he also drew value consequences related to the Semites' contribution to humanity. In 1855, in *Histoire générale et système comparé des langues sémitiques*, Renan evaluated the contribution of the peoples of the Semitic race to human culture in unequivocal words: "The absence of philosophical and scientific culture is due, it seems to me, to the lack of space and of diversity and consequently of analytical spirit which characterizes them. [...] One does not find in their midst either great empires or commerce or public spirit. [...] The true Semitic society is that of the tent and tribe. [...] The Semite knows almost no duties, except to himself" (Katz 1979, p. 121). Thus, Renan transferred the ancient accusations against Jews

---

[9] "[B]eauty, that of both male and female, deeply influenced anthropologists' assessment of the world's peoples. [...] Early modern anthropologists were as intrigued by beauty as their nineteenth-century *confrères* were by skulls. [...] For many Europeans skin color determined beauty (as it did political power and moral worth)." (Schiebinger 1993, p. 126)

from the theological level to the hereditary-biological level, and was one of the first authors to raise the explanatory power of race to social and historic levels. Racists derived much encouragement from his classification of the Aryan and Semitic languages, which he padded with a fair amount of prejudice. Yet apparently the rise of racist anti-Semitism in the early 1880s worried him. In accordance with the concepts of the period, which accepted that acquired properties may become hereditary, Renan acknowledged that certain conditions may affect racial properties and concluded that there is neither a Jewish race, nor one typical of Jewish appearance. "At most, self-isolation, endogamous marriage and the long periods in the ghettos had produced a certain Jewish type" (Sand 2009, p. 236). Apparently, also from his acquaintance with the people of the East, Renan expressed some doubts regarding the Jews of the present: "the Israelites of our days who descended in direct line from the ancient inhabitants of Palestine have nothing in the Semitic character and are no more than modern men, assimilated through that great force superior to races and destructive of local originalities which we call civilization" (Katz 1980, p. 138).

Renan's contemporary, Joseph Arthur Comte de Gobineau (1816–1882), went one step further. In his book, *Essai sur l'inégalité des races humaines* [An Essay on the Inequality of the Human Races], published between 1853 and 1855, he expounded a racial theory as an intellectual tool for the explanation of historic phenomena. Gobineau tried to explain all of human history – the rise and fall of kingdoms, governments as well as cultures – as the function of interaction among three extended races: the White, the Black, and the Yellow. The Semites were presented as one of the races of humanity, without any specific derogatory assertions relating to Jews. The ideological purpose of his book was to provide an explanation for the degeneration of human society that, according to the pessimistic opinion of Gobineau, was a fact. Following Gobineau, a comparison of the facial features of present-day Jews with those that look at us from the ancient reliefs in Egypt and Mesopotamia, provided evidence that environmental conditions do not possess the power to change the (physical) characteristics of the race (Katz 1979, p. 43;1980, p. 310). The degeneration is, according to him, due to the loss of racial *purity*. This is also what happened to the Aryans, the superior race who, in Gobineau's mind, had been the source of exceptionally high cultural productivity.

Contrary to claims of common historic roots of nations that were based on evidence from linguistic associations, other anthropologists emphasized the need for developing physical parameters to establish the biological relationship between populations. These researchers argued that linguistic anthropologists who tried to detect *biological* origins by following *linguistic* relationships were mixing apples and oranges. For example, as early as 1844, the Swedish anatomist, Anders Retzius (1796–1860) introduced the cephalic index for the identification of races, based on the height of the skull and its width. One of the early investigators of the physical anthropology of the Jews was Felix von Luschan (1854–1924). He asserted that "all those who speak of the Semitic race confuse the concepts in a distorted fashion, as if they were talking of a language of the dolichocephalics [long skulled]" (see

2.2 The Biologization of Race

Ruppin 1930a, p. 21, ftn. 1).[10] Nevertheless, the criteria developed by the physical anthropologists for the "objective classification of populations" and for tracing their origins were, in retrospect, no less biased than those of their colleagues, the cultural and linguistic anthropologists, whose criteria also reflected the prejudices of their authors.

As noted, the very fact that Jews were often considered the nearby "other" (presumably living under similar conditions) in Europe contributed to many physical anthropological studies referring to Jews. One of the early physical anthropological studies directly treating Jews was that of Richard Andree (1835–1912). In 1881 he claimed the existence of a Jewish prototype, and declared that environmental conditions are irrelevant in matters of race: "We all recognize the Jewish type. We immediately distinguish him by his face, his habits, the way he holds his head, his gesticulations or when he opens his mouth and begins to speak. And it is always possible [...] to recognize even the most assimilated, because he always bears some of the characteristics of his race" (Efron 1994, p. 22). Alfred Russel Wallace (1823–1913), Darwin's co-explorer of the theory of the origin of species, also asserted that Jews, spread over all continents, maintain the same facial features. Other anthropologists, who emphasized the unique biology of the Jews, accepted the existence of more than one type. Thus, one of the pioneers of the modern science of physical anthropology, Carl Vogt (1817–1895), argued in 1864 that there were two Jewish types: the one in Russia, Poland, Germany, and Bohemia, who has hair that often tends to be somewhat red, a short beard, a flat nose, small, grey and cunning eyes, a full-figured body, a roundish face, and prominent cheek-bones; the other type lives in the East and around the Mediterranean where, according to Vogt, another Jewish tribe, which has thick and dark hair and beard, big almond-shaped gloomy eyes, an oval face, and a prominent nose. The surprise was, therefore, great when the famous pathologist-anthropologist, Rudolf Virchow (1821–1902), who conducted a widespread study of the physical anthropometric variables of more than 10,000 school children in Germany in 1886, found that about 10 per cent of the Jewish students were blond, which was significant, even though the rate of blond hair among the German children was as high as 31 per cent (Efron 1994, p. 25).[11]

---

[10] See, however, Doron (1980), p. 398, who quotes Ludwig Stein (1859–1930), a Hungarian born scholar who taught philosophy at the Theological Seminar in Berlin. Stein denied that the Jews were a race, either in the physical sense or even the national sense: "As far as I am concerned, an ethnologist who talks of an Aryan race, Aryan blood, Aryan eyes and hair, is a sinner as much as a researcher of languages who speaks of a dolicocephalic dictionary, or a brachycephalic grammar."

[11] Blond hair was considered an important and significant indicator of racial origins. See Chap. 5, for suggestions of Salaman (1925) and others, for the origins of red/yellow hair among Jews. See also Adolphe Bloch's (1913) "Origin and evolution of the blond Europeans." Bloch rejected "The Aryan Hypothesis" that the first Indo-Aryans came from Central Asia, and insisted that the first blonds in history were the Celts: "The first habitation of the Celts, and consequently, [...] all the blond Europeans were, then, natives of Europe itself, and did not come from Central Asia. [...] As predecessors of the blond stock in Europe there is therefore none other to be considered than the Quaternary race of Neanderthal. [...] I think that under the influence of the cold climate of the epoch the production of cutaneous and capillary pigment in the human organism was so weak that the skin bleached, and their hair and beard became lighter [...]".

Fig. 2.2 Characterization of Jewish facial patterns using Galton's technique of superimposing film negatives (Efron 1994)

At Jacobs's request, Francis Galton took a series of photographs at the Jews' Free School in London in 1891. Galton then superimposed the ten original photographs to produce these four composite shots, which he claimed were representative Jewish types. The notion that essential Jewish features were to be found in all Jews was a central belief of race science.

A study by Francis Galton, the founder of eugenics, offers an illuminating example of studies that insisted on the (recognizable) Jewish type. Galton, who believed in the existence of essential types, devoted much effort and statistical ingenuity to calculating their parameters. Among others, he developed a technique of superimposing photographic negatives of several faces in order to identify the common denominator of the "type" (Galton 1878). In 1891, upon the request of his colleague, Joseph Jacobs (1854–1916), Galton took a series of photographs at the Jews' Free School in London. He superimposed them to produce four composite faces, which he claimed were representative Jewish types. Both men agreed that these visual images revealed something fundamental, even essential, about the nature of modern Jewry. To Galton, "every one of them was coolly appraising me at market-value, without the slightest interest of any kind." Not surprisingly, Jacobs' reaction was remotely different, "I fail to see any of the cold calculation which Mr. Galton noticed […]. There is something more like the dreamer and thinker than the merchant." And for Jacobs too, the overall significance of these photographs was that they were evidence of "a definite and well-defined organic type of modern Jews. […] There has been scarcely any admixture of alien blood amongst Jews since their dispersion" (Fig. 2.2) (Hart 1995, p., 165; Efron 1994. See also Glad 2011, pp. 121–122).

Ruppin, in the first publication of *The Jews of Today* in 1904, criticized the numerous studies of the issue of the Jewish race, many of which proved not to be

2.2 The Biologization of Race

scientific. He noted, however, the 1881 study of Richard Andree, mentioned earlier, as a good though old-fashioned study. He also favourably mentioned Galton's colleague Joseph Jacobs, the anthropologist and Jewish folklore researcher. Jacobs was born in Australia and immigrated to England. He was later called to the United-States to edit the *Jewish Encyclopedia*. Jacobs consistently complained about those who claimed that the Jews were a spiritually degenerated race, and he responded aggressively to those who declared that the Jews' religion was outdated. Like many others, Jacobs built his arguments on the ancient texts, but at the same time he tried to support these arguments with statistical demographic data that he collected in the field. Likewise, Jacobs insisted that there was no linkage between Jews as a distinct biological entity and the properties attributed to them. These properties were, presumably, the consequences of the persecutions and living conditions in the ghettos. Jacobs' research method was to sort out which properties attributed to Jews were the result of the social conditions of the Jews and which were hereditary. As mentioned, he even managed to convince Francis Galton to examine whether it was possible to discern typical Jewish faces. Jacobs first claimed that he had successfully identified four facial patterns using Galton's technique of overlaying photographs of facial patterns. Later, however, based on his statistical calculations, Jacobs acknowledged that he could not identify "average types" or that all Jews appeared to belong to one type, although he claimed to recognize the existence of distinct anthropological types for Ashkenazi and Sephardic Jews (see Efron 1994, pp. 58–90).

> Scientific studies of racial character and continuity that utilized visual imagery as evidence rested on a fundamental assumption about the nature of iconography: images offer direct, unmediated access to objective reality. [...] In this crude positivist approach the artist all but disappeared. The image was equivalent to the numerical measurement in that both provided the scientist with a body of facts with which he could then construct a narrative about race. [...] Just as time was made irrelevant by the assumed continuity of racial essence, so too artistic and technological developments became meaningless in these texts. [...] Photography, it came to be widely believed, had replaced painting as the "mirror of nature." (Hart 2000, p. 176)

Toward the turn of the twentieth century, when the Zionist movement granted a kind of approval to the national social alliance of Jews, rather than merely to their traditional religious or cultural uniqueness, the flood of studies that ascribed to Jews a biological essence as a race swelled. In the 1911 edition of the *Jews of Today*, Ruppin referred approvingly to Ignaz Zollschan's (1877–1944) study, published in Vienna in 1909, as well as to studies by other Jewish scholars, such as Weissenberg, Judt, Elkind, Auerbach, Fishberg, and Sofer, and his (non-Jewish) teacher, the German anthropologist, von Luschan (Ruppin 1911, p. 213).

In contrast to authors who presented anthropological evidence of the biological distinctness of the Jews; others, like the German-Jewish philosopher and psychologist Moritz Lazarus (1824–1903), made enormous efforts to deny any racial or national distinctness of the Jews. He was one of the advocates for the preservation of Judaism as the religious-moral tradition of a distinct Jewish congregation within the German nation. In a speech that he delivered in December 1878, entitled "What is the meaning of nationality?" Lazarus analyzed the concept of nationality in general and Jewish nationality in particular, and rejected outright any claims of Jewish nationality (Lazarus 1925). In his book, *The Ethics of Judaism* (Lazarus 1900–1901), he made an effort to convince his readers that Judaism is nothing but a religion and that the Jews were "Germans of Mosaic Faith": The Jews are not members of a separate race, but it is rather their religion that distinguishes them. Their religion has universal significance; it provided the base upon which Christianity was founded, and it continues to provide a foundation for Western ethics. Lazarus tried to fight anti-Semitism with his conviction that non-Jews would eventually embrace those who contributed and continued to make such a basic contribution to Western culture. Lazarus's movement formed the beginnings of the *Reform* movement of Judaism which eventually flourished, mainly in the United States. It is an irony of fate that Lazarus's book, *The Ethics of Judaism,* was translated into English in the 1920s by Henrietta Szold, the founder of "Hadassah" in the United States and the head of "Youth Immigration" to Palestine.

It was only toward the 1890s that organizations appeared in West Europe that not only disengaged from denying being Jews, but explicitly stressed their distinctness without excuses, demanding that they be given respect and consideration. These groups did not favour the assimilation of communities or the elimination of Judaism *per se*, but rather a return to Judaism with confidence in its essence. *Zentralverein deutscher Staasbürger jüdischen Glaubes* [Central Union of German Citizens of Jewish Belief], established in December 1893, and the Zionist Movement, established in September 1897, are two examples of such organizations, which had diametrically different aims. Whereas the former engaged in a "thirty years war" – that became a forty years war until its complete annihilation – for recognition of the Jews as a cultural and religious minority within the German nation, the later engaged in a political, cultural, and colonization campaign for the admission of Jewry as a national independent entity. Like the legendary phoenix, the realization of the aspirations of one movement rose from the ruins of the other (See Mosse 1970, for a detailed discussion).

Thus, the claims that the Jews were immanently different, namely, that they were distinctly a biological rather than merely a religious, cultural race, evolved hand in hand with the emancipation of the Jews. However, the variable paths that led different researchers to such conclusions – often one contrary to the other – strongly indicate that, by and large, the ends justified the means. No wonder that against such a background, when the Nazis came to power, they had to mobilize their best anthropologists to identify – in vain – Jews in order to discriminate against them. Of course, soon they had to fall back on more straightforward devices to label Jews, such as the Yellow Patch.

2.3 Anti-Semitism

The years of 1850–1871 marked the end of the struggle of the Jews of Germany for political emancipation. With the union of the northern German states in 1866 and the establishment of the German Reich in 1871 under the leadership of Otto von Bismarck (1815–1898), a new constitution was established that included the rule of the equality of all citizens before the law, independently of their religious beliefs. Already from the beginning of the nineteenth century a new type of a Jew appeared in Western Europe alongside the traditional, stereotypic Jew who was distinguished by his or her language, dress, education, and the limited range of occupations that he/she was allowed to hold. Although this new type of Jew shook off all the traditional identification marks of his congregation, Jew-hatred did not diminish. Disapproval of "The Jews" by members of the society at large – who were by now increasingly liberated from the biases attached to Christian dogma – was now directed at the economic, the social, and the political roles that Jews had filled in the past, or might fill in the future.

In 1861 the book *Die Juden und der deutsche Staat* [The Jews and the German State] was published by an anonymous author, soon identified as Johannes Nordmann (1820–1887). The book gained immediate popularity and was published in several editions, the last one as late as in 1920. Its overt intention was to prevent Jews from occupying public offices. Nordmann's argument was not based on the Christian religion, but rather on Christian morality, which endowed society and the state their specific character; and since the Jews were not able to rise to that level of morality, they were inadequate to be candidates for state office. Although there was nothing new in this argument *per se*, a new dimension entered the discourse, namely the biological foundations of the moral inferiority of the Jews. According to Nordmann, a detailed analysis of Biblical and Talmudic texts revealed an inherent Jewish mentality. As a consequence of the continued cultural isolation of the Jews over many generations, he claimed, these definite Jewish traits prevailed: "Seclusion and inbreeding over many thousand years strengthened the thorough domination of the race type and made the way of thought a part of it. Jewish blood and Jewish sentiment became inseparable and we have to conceive Judaism not only as a religion and congregation [*Kirche*], but also as the expression of racial peculiarity." To complete the picture, Nordmann declared that "Jews, in contradiction to the Germanic tribes, possess the deficiencies of the Southern races without their merits" (Katz 1980, p. 212).

Nordmann may have been the first in Germany to make immanent race the pivotal concept of his anti-Jewish ideology, but he was not alone. Others who could not overcome their sentiment were content to exchange the theological elements of their anti-Judaism for seemingly rational arguments and ostensibly verifiable claims. The Jews' characteristics and their social inferiority were interpreted as imprinted collective properties and as the unavoidable acquired outcome of a unique historical process. In the 1870s, writers, poets, and other intellectuals came up with so many varied accusations against Jews that a new trend was established with the sole, unequivocal purpose of rejecting the Jews. The leaders who carried this banner denied being the successors of Jew-haters of the past. According to them, the former justifications for opposition to Judaism and to Jews were obsolete: In the past religion formed the chasm between Jews and Christians that nurtured alienation and hatred; in the present, religion was no longer a factor in Jew-hating, or at least it was much weaker. And although there were still many Christian believers among them, they all insisted that they adhered to the principles of religious tolerance.

Modern anti-Semitism acquired its full-fledged ideological significance only in 1879, when the Berlin preacher, Adolf Stoecker (1835–1909), started distributing his anti-Jewish propaganda. That year, in Bern, Wilhelm Marr also published a pamphlet entitled *Der Sieg des Judenthum über das Germanenthum* [The victory of Judaism over Germanism], which aimed to unite all societies embittered by purported Jewish domination. Marr introduced the term *Antisemitism* in this pamphlet (see Chap. 5), although he had already sharply expressed anti-Jewish sentiment, which would be the prime motivation for his anti-Semitic outburst in his writings in 1862. Marr's proposal for European order was based on the principle of the partition of Europe into three racial spheres of influence: the Latin, the Slavic, and the Germanic. According to Marr, these three races are the three nationalities that were defined by "language and custom." This system pits the European races against Orientalism or Asianism, a radical arrangement that included the Jews. At first, Marr favoured intermarriage as a solution to the Jewish issue. But when asked what method was to be employed when the belief in racism rejected the old solution of assimilation and intermarriage, he replied: Jews to Palestine. "Palestine had to be the ideal location, since the Jews were racially close to the Moslems" (Zimmermann 1986, 1988). Marr repeatedly made the point that it was not the religion of the Jews that incited resistance, but rather the traits of their racial essence, imprinted in their characters and expressed in their behavior. Accordingly, his anti-Semitism was aimed at rescinding the emancipation of the Jews, or at least denying its social consequences. Shortly thereafter, in 1879, the historian and liberal-nationalist politician, Heinrich von Treitschke (1834–1896), attacked the Jews for their refusal to assimilate into German society and to integrate into its culture, in his famous article *Ein Wort über unser Judenthum* [A Word about our Judaism]. The fact that Treitschke used the term anti-Semitism undoubtedly legitimized it (Zimmermann 1986, p. 94).

These developments were closely connected to the social and economic events of the period. After a century of economic prosperity in Europe, especsially in Germany, the year 1873 is remembered as the year of the great bankruptcy and the collapse of financial enterprise. With that collapse, Bismarck's "period of economic depression" set in, with its ups and downs, which lasted until 1896. Politicians diverted public opinion from the fundamental problems of society by referring to the Jewish issue. Adolf Stoecker (1835–1909), the head of the Christian-Socialist Workers Party, described the Jews as a camp of strangers who do not belong to the German nation and still invade the cells of its internal life. In 1881 Eugen Dühring (1833–1921) instigated a frontal assault against Judaism in his pamphlet *Die Judenfrage als Frage der Rassenschaedlichkeit* [The Jewish question as an issue of racial damage]. According to Dühring, the Jews were a kind of human being who

2.3 Anti-Semitism

had a special body and spirit, which of course, was negative. He honestly believed that he was able to discern in the faces of contemporary Jews the trait patterns that gazed at him from Biblical tales. Claiming that Jewish inferiority was a stable, unchangeable property, Dühring affirmed the universal character of Jew-hatred. Modern-day anti-Semitism, according to his conception, was nothing but a new stage in the permanent defense of the entire world against the eternal Jewish danger (Katz 1980, pp. 266f).

Anti-Semitic attitudes were not confined to Germany. In Austria the racist Karl Lueger (1844–1910), the representative of the Christian-Socialist party, was elected in 1897 as the mayor of Vienna on the basis of an explicit anti-Semitic agenda. In France, Édouard Drumont (1844–1917) first attacked Jews in 1886 in his book, *La France Juive* [Jewish France], adding racial slander against the Jews to the traditional Christian accusations and arguing for their exclusion from society. Within a short time, up to a hundred editions were issued. Drumont was elected to the Senate as a representative of Algiers. Other virulent anti-Semites were also elected, such as Maurice Barrès (1862–1923), a well known author, who claimed that anti-Semitism comprised a foundation for "national union" as a barrier against the domination of foreign Jews and cosmopolitans. Profesor Jacob Katz claimed that it is difficult to exaggerate the contribution of Drumont's newspaper, *La Libre Parole,* to the anti-Semitic atmosphere surrounding the Dreyfus Affair. Drumont's paper bluntly accused Jewish officers in the army of disloyalty to the State, and even of espionage, two years before Dreyfus was arrested (Katz 1980, pp. 297f.).

The German Theodor Fritsch (1852–1933) may be considered a living bridge between the inception of the anti-Semitic movement and its apocalyptic climax during the Nazi regime. Fritsch represented the school that essentially relied on the impact of propaganda and attempted to integrate anti-Semitic ideas into wide circles of society. He tried to prove that Jews were a corrupt and spoiled people. Fritsch was one of the pioneers who attempted to reestablish racial-biological anti-Semitism on Christian-religious foundations by declaring that Jesus was of Aryan origins: "Surely Christian teaching arose as a protest of the Aryan spirit against the inhuman Jew-spirit" (Katz 1980, p. 306). In 1899 this type of interpretation of Christianity was also a major theme of *Die Grundlagen des neunzehnten Jahrhunderts* [The foundations of the nineteenth century] by Houston Stewart Chamberlain (1855–1927), an Englishman by origin, a German by choice, and an admirer of the Wagnerian cult. He married Wagner's daughter, Eva, in 1908. Chamberlain rejected Darwinism, evolution, and social Darwinism and instead emphasised *Gestalt* [form, being a complete and unanalyzble whole]. Considering the virtues or the deficiencies of a race to be immanent properties, the "survival of the fittest" was nothing but the consequence of the process of nature which determined the combination of racial properties that happened to be successful (Katz 1980, p. 309). The antithesis of the heroic Aryan race with its vital, creative life-improving qualities was the "Jewish race": every positive quality the Aryans had, the Jews had the exact opposing negative quality. The American historian Geoffrey Field wrote: "To each negative 'Semitic' trait Chamberlain counter-posed a Teutonic virtue".

2.4 Judaism as a Historic Entity

When the concept of progress came to a halt during the counter-revolution and the Restitution (1815–1848) and crises following industrialization and modernization began, there was a general abandonment of the values of the Age of Enlightenment and its idea of universal progress. Yet at the same time, from the beginning of the nineteenth century the notion of history increasingly became an organizing principle. As a scientific discipline, history eliminated the traditional disciplines – theology and jurisprudence – and embraced the belief that it was possible to know the past: A critical-historical approach may achieve an "objective and bias free" research. The founders of a Science of Judaism adopted this ethos (see Livneh-Freudenthal, "The historical perception of the founders of the Science of Judaism" [in Hebrew]. Talk at the 2001 World Congress for Jewish-Studies).

Religious definitions and, to a large extent, also social ones, which had previously influenced the living patterns of many Jews in Central Europe and distinguished them from their non-Jewish surroundings, began to fall apart. Attempts to find an "enlightened" and "scientific" definition of the Jews at the beginning of the Emancipation drove Jews to engage in research on their anthropological status as they attempted to integrate into society while still maintaining their uniqueness. The Science of the Jews [*Wissenschaft des Judentums*] that was born as a discipline in the 1820s within the framework of the Union for Culture and Science of the Jews [*Verein für Kultur und Wissenschaft der Juden*] appeared to provide a proper foundation on which to construct a coherent, continuous narrative that would be meaningful for Judaism. Leopold Zunz (1794–1886) and his colleagues, in line with the spirit of the time, tried to present Judaism as a "collective I," conscious of itself, acting within history according to a plan and a purpose. He supported his claim for the existence of the culture and science of the Jews by writing its history (Livneh-Freudenthal, 2005). In 1822, Zunz published a paper on the "Outlines for a future statistics of the Jews" [*Grundlagen zu einer künftigen Statistik der Juden*] in the first volume of the *Zeitschrift für die Wissenschaft des Judentums*. He called for examining all aspects of Jewish life – language, religion, customs, occupations, and the life of the individual and of the community – in accordance with traditional, objective scientific research. He pressed for a study to discover the sources of the physical properties typical of Jews and to draw the necessary conclusions. Zunz was convinced that the inferior status of the Jews stemmed from the fact that non-Jews were not aware of the cultural richness of Judaism. He thought that presenting this knowledge in a palatable form to intellectuals and rulers would bring about the elimination of Jew-hatred. However, Zunz, like many of the Jews who contributed much to the *Wissenschaft des Judentums*, did not discern cause from effect. Deplorably, preoccupation with Judaism was for many non-Jews a cause for discrimination against Jews rather than result of its effect. In spite of the Emancipation, knowledge about

Judaism served only to increase Jew-hatred rather than to provide a means for knowing and understanding the Jews.[12]

The two main paradigms that dominated Jewish historiography were that Jews are a nation with a unique character and a common biography, and that history is an objective science, hence a guarantee for uncovering the truth. Thus, the task of the *Wissenschaft des Judentums* was to restore the existence of the Jewish nation by recounting its history as a national narrative. Indeed, following the philosophy of Herder, Zunz and his colleagues applied the notion of a *Kulturvolk* [culture-nation] to the Jewish people. Such a halo appealed to some intellectuals who were attracted to the romanticism of a Jewish national culture. However, it should be kept in mind that the researchers of the *Wissenschaft des Judentums* actually imposed values and categories taken from their non-Jewish milieu on the Jewish national narrative (as did the Zionists). In hindsight, it is difficult to ignore that the very same romantic spirit that drove the German *culture-nation* was also the one that soon propelled the Germans further towards the *national-culture* of the Nationalist State: The state that many Jews now considered their fatherland soon defined itself by excluding the "foreigner," who turned out to be primarily the Jew. Soon, many Jews were searching for a different rational for the vindication of the continuous existence of the Jewish nation. They found it in the idea of liberty, which ever since the French Revolution, was a pivotal idea of political discourse and a central theme of Hegel's philosophy. In Zunz's words, the realization of liberty is a process that reveals the "world spirit" at different stages of history. The notion of liberty was at the core of the ambition to establish a "civil society" in which each individual (and nation, which is also a kind of an individual) maintains its uniqueness in the framework of the wider culture. The definition of Judaism as a *culture-nation* was supposed to sever Jewry from the traditional religious framework that isolated it, allowing Jews to integrate into the society at large. The conception of the people of the *Wissenschaft des Judentums* was to the represent the Jewish culture in a space of culture *per se*, which would confirm the national existence of Judaism (Livne-Freudenthal 2005).

This explains why the agenda of the Jewish scientists was, as a rule, different from that of their non-Jewish colleagues, although Jews and non-Jews alike engaged in the new science of the history of the Jews. Whereas non-Jewish scholars took advantage of the scientific framework to establish claims of the inherent *racial separateness* of the Jews, most Jewish researchers endeavored to prove that the Jew was a product of a *cultural-social isolate*, rather than a racial-biological one, and that an enlightened process of emancipation would allow the Jew to integrate – to assimilate – into the local population. Obviously, such assimilation had different meanings to different researchers. There were those who expected the Jew, as a

---

[12] Meira Yifat Weiss (2002) comments: "The duality of perspectives of both an observer and a participant illuminates the internal paradoxes of the process of the integration of the European Jews in the scientific deliberations of the time. As scientists they are observers, however as Jews they serve simultaneously and permanently as subjects for deliberations on heredity and environment, origins and culture, assimilation and essence" (p. 136). Therefore, "the main effort is directed not at an attempt to formulate essential conclusions [...] but rather at an attempt to write a cultural-Jewish history that is neither aloof nor without a political-cultural context" (p. 139).

unique being, to assimilate completely into his environment without leaving any "fingerprints"; others claimed that the assimilation of a Jew as an individual is not in conflict with the maintenance of a distinct cultural (religious and moral, rather than national or socio-economic) identity in a multi-national society of Europe. Among the latter were those who adopted the national-cultural idea and actually joined those who claimed that the Jews are by nature, i.e., by biological inheritance, different from the Gentiles among whom they lived. Many of these eventually arrived in the Zionist camp. Others, some of whom became convinced proselytes, were leading ideologists of Jew-hatred who took advantage of their intimate acquaintance with Judaism to provide ammunition for their hatred. Evidently, many Jews, including the adherents of the *Wissenschaft des Judentums*, conceived it to be part of their duty, part of the effort to facilitate their acceptance in the society and the culture at large, to explain the phenomenon of the Jews who maintained their separate existence for so many generations. The leading argument was, undoubtedly, that Jews were *not* a biological entity, different from the non-Jews around them, or at least were no longer such a distinct biological entity. For example, Abraham Geiger (1810–1874), one of the leaders of the liberal German Jewry and a founder of the school for the Science of Judaism in Berlin in 1872, supported the universal perception of Judaism and dismissed any national element from his teaching. Samson Raphael Hirsch (1808–1888), the founder of the modern orthodox community in Germany, was no less extreme in his rejection of any national Jewish element.

Chapter 3
Heredity or Environment?

> Groups commonly evaluate their characteristics in comparison with others. Racism falsely claims that there is a scientific basis for arranging groups hierarchically in terms of psychological and cultural characteristics that are immutable and innate. In this way it seeks to make existing differences appear inviolable as a means of permanently maintaining current relations between groups. (UNESCO Statement on race and racial prejudice, Paris, September, 1967)

Similarities and differences between parents and their progeny have always intrigued humans. Already in the fifth chapter of the Book of Genesis it is written: "And Adam lived a hundred and thirty years, and begat a son in his own likeness, and after his image; and called his name Seth" (Genesis 5:3). Primarily, however, the concept of inheritance did not refer to similarity of features between parents and their progeny, but rather to the transfer of property: "And Abram said, Behold, to me thou hast given no seed: and, lo, one born in my house is mine heir" (Genesis, 15: 3); "And Ahab spake unto Naboth, saying, Give me thy vineyard, […]. And Naboth said to Ahab, The Lord forbid it me, that I should give the inheritance of my fathers unto thee" (1 Kings 21: 2–3). Only later was the concept of inheritance applied to the properties of living creatures, man included, to indicate the constancy of the patterns of family relations against the background of the variability of properties, and especially those of populations with sequential generations.

As mentioned in the introductory chapter, the institutionalization of Linnaeus' system of classifying animals and plants expressed the eighteenth century concept of hereditary fixity of species. Linnaeus strove to impose on the richness of nature the order that had been discovered over previous centuries by classifying living beings in a hierarchical system, a series of separate boxes of different sizes that fit one into the other, as designed in the Beginning, according to God's premeditated plan. According to this notion, all individuals of each species may be represented by a prototype of the essence of that species. There might be considerable variability among individuals of a species, but such variation, which for the classifier is merely "noise," is unavoidable, considering the diverse environmental conditions in which individuals of the species dwell. Linnaeus intended his system of classification to demonstrate that the spirit of God is imprinted on each living creature, and the most significant expression of it is hereditary continuity from one generation to the next. One aspect of this conception of variability and similarity in nature as an expression of God's determinism was the predictable certainty of embryonic development. According to this conception of preformation, the essence of the developing embryo is already present in the ovum (others say: in the sperm), and embryogenesis is nothing but the unfolding of that essence.

In contrast to these teachings, which emphasize the constancy of plant and animal life with respect to the transmission of traits from one generation to the next, other scholars accentuated variation – the flow and disparity in nature – with reference to the unfolding of the traits in embryonic development. These scholars wished to stress the effect of environmental or circumstantial conditions on embryonic development, as well as on the variation of species ("beyond the inherited," thus, epigenesis). The character of the embryo is determined by the food it obtains and the environmental circumstances in which it grows. The apparent constancy of the species from one generation to another and of the invariable patterns of development is simply a consequence of the similarity of the conditions under which they grow.

Linnaeus's contemporary, Georges-Louis Leclerc, Comte de Buffon (1707–1788) rejected essences in the classification of living creatures, and was more liberal with respect to the mutability of species. For him, all classifications were man-made constructions rather than immanent reflections of nature. It was his associate and pupil, Jean Baptiste Lamarck, who drew the conclusions and developed a theory of the evolution of species (Lamarck, 1984[1809]).

According to Lamarck, the difference between living and inanimate matter lies in its organization. "Spontaneous" production of life is a normal process, and an inborn inclination toward progress is an essential property of life; consequently life is becoming more varied and increasingly complex. Since living creatures are capable of responding to their environment, and since environments are many and varied, living forms diverge into many varied forms. This is the process of evolution (Bochard and Lohlin 2001; Gissis and Jablonka 2011; Jablonka and Lamb 1995). Thus, Lamarck conceived of life not only as a progressive process, but also as an evolving process that leads to continuous differentiation into many, varied, and increasingly complex forms of life. Frequent and continuous use of an organ or a property enlarges and invigorates it, whereas disuse leads to its degeneration. Acquired changes may with time become inherited. Such notions of the inheritance of acquired properties were common at the time and for many years to come. Lamarck's contribution was to propose a theory of the evolution of species through the acquisition of increasingly adaptive traits that could be hereditarily transmitted to future generations. Lamarck reached the conclusion that an inbuilt tendency toward progress and increasing complexity were enough to secure the continuity of acquired adaptive properties; thus he initiated a path that led to undermining the idea of the species as a constant essence brought into being by the unique act of Creation.

During the nineteenth century, there were intense and often vehement discussions about the role of inheritance relative to the role of external environmental conditions in shaping the intrinsic properties of living beings, their forms, and their behaviors. Charles Darwin's crucial contribution to the notion of the evolution of species was that he liberated evolution from the need for any intrinsic property of life (such as "progress") for living beings to evolve.[1] Evolution is simply the result of the confrontation of divergent organisms with the environments in which they live. As long as there are individuals who have properties that allow them to live and produce progeny (or are capable of developing them), they may survive as a population. If they lack such properties, or if their progeny are not effective competitors in the circumstances of their lives, they will not leave their impact on future generations. Evolution, according to Darwin, is the natural selection of some extant living forms from the great variability available in nature. If the properties selected are inherited, then the structure of the population may change over generations. There is nothing intrinsically progressive in evolution. Thus, in dark breeding sites like caves, it may even be a selective advantage from the perspective of the effective breeding population to be eyeless, although from the perspective of the development of the wider category (say, the family), the loss of eye sight is regressive, considering the long and painful (in hindsight, progressive) path of their ancestors in adopting eyes. According to Darwin, hereditary variability is not merely "noise" about the prototype, but the essential raw material that enables life's continuation and evolution. The hierarchical system that describes how species converge into genera, which then converge into families, and so on, is merely a reflection of history and the human mind that confines life to categories.

The establishment of the Darwinian theory of evolution turned attention to the forces that shaped species (or, for that matter, varieties), essentially to the struggle between the forces that maintain so-called biologically meaningful entities and those that antagonize them. Borrowing from Shakespeare's *The Tempest* (Act IV, Scene, 1),[2] Francis Galton called this struggle "Nature versus Nurture," whereby conservancy is the Nature of living beings and Nurture is the source of variation caused by circumstances. This juxtaposition of contrasts or internal struggle may be conceived as an extension of the efforts of the ancient Greek philosophers, Democritus, Plato, and Aristotle, to expose the lawful order of *nature* in the face of the impact of the immediate constraints (*nurture*) of the unpredictable here and now.

---

[1] Although Darwin also used the term, "progress," occasionally, Stephen Jay Gould claimed that there was no relation between the terminology of Darwin and his predecessors, including Lamarck. Darwin used the term, "progress," to describe events (from a post factum perspective), whereas Lamarck considered progress an essential, given property of life. See e.g., Gould (1996), pp., 19–21. Notwithstanding, some scholars, such as Michael Ruse (1996) and Robert Richards (2013) read Darwin as a progressionist.

[2] Prospero: A devil, a born devil, on whose nature

Nurture can never stick; on whom my pains,
Humanely taken, all, all lost, quite lost:
And as, with age, his body uglier grows,
So his mind cankers: I will plague them all.

During the nineteenth century, researchers in the life sciences, medicine, physiology, and anthropology increasingly applied the methodologies of the physical sciences; namely the analytical-experimental approach that reduced phenomena to the laws of physical causality. They studied phenomena by following one variable at a time, keeping all others neutral or, as far as possible, randomly distributed. By adopting the laws of the physical (and chemical) disciplines, researchers in the life sciences tried to shake off any mystical or metaphysical explanations. In *On the Origin of Species* Darwin took a crucial step by freeing biology from having to resort to attributing irrational concepts, such as "will" or "need" to living beings. However, Darwin did not comprehend the nature of hereditary variability, its origins, and the regularity of its transmission that his contemporary, Gregor Mendel (1822–1884) postulated. In spite of it, Darwin turned his attention to the evolutionary significance of the existing variability among individuals of any given species, without discriminating between inter-specific variability and intra-specific variability. As a matter of fact, much of his argument for the role of inter-specific variability in the evolution of species was based on his and others' observations on intra-specific variability. His deduction of speciation in nature from breeders' artificial production of varieties is an example of his insight that the classification into Linnaean hierarchical categories is artificial and man-made.

In reality, Darwin's deduction of the natural processes of the creation of new species based on the artificial selection of plant and animal varieties was also used in the opposite direction: If species are immanent "real" entities of nature, then varieties and races should also be recognized as natural entities. These deductions had immediate consequences as did two other consequences of Darwinism, which fed ideas that reached far beyond biology and science. Although Darwin was careful to avoid discussing the evolution of man in *On the Origin of Species*, he eventually extended his theory in *The Descent of Man* (1871) and *The Expression of the Emotions in Man and Animals* (1872). And no less significant, Darwin's contemporaries, notably Herbert Spencer, elaborated and extended the consequences of the theory of evolution by natural selection to man's psychological traits and his social structure (sociology) (see, e.g., Paul 2009, Richards 2013).

Evolutionists conceived of social processes as resulting from exposure to slow but consistent Darwinian change of increasing complexity; namely the elimination of "unsuccessful" traits and the establishment of "successful" ones in response to the living conditions of the communities. The progress of the human species is expressed by an increase in social complexity, or inter-subjective relationship, to use Yuval Noah Harari's (2015) language, at least some of which is due to "soft inheritance." In other words, social and other properties are purportedly inherited.

Darwin was, of course, aware of the crucial role of inheritance in the process of evolution, but he had no sound foundation for his theory of inheritance. The "provisional hypothesis of pangenesis" that he eventually proposed was actually "neo-Lamarckian" in character as it allowed for the inheritance of acquired characteristics. His contemporary, Fleeming Jenkin, demonstrated that this hypothesis was inconsistent with that of evolution by natural selection (see Fleeming Jenkin, 1973, pp. 302–350).

Yet, both the burgeoning medical sciences and anthropology increasingly accepted the notions of Darwinian evolution, although in the framework of the determinism of the laws of nature. As I will argue below, people accepted that "social deviations," whether crime, poverty, or disease (or for that matter, genius), were inescapable facts of life. The only way to change society was by way of natural selection. This is how the evolution of species occurred, and this is the way it must act in human society. Little or no thought was given to the possibility that a biologically acquired (Darwinian) advantage of the human species might have been an insight that allowed humans supervening elaborations in ethical notions.

Although many biologists were inclined to accept the inheritance of acquired characteristics, for most experimental biologists this conundrum ended around the 1880s upon acceptance of the argument of Francis Galton in England and later that of August Weismann (1834–1914) in Germany that acquired traits are not inherited. Weismann further elaborated a theory that distinguished between germ plasm, which is continuous from one generation to the next, and soma, which in animals is created by, but does not contribute to the germ plasm (Weismann 1893. See also Buss 1987 and Churchill 2015). Accordingly acquired properties of somatic cells in animals have no way to directly affect inheritance, which is restricted to events that might occur in the germ cell-line. Spontaneous hereditary changes that occurred in the germ line accounted for evolution by natural selection. Johann Gregor Mendel's 1866 theory of inheritance based on the notions of reductionist particulate principles, upheld the distinction between germ plasm and soma, or nature and nurture. As formulated in 1909 by Wilhelm Johannsen (1857–1927), the genotype was the hereditary potential and the phenotype its somatic expression (Johannsen 1909). The achievement of reductionist genetics was highlighted by the juxtaposition of the gene was its material and functional essence, like the atoms of physics and the elements of chemistry (Muller 1922). Eventually, R. A. Fisher (1890–1962) settled the apparent inconsistency between continuous, phenotypic Darwinian evolution and particulate Mendelian genetics. Using statistical methods, he distinguished the inputs of nature and those of nurture, consistently reducing Darwinian evolution to particulate genetics (see Fisher 1930).

The genetic reductive determinism of the first decades of the twentieth century accorded perfectly with the conceptions of the upper classes of society and those in nationalist circles who were convinced that they were the true representatives of a long and painful process of evolution by natural selection. They found in the theory of evolution justification for rejecting the "other." This "other" took on different faces in different places. In England, the "others" were usually those of the lower social classes; in the Unites States, they were usually the immigrants from the Far East and from eastern and southern Europe (not to mention the Blacks from Africa!); and in continental Europe, these were usually the Jews (and the gypsies). Evidently, besides ethnic minorities, the sick and disabled were also considered "others," who had to be isolated and whose reproduction had to be regulated in order to allow natural selection to maintain a sane and socially balanced human species. The failure of natural selection was the consequence of improvements in hygiene, medicine, and welfare.

These issues gained new significance in the 1940s, when the criteria for defining species became gene frequencies rather than morphological and physiological criteria. Speciation became a process of population genetics, and species were defined as genetic populations whose members are capable of (regularly) producing fertile progeny with each other, but not with members of other species.

The increasing awareness of the role of the individual in society changed the perspective of science. Uncovering biological variability undermined social and racial discrimination (see Barkan 1992). This trend took a surprising turn with the discovery that, at the molecular level, the variability of populations, humans included, was orders of magnitude higher than had been previously observed (Lewontin and Hubby 1966; Harris 1966). In 1972, Lewontin and Hubby's analysis showed that less than 10 per cent of human population variability is *between* populations, whereas more than 90 per cent is *within* populations, which provided clinching evidence that, at the biological level, there was no support for social or political segregation of humans into races (Lewontin 1972). Yet, during the past twenty years, the use of the category of race in the social sciences has increased significantly (Gissis 2008).

Improved methods of molecular genomic analysis and of DNA screening of the whole genome in ever more species has generated a growing industry that utilizes these data to reconstruct evolutionary history, or the phylogenesis of species, as well as other associations (including individual families), on assumptions of common roots and increasingly divergent trees of descent. Somewhat unexpectedly, however, this trend seems to be reviving the notion of biological races. A population's variability should not be reduced to isolated genes; such a reduction disregards the fact that isolated genes are artifacts of our analyses. Once we consider the aggregation effect, a clear group structure emerges (See, e.g., Sesardic 2010a and b. See also Wade 2014). It must, however, be kept in mind that another development in the study of the phylogenesis of species is the increasing awareness of the role played by lateral gene exchange (i.e., between individuals who are not related as branches of one – diverging – phylogeny, but rather by gene exchange between phylogenies, whether by "intermarriage" or, like in bacteria, by asexual means of gene transfer). Thus, geographic or cultural and social contacts may imitate historical, ancestral phylogenetic trees; sharing the "same genes" may be due to lateral transfer rather than to vertical transmission. As we shall see, such lateral gene transfer undoubtedly played a role in the history of the Jews. This should be considered when we make efforts to find support for the linear historic story in the biological data, rather than follow the old convention of evaluating the reticulate biological data in light of our historical contentions.

3.1 Heredity or Society?

Almost precisely a century after the publication of Lavater's *Physiognomic Fragments for the Advancement of Knowledge and Love of Our Fellow Man* – which eventually contributed mainly to fostering discrimination among our fellow man – two books were published on both sides of the Atlantic that enormously affected scientists and non-scientists alike. In the name of social liberalism, these two books spread the idea of biological fixity and, thus, contributed much to a somber atmosphere of despair and worry about the future, which the progress of science and technology had promised to enhance. One book was *The Jukes: A Study of Crime, Pauperism and Heredity* by R. L. Dugdale (1841–1883), published in New York in 1877. The other was Cesare Lombroso's (1835–1909) book, *L'uomo delinquente* (The Criminal Man), published in Rome in 1876. Dugdale, a merchant, who was a member of the Committee for Social Correction in New York State, was deeply affected by his visits to state prisons, and especially by the fact that many prisoners were related to one another. Consequently, he conducted a research project to follow 709 members of the "Jukes" family – a fictitious name – over several generations. One hundred and forty of them were prosecuted for criminal offenses; 180 were residents of almshouses or were persons who needed other welfare assistance and much more. Dugdale calculated that over the years, the "Jukes" tribe cost New York State taxpayers $1,308,000 (according to the value of the dollar at that time). Trying to appeal to the mentality of his countrymen, he argued that this heavy financial burden was due to the living conditions of these miserable persons, and that it would be more profitable to improve their living conditions by applying the achievements of science, medicine, and technology, rather than bear the damages of their crimes and the burden of their incarceration in prisons and almshouses. However, Dugdale's arguments were utterly misconstrued by researchers of human behavior, and were interpreted as evidence for the hereditary fixity of human characters and the insufficiency of environmental (i.e., social) conditions to change this fate. The misinterpretation of Dugdale's observations was considered for many years evidence for human cultural-behavioral determinism (see e.g., Kevles 1985, p. 71; Paul 1995, pp. 44–54).

Cesare Lombroso came from the liberal, medical, scientific community, and in the beginning supported the extension of socialized medicine. However, somewhat later, he became an avid campaigner for the idea that behavioral properties were hereditarily fixed. He claimed that it was possible to discern the character of compulsive criminals by their appearance. This was, actually, nothing but an extension of Lavater's old idea. Although influenced by Darwin's theory, he developed the notion that criminality, even though not "natural," was hereditary: Human beings went through a process of evolution, at every step of that evolution their corporal as well as their mental characteristics became fixed. The normal human being, a member of the culture in appearance and in socio-moral behavior, is the product of evolution. The criminal is nothing but a person in whom an "evolutionary regression" occurred at a stage at which most people proceeded forwards. The contemporary sick and criminal reveal signs that still exist in today's more primitive tribes, and also in animals from which humans evolved. Lombroso's theory, which had explicit socio-political sources (the struggle between northern Italy's European bourgeoisie and southern Italy's poor Mediterraneans), became the cornerstone of new social theories in the disciplines of criminology and psychiatry, as well as in anthropology. His theory profoundly influenced social and political reformers at the end of the nineteenth century.

3.2 Racism

> Race classification [...] is the most difficult of the tasks of the anthropologist. [...] but in this instance the inability of the anthropologists to clarify opinion by a generally accepted definition of 'race', and their failure to arrive at an agreement as to its implications have had serious repercussions in the outside world, as is now familiar to everyone. Extreme and biased views have been accepted as endorsed by science because no voice is raised to contradict them with full and unquestioned authority. [...] It is still more unfortunate that, so far from the disparity between the hesitancy of science and the dogmatic certainty of popular opinion being a matter of merely academic interest, up to the present the popular judgment has prevailed in practical affairs. (Editorial, April 18, 1936. *Nature, 137*[3468], 635-637.)

Population geneticists define species as more or less closed entities, whose members may mate with each other and produce fertile progeny. But races are, by definition, entities *within* species. What then empirically defines "race"? Lawrence D. Bobo notes that "like most social scientists, I have grown comfortable in the thought that we study race without actually believing in race" (Bobo 2008, p. ix). As has already been noted, the concept of race is an ancient one that served to designate (and identify) groups of people according to some external criterion such as skin color, hair texture and color, domicile, language, etc. and to attribute other properties, either physical, mental, or social, to them. Such designation implicitly indicated a belief in some common roots. The Israeli educator, Zvi Rudi, suggested that the term race (*Rasse* in German), was introduced by the Arabs, and entered European languages via Spain: "In Arabic, *râs* means "origin" (Rudi 1927. See also Boyd 1950, p. 186). Thus, the term could signify the "human race," the "black race," or the "Jewish race." However, in the, eighteenth century "the discourse on society collapsed into the discourse on nature" (Gissis 2011).

Notwithstanding, over the next two centuries in the "discourse on nature," race could not be defined without falling back on sociological and historical terms. Still, with the increasing role of science in society, attempts to impose a biological concept of race on the social notion have been as popular as those to impose biological notions on politics and society. This imposition of meaning from one discipline on another, such as the attempts of Nordmann, Stoecker, Marr, and other anti-Semites (see Chap. 2), is racism. Racism is the claim that socio-geographic variables reflect biological classification. Of course, people on different continents may look different: Mating is not random – even today in the era of fashionable jet-flights. Partners have always been determined primarily by geo-sociological variables. This is not to say that in practice such variables may not be helpful guides and indicators of genetic differences. It would be more reasonable to suspect a person of Ashkenazi origins from Lithuania, where the hereditary Tay-Sachs disease is at relatively high frequency of being a carrier of Tay-Sachs disease, than a member of a Sephardic community, born in Iraq, where the disease is practically unknown.

Such issues become increasingly problematic with developments in pharmacogenomics. It is not (yet?) practical to adapt a pharmaceutical product to each individual; therefore, classification of people by proxy markers, whether skin color or cultural-religious background, seems to be indispensable. Whether "Jewishness" is a racist indication depends on circumstances. Concerted attempts to find common genetic roots of Jewish communities and link these common roots to other Mediterranean populations may, therefore, be conceived as an attempt to find evidence in biology for historical roots of socio-political conceptions. At the same time such classification can serve as a foundation for a possible data base that lumps people together for the practical application of the achievements of the genomic age.

Slightly modifying the quotation from the Foreword to Koening et al. (2008) *Revisiting Race in a Genomic Age*:

> Advances in biological science, for the most part, present something of a double-edged sword. First, most scholars recognize that the very concept of race "reflects a marriage of the social and the biological," rather than something obviously given in the nature of things. Second, there is a general acceptance of the fact that the great bulk of genetic variation occurs within conventionally recognized racial or continental groups, not between them. Third, "variation is continuous and discordant with race."

> On the other hand, there is much unreflective racialized thinking still going on, including among those working on mapping the human genome. Three trends should concern us all: 1. Conventional racial categories still appear to influence the work and thinking of many who are developing, for example, the DNA repositories used by researchers who continue to be catalogued by racial and ethnic identifiers. 2. There is plenty of evidence that clinical practitioners rely on lay or commonsense race categories and racial cues in their assumptions about needs and risks for certain conditions and health outcomes. 3. The pharmaceutical industry sees real market potential in developing lines of race-based drugs. (Bobo 2008, pp. x-xi. Modified).

3.3 Eugenics

Francis Galton coined the phrase 'Nature versus Nurture,' to formulate the conclusions that he drew from Darwin's theory of evolution. According to him, modern life jeopardized the future of the human race by relaxing selection against the 'unfit.' The conclusions that Galton drew from Darwin's theory with respect to the changes that occur in human populations as consequences of the technical and social changes of his time were no less threatening than those described by Nordau (*vide infra*). Galton was convinced of the determinism of human traits – primarily, but not limited, with respect to mental properties. In his book, *Hereditary Genius,* published in 1869, he noticed the noble characteristics of Supreme Court judges, scientists, clergymen, politicians, as well as of oarsmen, that had been transmitted from one generation to the next in British society. These admirable properties were allegedly the outcome of natural selection (or more appropriately, artificial selection) and, according to Galton, were the rational behind the classes of British society. Galton believed that in his own society, "indiscriminate charity" allowed the least capable to have many children, whereas the most able married late and had fewer progeny. Given the complexity of modern society, reproductive restraint by the intellectual elite was a particularly urgent problem. Galton's principal concern was to increase the proportion of very high-functioning individuals in the population rather than to prevent biological delinquency. He hoped it would be addressed through education and policies, such as subsidizing the marriages and procreation of the especially gifted.

At the end of the nineteenth century and, even more so, in the first decades of the twentieth century, the eugenic movement gained popularity among social scientists and natural scientists alike, and especially among politicians and various social reformers. There was concern about the consequences of relaxing the effect of natural selection in human populations – or, to be precise, in the European cultural sphere. Quite a few scientists and social reformers hoped to promote Galton's notions in their innocent sense. Many of the liberals and social reformers were worried that social reforms would relax natural selection in human societies. They enthusiastically supported the introduction of modern biological methods to select against so-called "undesired" traits and in favor of "desired" traits. But it soon turned out that these concerns were exploited by those who harbored prejudices against different nations and different classes of society, and eugenic arguments were put forward to uphold their prejudices. Not only discriminatory immigration laws, like those of the 1920s in the United States, which served to maintain racial discrimination and to protect the labor market during economic crises (see, e.g., Reuter 2006), but also the laws of Nazi Germany against Jews and gypsies, were based on biological-eugenic claims. Although not racial, sterilization laws that targeted mentally defective patients that were included in the social security policy of Scandinavian countries were also based on eugenic arguments (Broberg and Roll-Hansen 1996; Roll-Hansen 2000). And as I will show further on, many of the supporters of the Zionist project to settle in Palestine conceived of it as the return of the Jews to conditions of "normal" natural selection, replacing the degenerating life in the ghettos and in big towns; in other words, the implementation of an effective rate eugenic program.

***

Shortly before his death in, 1911, Galton was interviewed for *The Jewish Chronicle* (Galton 1910). Asked how he would define eugenics he answered: "It's the study of the conditions under human control which improve or impair the inborn characteristics of the race."

- Do you think that the hygienic regulations of the Mosaic Code have contributed to the fitness of the Jewish race?
- I am willing to believe that their indirect influence has been great.
- Are they more responsible, in your opinion, than the vicissitudes which the Jewish people has had to go through?
- Both have played a part in producing the fitness of the Jewish race. The wish of the Jewish woman to be married and have children is an important factor. It is one part of eugenics to encourage the idea of parental responsibility the other part is to see that the children born are well born. It is a praiseworthy feature of the Jewish religion that, as a religion, it enjoins the multiplication of the human species. But it is still more important to determine that children shall be born from the fit and not the unfit.

The reporter noticed that Galton developed "the modern but somewhat startling view that the environment had only little effect on human development."

- What effect do you think persecution has had on the Jewish race?
- So far as persecution weeds out those who are unfit so far it tends to evolve a race suited to meet hard conditions. [...]

Fig. 3.1 The title page of Theodor Herzl's *Altneuland* (1902)

- Is it not rather immoral to look with satisfaction to persecution as an aid to race culture?
- It is not immoral but unmoral – it has nothing to do with morals. Persecution does not always produce good results. [...]

Thus, although the notion of eugenics is as old as Plato's *Republic*, it was Francis Galton who founded the modern eugenic movement and gave it its name. It became a movement for combating the biological degeneration of the human species, 35 years before Theodor Herzl founded the Zionist movement, which sought refuge for the Jewish people from the persecution of the Diaspora; both were utopian movements that sprang from similar intellectual and emotional background. Eugenics aspired to save the human species by confronting its biological realities. Zionism aspired to save the Jewish people, who were suffering physical (though not mental?) degeneration by exile and persecutions, by forcing its members to confront its biological essence. Zionism, just like eugenics, was influenced by Darwinism. Both movements tried to establish order in human communities based on notions of the struggle for survival as understood at the time in relation to the evolution of species.

Galton and Herzl were visionaries who struggled to realize their utopian dreams. Herzl presented his vision for Palestine, the Land of Israel, in his book, *Altneuland*, published in 1902 with the epigraph, "If you wish, this is no fairy tale" (Fig. 3.1). Sections of Galton's *Kantsaywhere* were published many years after his death. It is a tragic irony that both utopias were carried out in some sense – one with the founding of the State of Israel and the other in the racial purge perpetrated by Nazi Germany.

Chapter 4
The Response: Zionism

> The use of the term "nation" among the founders of Jewish secular nationalism [...] was almost always in the ethnic-biological context, and not the cultural-territorial.
> [...] Zionism almost always interchanged in its historic imagination the principle of the religious Jew and the ethnic-Jewish principle. (Sand 1999, p. 343)

Zionism is a political movement for the return of the Jews to their homeland. It was formally established on August 29–31, 1897, at the First Zionist Congress in Basel, Switzerland. To most Western European Jews this was an odd idea: Although identity crises were quite common in Jewish history, this alliance that had discarded many of the traditional formal religious symbols in an attempt to integrate into the non-Jewish world, now blatantly declared its Jewish national identity. Emancipation had presumably opened the gates for Jews to enter the world at large; however, they soon realized that even though they might have dropped all their Jewish identification marks to others, they were still Jews. As Jews in Western Europe learned, liberation from the restrictions of the ghetto did not lead automatically to integration into society. Even when Judaism as the religion of a minority community was grudgingly accepted by non-Jews, very few Jews were successfully assimilated. Thus, many Jews who would have preferred assimilation faced identity crises that they had never before experienced, and Zionism seemed to offer a gate to the world at large according to its rules. Instead of denying their Jewish identity, Zionists claimed that prevalent social customs should recognize their Jewish uniqueness. Until then, Judaism provided a common past; now it offered a common future (Katz 1986, pp. 131–132). Thus, in response to physical racial associations, political Zionism was coupled with the idea of the biological identity of the Jew. Discussions about the right of the Jews to their homeland in Palestine were actually discussions about their biological essence, which went above and beyond their religious, cultural, or even national arguments.

Obviously, the founding of political Zionism was more than a reaction to anti-Semitic racism. As noted with regard to the issue of the nature of the Jewish people [*Wesen des Judentums*] as a race, Jews who identified themselves as members of a nation positioned themselves on the other side of the barrier opposite liberal Jews.[1] Whereas most researchers of the "science of the Jews" directed their efforts to historical and cultural evidence and insisted on independence from the natural sciences, those who aspired to independence asked themselves what was needed for a community to become a nation. In Western Europe, it was the state that provided the cohesion needed for the establishment of a nation, whereas in Central and Eastern Europe, it was a common language and a common culture that defined nationhood. A Jewish national organization presumably had enough tradition and history to build on, but it was difficult to separate tradition and history from religious rituals. And with respect to language, here too there was a lack of consensus – Yiddish, Hebrew, or local languages? The conclusion was, therefore, that the essence of the Jews-as-a-nation lay in their *biological essence*, in their being a race in the sense of the life science, biology. Nathan Birnbaum (1864–1937), one of the early supporters of Theodor Herzl who invented the term, "Zionism," wrote[2]:

> The insights and mental dispositions of a nation may not be explained otherwise than by the natural sciences. "Race is everything," said our great race-member, Lord Beaconsfield [Benjamin Disraeli]. In the uniqueness of race, the uniqueness of the nation is enfolded. National variation is founded on racial differences. It is by virtue of race that the German or the Slav feels different from the Jew. In this antagonism one must look for the explanation that the Germans produced in the Song of the *Nibelungen*, and the Jews discovered in the Bible. It would have been absolutely wrong to state that the *Nibelungen* saga had produced the characteristic German spirit, and the Bible the Jewish spirit. This would be nothing short of a ridiculous exchange of cause and effect. (Birnbaum, cited in Doron 1980, p. 401. TRF)

Race for Birnbaum was a positive fact that did not need justification. Such an insight does not uphold racism, but rather refutes the possibility of assimilation. "Races and nations exist […] because nature develops and goes on to generate various human races, just as it generates different seasons and different climates" (Doron 1980, p. 403).

Those Jewish intellectuals who came to Zionist-political conclusions following the Enlightenment were, of course, not the first Zionists. The buds of Zionism were already evident in societies like "*Hibat Zion*" [Fondness for Zion] and "*Shivat Zion*" [Return to Zion] that erupted in the large Jewish centers of Eastern Europe and, no less intensely, among the Jews of North Africa and Near East Asia. However, living conditions in those countries, both in terms of internal organization and outside pressures, did not allow the establishment of a movement founded on the concepts

---

[1] "The position of national Judaism prior to 1914 in relation to the issue of race differed fundamentally from that of liberal Judaism. […] Whereas the liberal *Wissenschaft des Judentums* oriented itself according to the humanity-sciences, the national Jewish *Auto-emancipation* wished to establish its ideology on the natural sciences." In the framework of German culture, the Zionists were the main Jewish group that insisted repeatedly on essentialist positions to support its arguments (Doron 1980, p. 400). Essentialist interpretations of identity were not limited to biologization. "Essentialism assumes that there are defined characteristics that exclusively hold for all members of the group or class – which in our case – hold for all Jews" (Weiss 2002, p. 145).

[2] Birnbaum later changed his ideas. He went through several changes of faith until finally he became ultra-orthodox. As the secretary of "*Agudat Yisrael*," he became an extreme anti-Zionist.

of the new world that was formed in western Europe. It appears that an additional contemporary dimension of Western intellectual, social thought was required for Zionism to become a political movement.

Indeed, the changes that occurred in Western Europe brought about enormous economic and social changes thanks to the achievements of science and technology. However, these changes also carried deep political and, especially, economic and social crises. On the one hand, a kind of nostalgia for the ancient "cultural" values that concealed images of national uniqueness flourished throughout Europe, and liberation movements claimed immanent rights of national entities for political expression. On the other hand, the great social changes that accompanied the emigration of citizens to the big cities brought with them class solidarity and meta-national identification. At the same time, the opening of new markets outside of Europe demanded a very pragmatic approach to the new values of the West in order to carry out a policy of controlling the expansion of the colonies and the "natives" in accordance with the colonial needs of the West. Such was the socio-political and romantic-intellectual background of political Zionism that sought to find its poetical path (for an extensive review, see Mosse 1970).

To the Jews of Eastern Europe and to the Jews of North Africa, the Orient, and the Near East, the idea of a common (national) future appeared less revolutionary than it seemed to many Western European Jews at end of the nineteenth century; though even for the former, as a rule, the notion was not framed in formal political terms or in a plan of action, but rather as a wish for the day of the Coming of the Messiah. Thus, societies for settlement in the Land of Israel, like that of *Hibat Zion,* gained increasing support in Eastern Europe even before the Zionist Congress, and especially after the pogroms of 1870. Jewish intellectuals, whether insisting on the idea of national resurrection or not, rejected outright any ideas of assimilation. Still, the great majority of the Jewish public, including the revolutionary elements, dared not think of turning their backs on the non-Jewish society in which they lived. Most of those who towards the end of the nineteenth century preached for the revitalization of Jewish life, now conceived of their nationalist sentiment in secular and literary terms rather than in religious-halachic terms (Vital 1987).[3] Contrary to the Jews in Western Europe and in spite of the penetration of the ideas of Enlightenment, Eastern European Jewry was still inclined to turn inwards, thinking of "them" versus "us" (see Vital 1982, pp. 163ff.).

Many Jews in Eastern Europe sought a practical solution, i.e., integrating as a distinct entity into the local community. Thus, in 1897 the *Bund*, the Association of Jewish Workers in Lithuania, Poland, and Russia was established. This was the peak of the radical social democracy movement of Eastern European Jewry, that was founded one year before the Social Democratic Party of Russia – the predecessor of the Communist Party of Russia. The propagation of the *Bund* indicated the progress of the secularization processes of Jewish life in Eastern Europe, and the hope of the

---

[3] Halacha – Jewish law and jurisprudence, based on the Talmud and later rabbinic law. Historically in the Diaspora, Halacha served many Jewish communities as an enforceable avenue of civil and religious law.

Jews that when emancipation finally reached these countries, they would be able to occupy their proper place in the universal process without the need to regress into the "bosom of national romanticism." Therefore, Jews demanded emancipation from the Gentiles among whom they were living without taking part in non-Jewish life. Their demand for emancipation was to live as a separate congregation with equal rights. However, they were soon confronted by reality and awakened from the dream of emancipation, at least as far as the regimes in Eastern Europe were concerned.

As a matter of fact, the formal beginning of the Jewish national awakening occurred, of all places, within the circle engaged in the struggle for a meta-national class struggle in Western Europe. Moses Hess, who published his book *Rom und Jerusalem* in 1862, was a member of Karl Marx's inner circle (see Chap. 1). In his book, for the first time, a Jew made the explicit statement that Jews were a race, which in spite of climatic influences accommodated to all situations and maintained its cohesion. It is doubtful that Hess really understood the full meaning of this biological statement written only three years after Darwin's publication of *On the Origin of Species*.[4] Hess was influenced by the ideas of Giuseppe Mazzini (1805–1872), who tried to integrate the call for Italian nationalism with the idea of universalism: "By my very being a member of a nation, I am also a member of the human race." Thus, Hess concluded that the revival of the fatherland in Palestine, rather than assimilation among the Gentiles, was the proper answer to the plight of the Jews. Shlomo Avineri considered Hess one of the forefathers of Zionism because he discerned that the failure of the Jewish reform movement was that it presented the Jewish problem in terms of religion (Avineri 1986). By doing so, it distorted the historical essence of Judaism as an entity of common material and biological roots; namely, as a *Volk* in the philosophical terms of the German philosophers of that era, primarily Herder and Hegel.

In Eastern Europe, Moshe Leib Lilienblum (1843–1910) described the situation in Russia after the 1870 pogroms as a reversion to "prevailing conditions in the Middle Ages," with one significant difference: By now, it was no longer religion but rather nationality and race that became the Jewish trademark. The present-day Jew is not a Teutonic of the German nation, of the Magyar of the Hungarian nation, nor of the Slav, but rather a nameless being. There is only one way to escape the predicament – stop being strangers. Those who are impatient and narrow-minded suggest assimilation, complete absorption among non-Jews. This solution, according to Lilienblum, is not only unjustified and despicable in principle, but also impossible

---

[4] According to Doron (1980, pp. 392–394), the first article in a liberal Jewish publication that addressed the issue of race was that of Ludwig Philippsohn, published in 1865 at the time of Prussian liberalism and prior to the anti-Semitic wave of the 1870s. Philippsohn attacked the contemporary concept of "race." He rejected the claim of the natural sciences of the existence of race as well as the tendency to accept laws of the natural sciences in the social sciences and in historic arguments. History was for him a campaign for human liberty. There is no mention of "Jews" in the paper. Twelve years later, Philippsohn attacked Heinrich Graetz, who instead of outright rejecting Aryan racial claims, joined them by ascribing to the Semites "the real civilization" [*die wahre Zivilisation*].

and impractical. Jews must leave Europe, he insisted. It is possible that the exodus of Jews from Europe may last a hundred years, but we must start. Where to? Not to America. There too we will eventually be strangers. Rather to Palestine, "For we have not lost our historical rights to it with the loss of governing it. Just as the Balkan nations, when they lost governance, did not lose their rights in their countries." But how will this be realized? (Vital 1982).

A crucial step towards the realization of this idea was a change of conceptions. There is no way out except by self-reliance, as implored by Leon Pinsker (1821–1891) in his pamphlet *Autoemancipation: Mahnruf an seine Stammgenossen, von einem russischen Juden* [Auto-Emancipation: A Warning to His Brethren from a Russian Jew], which he published anonymously in 1882. This pamphlet presented the principal ideas that Herzl brought up 13 years later in his *Der Judenstaat*. Although Pinsker wrote his pamphlet while he was living in the West, his perspective was that of an Eastern European Jew. This gave Pinsker the authority to be explicit about the need for auto-emancipation, without falling prey to the auto-suggestion of Western European Jews who hoped for a change in the attitude of the surrounding non-Jewish society, a hope that many still clutched. Pinsker diagnosed the root of the problem as being the fact that the Jews were considered a strange element wherever they lived, an element that could not assimilate completely. Remaining unacceptable and unwelcome, they aroused fear and hatred that led to the deprivation of their rights and loss of status in society. The great achievements of the Emancipation in many places could not be denied, although it was only a legal emancipation, not a social one. The Jews remained strangers who, lacking a homeland, could not establish their own leadership; and in order to acquire a homeland, they had to take their fate into their own hands.

Still, even auto-emancipation – taking responsibility for freeing the Jews by their own hands – required a socio-cultural, rather than merely a political framework. In 1889, a couple of years after Pinsker's *Autoemancipation*, the publicist Ahad Ha'am (Asher Z. Ginsberg, 1859–1927) claimed in his article, "*Lo Zu Haderech*" [This ain't the way!], that the Lovers of Zion had gone astray: In their eagerness to save the lives of persecuted individuals, they missed the real, wider national task that they had taken upon themselves. Why didn't the awesome idea of national resurrection that inspired other nations take hold among the Jews? Why did the idea take a step backward since its initiation? Ahad Ha'am categorically rejected the claims that Jews ceased to be a nation like other nations and that "we have no right to fasten a national sentiment to Israel," and relating to the argument that only religious relationships connected us to the country: "Anyone who says that no national sentiment can be found in the sons of Shem, or more specifically in the sons of Israel – the kind of sentiments that according to one of the greatest contemporary sages have been observed even in animals – must bring forward a more convincing proof" (Ahad Ha'am 1930, vol., 1, pp. 10–13).[5] Ahad Ha'am believed that "national sentiment" is

---

[5] Ahad Ha'am refers to the physiologist, Emil Du Bois-Reymond (1818–1896), who was involved in the electro-physiological studies of nerves and muscles. He was one of the major opponents to vitalism that was prevalent at the time in German science, and wrote extensively on the relation between body and soul.

based on an inherited trait of living beings. Nevertheless, he doubted the so-called scientific claims about the nature of the racial properties of the Jews how they were acquired and the means by which they could be changed. He noted the discussions among scientists relating to the Jews' traits, whether acquired or inherited.

The following quotation reflects the attitude of the socio-political authors of that period toward the sciences in general, and to determinist biological notionss in particular:

> As a rule, the ethnological investigations concerning the properties of every nation are like mounts leaning on a single hair that provides nothing to be relied upon. [...] Who is the wise man that will draw the line between properties ingrained in us since the beginning of the history of our nation and those engendered by our living conditions in Exile? Who is the one who will enumerate one by one the varieties of suffering of one generation or another, in one country or another, and predict what other properties may or may not change with the change of our living conditions? Thus, one of my critics claims that the properties of the Israeli nation exist forever and everywhere, with no change whatsoever, contrary to one of the most famous wise men of our time who chose us from among all the nations to prove the truth of his belief that national properties depend on living conditions and on social order, rather than on ancestral heritage? By studying our migrations he proved how our properties vary in different countries and change over time, all according to our living conditions and the mentality of the nation mong which we live. (Ahad Ha'am 1930, vol. 1, pp. 11–12. TRF).

According to Ahad Ha'am, the real problem with the efforts of Jewish settlement in Palestine was not the lack of an inherited property of national identification, but rather the fact that the leaders, driven by the concept of Jewish national resurrection, "in their hurry to obtain great achievements in no time, neglected the long tedious road of natural unfolding, and artificially forced on the world of action a supple and young idea, still immature, before its vigor had unfolded properly." The Lovers of Zion erred in giving priority to attending to the needs of individuals who were hungry and needed immediate refuge. For this purpose, they described the situation in Eretz-Israel in misleading terms and gave promises that they were unable to fulfill. Admittedly, under the burden of the living conditions in the Diaspora, "national aspirations were confused: National devotion was no longer unconditional, [...]. Everyone's aspiration is nowadays his private individual success." Nevertheless, private interests alone would not maintain an orderly society. Consequently, "there is no doubt that we should have devoted our primary activity to the *revival of hearts*, to increasing love for public life" (Ahad Ha'am 1930, vol. 1, p. 6, emphasis in original). Put differently, nothing is wrong with our inborn properties of aspiring to become a nation like other nations; notwithstanding, what was needed first was the revival of the national consciousness of the Jews, which had been weakened by persecutions and years of wandering in the Diaspora (Vital 1982, pp. 28 ff.). To sum up, Ahad Ha'am called for an awakening from the romanticism of the Enlightenment and from the hope that Jews might be integrated into the non-Jewish world, whether through assimilation or as a recognized minority that considered itself an integral part of the world at large. It was the disillusionment with such a framework that prompted attempts to give the ideas of national revival practical validity in Europe at the end of the nineteenth century. But the publication of the *Judenstaat*, Herzl's utopia of national revival in Palestine,[6] and the assembly of the Zionist Congress with its strict, pretentious formalities, were at most, steps to "revive the hearts" (Vital 1982, pp. 358–359). This pattern was alien to that preached by Ahad Ha'am, and it is no wonder that he soon became one of Herzl's main opponents.

The intellectual internal conflicts of many of his contemporaries were factored into Ahad Ha'am's conception, which was first published in 1889. It combines respect for the achievements of science with suspicion of its social impact, and a profound conviction in the moral priority of the good of all humans with the need to sacrifice the good of the individual for that of the community. It takes into account the claim of scientists that national sentiment is inherent in all human beings, yet it relies on discussions among scientists on the meaning of inherited and acquired properties in an attempt to stress the priority of national needs over the needs of the individual.

The biological roots of human communities in general, and those of the Jews in particular, were major components of the socio-political thinking of the time and Zionism was no exception. There were, however, major differences in the educational and socio-political background of the persons involved, as well as in the perceptions of different Zionist thinkers of Zionism. The following is an examination of the extent to which Zionist leaders of the era were influenced, even absorbed, by a biological conception of the place of the Jews in the order of the world.

4.1 Theodor Herzl

Theodor Herzl was well equipped to found political Zionism since he was profoundly integrated in the cultural and political mood of neo-romantic *fin de siècle* Central Europe. This was a Europe that vacillated between the fears and despair of the *Bildung* of the post-Enlightenment era and the awe and respect of *Wissen* of the era of post-industrialization.[7] In this framework, biological determinism played a crucial role. Already as a young student, when Herzl learned in 1882 of the imposed isolation of Jews in the Middle Ages, he drew conclusions about the biological properties of the Jews: Isolation "prevented the physiological improvement of the race by mating with others"; therefore, Jews have "a physical and mental physiognomy" that is different and unique as they do not intermarry with members of other nations. "Mating among members of Western race with those called Eastern on the basis of a common religion-state is an important and desirable solution" (Ragins 1980).[8] While stationed in Paris in 1893 as a correspondent for the Viennese *Neue*

---

[6] Adi Zur (http://www.e-mago.co.il/e-magazine/altneuland.html) claims that Herzl's *Altneuland* is a utopian rewriting of Herzl's ideology in *Der Judenstaat*.

[7] *Bildung* may be conceived as education; *Wissen* as knowledge.

[8] I thank Michael Hyman for calling my attention to a conversation between Herzl and the Baron Hirsch in the spring of 1895, and a letter in which Herzl noted the need "to improve the race" and considered "the degeneration of our race" as the cause for the Jews' lack of political courage.

Freie Presse, he admitted in conversations with a colleague the substance of the anti-Semitic accusation that linked the Jews with money, and defended them as the victims of a long historic process. "It is not our fault, not the fault of Jews that we find ourselves forced into the role of alien bodies in the midst of various nations. The ghetto, which was not of our making, instilled into us certain anti-social qualities." According to him, modern anti-Semitism was the consequence of Emancipation, which had been "an error of doctrinaire libertarian thought, the illusion that men are made free when their rights are guaranteed on paper. We were liberated from the ghetto, but we remained, we still remain, ghetto Jews." Herzl went further and, in the spirit of the period, took a Darwinian biologist's position: "Anti-Semitism has been helpful in development of the Jewish character. It is the education of a group by hardening [*Erzogen wird man nur durch Härten*], and it will perhaps lead to the absorption of the group. [...] The Jews will adapt themselves through a process of Darwinian mimicry. They are like seals, [mammals] that have been thrown back into the water by an accident of nature. If they return to dry land and manage to stay there for a few generations their fins will change back into legs" (Bein 1934 [1974], p. 173. See also Gilman 2010, pp. 10–11). Interestingly, at that stage, he definitely rejected the Zionist solution. Discussing the drama *Femme de Claude* by Dumas the younger, he said: "The good Jew Daniel wants to rediscover the homeland of his race and gather his scattered brothers into it. But a man like Daniel would surely know that the historic homeland of the Jews no longer has any value for them" (Bein 1962, pp. 99–101). It appears that in the mind of the founder of the Jewish political Zionist movement, assumptions about the biological nature of the Jews combined humanist notions on the one hand, and nationality on the other. Thus, the foundations of Herzl's movement are similar in kind to the attempts made in the first decades of the twentieth century to establish the fate of Europe on a balance of given truths. Although there was an apparently uncompromising insistence on the rights of individuals, there was awareness of the reality of the community, the family, as well as the *Volk*. In the first part of Herzl's utopia, *Altneuland*, the story of the establishment of a Jewish state in Palestine is intentionally called "a young and desperate youngster." Indeed, Zionism with its ups and downs was always a boat navigating between the forces of romantic illusions concerning the Jewish individual and the rational operations in the name of the *Volk*. The history of Zionism is a constant struggle between ideas of universal humanism and ethno-centric nationality, or between an emphasis on the historical-cultural aspects of Judaism and an emphasis on the historical-biological aspects.

When Herzl established the Zionist movement, he was assisted primarily by youngsters educated in Austria and Germany who were acquainted with and integrated into Western society. His closest associates came from well-to-do, well educated circles: Max Nordau, Max Bodenheimer, Otto Warburg, and David Wolfson. However, as soon as Herzl established political Zionism, Eastern European Jewry

---

However, a couple of months later, Herzl denied the racial element and wrote: "I'll only say this: we are a historic entity, a nation of different anthropological components [...] no nation is racially uniform."

joined him: Chaim Weizmann, Leo Motzkin, Berthold Faibel, Menachem Usishkin, Yechiel Zlenov, Nahum Sokolov, and others. Soon the Eastern Europeans opposed as a bloc Western European conceptions and became the leading force in the organization. The First Zionist Congress in Basel comprised 200–240 men and women from 24 countries, about half of them from Russia, Rumania, Serbia, Bulgaria, Austrian Poland (Galicia), and Bukovina. As a matter of fact, also many delegates from Western countries were originally from Eastern Europe. For example, about half the delegates from Berlin were students from Russia who had settled in Germany, among them David Wolfson and Zvi Herman Shapira (who served as professor of mathematics in Heidelberg). Most were secular Jews, but to what extent did they also have a biological-racial self-image behind their national, cultural-historical credentials?

Opinions varied among youngsters, Jews and non-Jews, educated in Austria and Germany, with respect to the chances of assimilation. The Zionists among them grasped the bull by the horns and declared that Jews were not merely a nation with a culture and tradition of their own, but also – as the anti-Semites of a different ilk claimed – a distinct biological entity, a race.[9] Considering this singularity, claimed the Zionists, there was no chance of assimilation into the cultural and traditional environment in which they live. Consequently, they argued further, the solution for the Jewish problem was to assemble in their homeland, where they may bring to full expression their uniqueness without infringing on the feelings or the interests of people of other nations, who themselves constitute a unique race and culture. This solution would allow the Jews to contribute essentially to humanity at large and, more specifically, to the local backward population living in the Land of Israel. Herzl's attitude toward the local Arabs in *Altneuland*, and especially to their prominent representative, Rashid Bei, was that they were a people who would be grateful to the Zionists for bringing "culture" to their primitive world. This attitude unequivocally exposes Herzl's humanistic, paternalistic approach.

4.2 Max Nordau

Max Nordau (1849–1923), a physician and publicist, Herzl's main supporter, was in many respects also Herzl's antithesis. Nordau, like Herzl, was born in Budapest, but spent his early years in Germany and eventually settled in Paris where he lived the rest of his life. Nordau was an avid supporter of "Social Darwinism," asserting that social phenomena express the biological essence of human beings, and are the result

---

[9] Chaim Greenberg, the editor of the *Jewish Frontier*, the paper of the Zionist Workers in New York, declared in 1942 that there were times when it used to be fashionable for Zionist speakers to declare that "to be a good Zionist, one must be somewhat of an anti-Semite." Even then, the Zionist Workers circles were affected by the notion that the return to Zion is a kind of a purification process from our economic uncleanness. Whoever doesn't engage in so-called "productive" manual labor was believed to be a sinner against Israel and against mankind (See Brenner 1983, p. 24).

of evolution by natural selection. According to his conception, not only the straightforward physical properties of people, but also their behavioral properties, their intellectual skills, and the moral social development that they adopt, were the outcome of their biology. Such Darwinism, based on Herbert Spencer's notions, appeared to resolve the conflict between inherited fixity of properties and evolutionary development, which claimed that recurring acquired properties, generation after generation, become finally – by an accumulative procedure – hereditary. As noted in Chap. 3, Nordau was a great admirer of Cesare Lombroso, whose notions of crime were diverted from juridical causes to biological ones. He became well known when he published his early work, *The Conventional Lies of Our Civilisation* (1883/1895). In 1892, he expressed his pessimistic views in his book on the *Degeneration* of human society. When he met Herzl he was infected by Herzl's (optimistic) ideas about the solution to the Jewish problem. One must, therefore, examine Nordau's Zionist conception of a deteriorating European industrial society in the context of the spirit of the *fin de siécle*. Nordau, the pessimist, warned of the threat of the degenerative processes of modern society.

> A race which is regularly addicted, even without access, to narcotics and stimulants in any form (such as fermented alcoholic drinks, tobacco, opium, hashish, arsenic), which partakes of tainted food (bread made with bad corn), which absorbs organic poisons (marsh fever, syphilis, tuberculosis, goitre), breeds degenerate descendents who, if they remain exposed to the same influences, rapidly descend to the lowest degrees of degeneracy, to idiocy, to dwarfishness, etc. [...] To these noxious influences, however, one more may be added, residence in large towns.
>
> [...] Humanity can point to no century in which the inventions which penetrate so deeply, so tyrannically, into the life of every individual are crowded as thick as in ours. [...] In our times, on the contrary, steam and electricity have turned the customs of life of every member of civilized nations upside down, even of the most obtuse and narrow-minded citizen [...]. The humblest village inhabitant has to-day a wider geographical horizon, more numerous and complex intellectual interests, than the prime minister of a petty, or even a second-rate state a century ago. [...] All these activities, however, even the simplest, involve an effort of the nervous system and a wearing of tissue.
>
> Its own new discoveries and progress have taken civilized humanity by surprise. It has had no time to adapt itself to its changed conditions of life. [...] It had not quite grown to this increased effort. It grew fatigued and exhausted, and this fatigue and exhaustion showed themselves in the first generation, under the form of acquired hysteria; in the second, as hereditary hysteria. (Nordau 1895, pp. 34–40)

Nordau's description of the processes of degeneration of society relates not only to the needy, who were barred from the working force, immigrating into big cities and creating slum neighborhoods, but also to those artists and writers who did not follow the traditional path and whose works were in his opinion rotten fruits of degeneration, as well as to societies and nations that deviated from the traditional patterns of old Europe.

Thus, Nordau, perhaps more than any other Jewish leader, expressed the mood of the time, the dissatisfaction of the place of man in the materialistic, capitalist society at the peak of the era of industrialization. He urged escaping from the social structures and the traditional politics of the time, and expressed his support for absolute values and the pursuit of idealistic commitments above and beyond daily realities.

4.2 Max Nordau

As noted, already towards the end of the nineteenth century, the German philosopher, Herder, conceived of the nation as an eternal frame of mind extending throughout history. This encompassed the *Volksgeist* just as much as the disposition of the individual. In the mood of the period, Nordau conceived of nations as entities existing in their own right, rather than entities whose task was to serve the needs of the individuals who comprised them. He rejected modern materialism and the longing for nationalistic romanticism, including the defamation of modern art as "degenerate art" and the reverence for nature. Equating urban life with animal domestication, a comparison made by Nordau, also characterized – it must be admitted – the cradle of Nazi nationalism.[10] The spirit of nationalist *Volkism* at the turn of the twentieth century was also shared by several Zionist leaders in Western Europe for whom Nordau was the spokesman. The "image of the Jew" was the antithesis of this *Volkist* image and for many, the Jewish image symbolized a lack of roots and a people lacking "blood and soil" [*Blut und Boden*]. Thus, Zionism, that borrowed many concepts from the same sources (even though explicitly defying them) as German nationalism, often adopted the language of "blood and soil," as well as did other naturalist *Volkisms*, long before their twisted deviations were introduced by Nazism (Mosse 1964. See also Mosse 1970, pp. 77–115).

Nordau's interest in neo-Lamarckian evolution, in the inheritance of acquired characters, indicates that he was less interested in the anthropological origins of the Jews, but rather in the conditions that brought contemporary Jews to their condition and in the circumstances that might release them from these. In the spirit of the *Volkist* ideas of "blood and soil" that were required for the existence of a nation, he accepted the claim that the life of the Jews as a separate and persecuted community had been the cause of their biological degeneration, of the debilitation of their body and their spirit. Accordingly, Nordau assumed that changing the Jews' living conditions would be followed by a biological resurection. In 1896 the Berlin Jewish weekly *Allgemeine Israelitische Wochenschrift* asked seventy Jews: "Are the Jewry and Judaism in the process of decline and, if so, what means could arrest this process?" Nordau answered without hesitation that, concerning the first question, indeed Jews exhibit symptoms of degeneration, both at the mental and at the physical levels. Millennia of oppression and persecution had left their mark – a depressive frame of mind. For two thousand years joy, self confidence, hope and élan did not reinvigorate them. They had lost their self-respect. As a people, they had lost contact with the soil. A people cannot, in the long run, remain healthy and strong if it does not again and again, at least temporarily, returns to the rejuvenating soil. He compared the situation of the Jews to the legendary Greek hero, Anthaeus, whose disconnection from the physical soil meant death. The growing rate of Jews who abandoned Judaism was the eternal indication of this decline. The intelligent Jew

---

[10] Urbanization conceived as a kind of animal domestication was a favorite theme in Nazi Germany. The clinching proof for such a degeneration process was, of course, the Jew. See, e.g., Konrad Lorenz (1940). Lorenz was a well-known Nobel laureate of animal behavior research. See also the collection of papers published by Zeiss and Pintschovius (1944) in Germany in, 1944 on the damages of civilization to humans.

leaves; only those who are too dull to feel persecution and insult remain, as do those stubborn enough to resist any enemy. Nordau warned that if this selective process continued, the remnant Jews would likely be intellectually inferior.

To withstand all this, he recommended first of all a return to agriculture. As many Jews as possible should become farmers: "Life in nature will rejuvenate their bodies, the secure possession of the soil will resurrect their self-esteem." Second, to hasten the process of regaining self-respect, Jewish education must make the name 'Jew' a title of honor. Third, internal Jewish solidarity must be strengthened, since so far "only persecution keeps modern Jewry together with its iron-ring. Without the enemies [...] we would fall apart like an untied stack of hay" (Ben-Horin 1956, pp. 179–180). In order to change the hereditary properties of the Jews, they have to adopt a different life style, abandoning the one they were accustomed to for generations. In order to foster "muscle Judaism" it is important to support physical training. David Biale in his *Eros and the Jews* (1992) noted that "one of the central arguments of Zionism was that the living patterns of the Jews in the Diaspora was like that of an entity without a soul, and that only healthy national life may reinvigorate them enough physically and materially. Such a political ideology was founded not only on the perception of the body metaphorically; it rather strove to bring about changes in the Jewish body proper, especially the sexual one" (Biale 1994, p. 231).

> Zionism meant both the physical rooting of the "impractical dreamers" (*Luftmenschen*) in the soil of Palestine and the reclaiming of their body. [...] The new nationalism was accompanied on the one hand, by a strong sense of respectability, inherited from European bourgeois culture, and driven, on the other hand, by the powerful asceticism of a national movement dedicated to goals that transcended the happiness of the individual. [...] Like other nationalists of the end of the nineteenth century, the Zionists were preoccupied by the physical and emotional degeneration of the nation and by the threat of demographic decline. [...] Nordau's Zionism reflected this diatribe against degeneration. [...] and the Jews, thought by many physicians to be the quintessential neurasthenics, could overcome their hereditary nervousness by developing their bodies. Jews, according to Nordau, must become men of muscle instead of remaining slaves to their nerves. (Biale 1992, pp. 176–179)

As Nordau phrased it:

> Gymnastics is of the highest importance for the Jews. Since our biggest fault is obstinacy, rigidity, and an aversion to respecting a fellow race-member [...]. It is not surprising, we used to say, that we lack muscle power and physical capacity. It was during the thousand years of life in the ghetto that inadvertently, because of lack of practice, we lost all physical skills. Now it is clear to us that we are lacking these and we must make an effort to regain them. (Nordau, 1902, *Was bedeutet das Turnen für uns Juden*, in Nordau 1909, pp. 382–384. TRF)

As for the origin of the Jewish characteristics, Nordau claimed that it would be difficult to formulate a satisfying and convincing scientific explanation, since the anthropology and ethnology of the Jewish race were a resource that had been hardly addressed. "We do not know if the original Jews were more corpulent, whose development was inhibited only as a consequence of their abnormal living conditions, or whether they were a small race to start with." (Nordau 1909).

Nordau was an outstanding speaker; his lecture was one of the highlights at the First Zionist Congress. Also the speeches he made in the subsequent congresses made the headlines. At the eighth Congress in The Hague, on August 14, 1907, Nordau explained his Zionist conception: "Zionism, like any other movement in history, stems from a strong and clear sense of need – the need for normal existence under natural circumstances." In his talk, he referred to several theories that were being elaborated at the time in Germany, noting especially race psychology – a pseudo-science, a pseudo-ideology – that elevated the *Volk* and the race to the level of entities with their own soul (and which later contributed significantly to Nazi thinking). He claimed that ethnic studies of race psychology and ethnology lead to the insight that nations, just like individuals, develop their fundamental individuality as it unfolds historically.

> Zionism is a beneficial movement, not a benevolent one [...] Zionism can give nothing to the Jewish nation and will provide nothing to the Jewish individual. It aspires to illuminate, develop, direct, and collect the resources of the Jews by which they will, to awaken them from the slumber of two thousand years, to produce for themselves natural circumstances for living on their land that will help them to carry out their own redemption [...]
>
> The claim that Zionism is a counterbalance to anti-Semitism and that all which can be learned from it is the refusal to surrender and to despair [...] is a surprisingly superficial claim. Anti-Semitism was at most a push toward Zionism, but not its cause. Jew hatred only awakened the slumbering conscience of the race in many Jews and gave them back the sense of their uniqueness. (Nordau, 1907, *VIII. Kongressrede,* in Nordau 1909, 174–187. TRF)

Thus, Nordau represented most faithfully the outgoing century, the spirit of scientific thinking in the service of neo-romantic socio-politics at the end of the nineteenth century. His Zionism aimed primarily to support the redemption of the nation, the *Volk*, in which the person, the individual Jew, is the servant, whether as a victim or as a leader.

4.3 Zeev Jabotinsky

Zeev Vladimir Jabotinsky (1880–1940) was born in Russia and studied in Italy at the turn of the century. A journalist and a prolific author, he was one of the most resolute intellectuals affected by the nationalist, futuristic notions prevailing in Italy at the time. He adopted the notion of "blood and soil" as did many intellectuals with a wide scope of opinions. Following the Kishinev pogrom in 1903, Jabotinsky got involved in Zionist activity. He was a member of the Zionist leadership who volunteered to serve in the Jewish battalions during the World War I. In the 1920 Arab revolt in Palestine he was active in the Jewish defense. Jabotinsky was arrested by the British authorities and later exiled from Palestine. In 1925, he founded the Alliance of Revisionist Zionists, withdrew from the World Zionist Organization, and in 1935 established a separate New Zionist Organization. Jabotinsky adopted an uncompromising, democratic-liberal, deterministic view. According to him, only distinct and clearly identified national entities would productively contribute to the human soul. In order to exploit its full potential, a nation must achieve the necessary conditions, primerily a unique language and a definite territory. In order to survive, a nation-race must develop a distinct national culture.

> A few years ago I asked myself: Where does our deep-rooted feeling of national self-identity originate? [...] The answer that first occurred to me was that the source lies in our individual education [...]. But then I realized that this answer is wrong [...] it is clear that the source of national feeling should not be sought in education but rather in something that precedes education. In what? I studied this question in depth and answered: in the blood. [...] The feeling of national self-identity is ingrained in the man's "blood," in his physical-racial type, and only in it. We do not believe that the spirit is independent of the body; we believe that man's temperament is dependent first of all on his physical structure. [...] The psychic structure of the nation reflects its physical type even more fully and completely than the temperament of the individual. The nation shapes its essential spiritual character, in that it adjusts its physical-racial type, and no other temperament based on this physical type is conceivable. (Jabotinsky, "Letter on Autonomy," quoted in Bilski Ben-Hur 1993, pp. 91–92)

Jabotinsky insisted that a precondition for a nation's contribution to humanity is cultural isolation. A national, cultural ethos is vital to every nation. "For this end a creative nation needs isolation, enclosure, just as is needed for the individual creative personality" (see Bilski Ben-Hur 1993, p., 100). Physical isolation ensures cultural isolation, though at the same time, physical isolation is affected by cultural isolation. Physical isolation and cultural isolation must undergo an integrated evolution. In his articles, "The Race" (1913) and "A Lecture on Jewish History" (1933), Jabotinsky claimed that any race that has overt properties aspires to become a nation, that is, to create for itself an economic, political, and intellectual environment unique unto itself. The existence of races, of nations, is a precondition of human creativity and progress. The fact that Jews are a distinct and essential race is the key to understanding most of their history, and this supports the argument that they are still vigorous enough to contribute significantly to humanity also in the future. Thus, the Zionist idea must be appreciated not in the narrow sense of Jews as a race, but rather as a matter of universal significance. But was Jabotinsky's concept of race 'biological' or merely cultural? To what extent did Jabotinsky make the distinction?

According to Jabotinsky, the characteristics of a race were molded by natural conditions prevailing in a country at the time and in the place that the race emerged. The impact of other factors, such as economic conditions or social structure and so on, is temporary and has no effect on the natural factors. The Jewish religion was not the treasure that the Jews protected in the Diaspora, but rather the treasure that protected Jewish national uniqueness. Thus, only a return to the soil and to the natural environment in which the specific Jewish nation was shaped in the past will allow the extension of national uniqueness and its universal contribution. Hence, the real destination of the Jewish nation in the Diaspora, namely the maintenance of its national uniqueness, is none other than to return to the fatherland to pursue its natural course, which cannot occur anywhere else (Bilski Ben-Hur 1993, pp. 124–126, 134).

Jabotinsky expressed a para-scientific, "environmentalist" notion about the indefinite preservation of the formative characteristics of a race:

> Another climate, other flora, other mountains would certainly distort the body and soul that were created by the climate, the flora, and the mountains of the Land of Israel; for the racial body and the racial soul are nothing but the product of a particular combination of natural factors. (Jabotinsky in Bilski Ben-Hur 1993, p.125)

The identification of the people of Israel with the Land of Israel is complete. Eventually Jabotinsky even accepted that "a Jew educated among Germans may acquire German customs, German words, to be entirely stuffed with German spirit. However, the kernel of his psychic structure will remain Jewish, for his blood, his body, his physical-racial type are Jewish." Yet, in order to maintain the "blood," the national uniqueness that secures the potential for universal, cerebral creativity, it is necessary to insure that there will be no assimilation or absorption. The natural and best niche for seclusion of any nation is its national territory (Bilski Ben-Hur 1993, p. 127). For Jabotinsky, being Jewish was obviously a biological identity.

4.4 Martin Buber

Martin Buber (1878–1965), a philosopher active in the Zionist organization who, like many, was deeply influenced by Nietzsche's teachings, adopted the *Volkist* conception that "blood" – heredity – was the ultimate essence of Jewish identity. Buber joined Herzl at the beginning of his Zionist activity and was for a while the editor of *Die Welt*, the newspaper of the Zionist organization. Buber became known as the student of the Hassidic Movement and as the translator of the Old Testament into German (together with Franz Rosenzweig). Max Brod and Franz Kafka, on the one hand, and Hugo Bergman, Robert Weltsch, and Arthur Ruppin, on the other, were among the participants in his "Bar Kochba" circle in Prague. Buber immigrated to Palestine in 1938 and was active in "*Brit Shalom*," an organization for Jewish-Arab coexistence.

Buber took exception to the emphasis that Herzl put on diplomatic activity. The Jews of the West were unfamiliar with the reality in Palestine – the language, the traditions, and the mentality – all these were foreign to their very essence. In short, they did not relate to the "blood community" [*Gemeinschaft seines Blutes*] to which they belonged. Therefore, for Buber, the Zionist project was primarily the means to an end, the spiritual resurrection of the Jews, mainly through the "soil." That is, the agricultural settlement in Palestine, consistent with the values of the European national movements: "A *nation* preserves its cohesiveness by the primary common factors of blood, fate, and a creative cultural force." Racial variables were, according to Buber, important for both the appreciation of the cultural capacity of nations, as well as of their limitations. However, although such factors define the personality of a nation, they are liable to change: "Racial characteristics are nothing but the products of the soil and the climatic conditions, of the economic and social constructs, of the patterns of life of communities and of the common historic fate"

(Buber 1920, p. 245). Therefore, Jewish nationalism must be nurtured to maintain a continuous reality.

Buber was not thinking in terms of heredity and environment (or Nature versus Nurture) in the sense of natural sciences when he spoke of "blood and soil." It is, however, remarkable how he and other intellectuals and philosophers of the humanities made use of terms like "race" as if they were well-defined concepts of science that may be borrowed as such and applied to the humanities. Buber's reference to "blood and soil," which was taken, of course, from the German philosophers, was more overt. When a person discovers the chain of generations and sees the series of mothers and fathers that extend to him such sentiments risk becoming realities:

> [H]e perceives in this survival of the generations a blood partnership, and conceives it, in a sense, as the prior life of his own ego, as a kind of perseverance of his ego from the infinite past. To this feeling is associated the revelation [...] that the Blood is the source and the force that nourishes the being of the individual [...]. The world around him is one of impressions and influences, whereas the Blood is the world of the Self that experiences these impressions and influences, and it is this world of the Self that absorbs all and reshapes it in its own image. (Buber 1984, p. 24. TRF)

Buber asserted that man is "like a link in a chain" and, therefore, "his status as an individual is dependent on his creation in relation to his nation." This relationship is not always actualized, "unless the homeland of that individual is also the homeland of his blood, and the language and manners in which he grew up." Buber repeatedly emphasized "the deepest layer of the capability [to shape impressions in his own image], of that which creates the type, the skeleton of the personality, is that which I called Blood, namely, that which had been implanted in us by the chain of Mothers and Fathers [...]. This is the large heritage of generations which is born with us" (Buber 1920, pp. 25–26).

Many years later Buber became aware of the meaning of the incautious use that he made of the concept of blood. Thus, he eventually attached a rejoinder, claiming that "the most evil persons distorted the notion of 'blood'." He claimed that when he used the term 'blood' he did not mean the "racial issue," which in his opinion is fictitious merely referred to "the continuity of begetting and becoming a nation." Obviously, Buber could not anticipate the future, but here again the magnitude of the Hebrew saying that "life and death is in the tongue," is revealed in the sense that the "tongue," or language, is shaped by its creator and his worldview. Whereas in 1909 he applied the term Blood to define nation, in his later writings he applid the term nation to define Blood.

Perhaps even more meaningful is the fact that it was Buber's circle in Prague that opposed the positivist attempts to institute explicitly spiritual claims on ad hoc material circumstances. They reasoned that "the racial theories that had been deduced from anthropological facts and arguments, derived from the natural sciences." Spiritual facts do not need natural science's support, and it is no longer necessary that "a mental claim should be founded on the natural sciences; it was a regrettable mistake to elevate the natural sciences to a Godly level, which endeavors to explain all the wonders of the human spirit in Zoological terms." (Doron 1980, p. 423).[11]

4.5 Arthur Ruppin

Contrary to many Zionist thinkers, Arthur Ruppin (1876–1943) was first of all a pragmatic Zionist. Even though he founded his Zionist perception on theoretical foundations, he wrote already in 1923:

> I think that Zionism is tenable only if it is provided with a completely different scientific foundation. Herzl's view was naïve and only makes sense in the light of his complete ignorance of the conditions in Palestine. [...] I think that Zionism is less than ever justifiable now except by the fact that the Jews belong racially to the peoples of the Near East. (Ruppin in Bein 1971, p. 205)

Still, it was clear to him that notwithstanding their Near Eastern roots, the common denominator that identified them as Jews was not their biological ties, but rather their cultural-social ties. Also, he could not ignore the emotional desire to return to the ancient homeland. Ruppin considered the preservation of the Jewish people of human interest, *per se*. He noted that the advantage of the Zionist idea was that it offered an ideal to those who had abandoned the Jewish religion and also lost faith in its noble ideals. Zionism thus provided many Jews with a solution to their integration in the social and cultural life of the developing Western World, which was diametrically different from the solution presented by assimilation.

Ruppin was born in Poznan, in the Polish district near the border with Silesia, which was heavily influenced by German culture. Although he grew up in a traditional Jewish home, he went to state schools where he was thoroughly integrated in the German intellectual life of the end of the nineteenth century. Ruppin earned a degree in law and political economy.

In 1899, during his first year of studies in Berlin, the student of economy and sociology responded to a call to participate in an essay competition addressing the question: "What can we learn from the theory of evolution about the structure of domestic politics and the constitution of nations?" Such essay writing competitions were common in the nineteenth century, and they may be considered the forerunners of the current calls for applications for research grants. That specific competition, as it turned out, was the initiative of a member of and industrialist clan and an admirer of Social Darwinism, Friedrich Alfred Krupp (1954–1902), whose main contribution was to the German arms industry. The prize committee was composed of the leading scientists in sociology, history, and biology of the time.

Ruppin tells in his memoirs that when he first read about the competition, he understood almost nothing of its contents. However, after thoroughly studying both

---
[11] Quotation from Gustav Landwer (1870–1919), pacifist and revolutionary, who cooperated with Buber in the Socialist Alliance in 1908. See Doron 1980, p. 426.

biology and the theory of evolution, he felt like an expert. Thus, in 1902, when a student in Halle, he again encountered the advertisement, he submitted his article *Darwinismus und Sozialwissenschaft* [Darwinism and social Science]. He won second prize (Ruppin 1903). The first prize was won by the physician Wilhelm Schallmayer (1857–1919) for his essay *Vererbung und Auslese in ihrer soziologischen und politischen Bedeutung* [Heredity and Selection in Its Sociological and Political Meaning] (Schallmayer 1910[1902]). This was the foundation of "racial hygiene," the German version of eugenics. The title of the competition as well as the works that won the prizes indicate that this was a call for the examination of the concepts of the social sciences in light of the natural sciences and, more specifically, of Charles Darwin's theory on the origin of species by natural selection. The work that won the first prize was aimed at fighting the common social habits and procedures which, according to believers, led to disease and the physical and mental degeneration of the German nation. Eventually, with the demand to "purify the race" from the mentally ill, invalids, Gypsies, and Jews, this movement was easily incorporated into the German nationalistic activity.

As is obvious by its title, Ruppin's manuscript was a social-Darwinist composition. Although Ruppin opposed the researchers of the nineteenth century who wished to impose the methods of the life sciences on the social and sexual life of the human species. "Dissecting society, or what is called the 'social body,' the way the anatomists dissect an organism into its parts, poses a problem. Even when we disregard the popular comparison of the single individual in society and the individual cell in a living organism, as being inadequate because the individual is conscious of purposes whereas the cell is not, the main reason for these efforts being futile is that they are methodologically wrong, because the form of argumentation in the natural sciences is not applicable in the social sciences" (Ruppin 1903, p. 7). According to Ruppin, what makes social sciences unique is not so much the subject of the investigations as the mode of observation. In the natural sciences there are explanations but no instructions; there is no reference to targets or purposes. "In the social life, on the other hand, the main point is precisely that of finding the norm of our actions." All that natural sciences can offer to the social sciences is the explication of the means that are available to human beings, of the natural limitations and constraints that act on social aspirations and inclinations. This is like "a traveler on a boat who decides on his own the purpose for making the trip but is practically constrained by and dependent on the physical means, the compass and the naval maps" (Ruppin 1903, p. 11). Ruppin thus conceived of the natural sciences as parallel or tangential to that of the social sciences. "Only by posing the question if at all and to what extent may society utilize the impact of its principles on the theory of the origin of species in order to change the human type to its needs, do we come to deal in the sphere of the social sciences" (Ruppin 1903, p. 41).

Ruppin was thoroughly influenced by the socio-biological teachings of Ernst Haeckel (see, e.g., Haeckel 1876). Haeckel's universe is monistic – namely, all its physical, chemical, organic, and human factors emanate from one primary vital force. According to his Darwinist conception, the environment acts directly on organisms to produce new stocks and races. Ruppin presented his Haeckelian

4.5 Arthur Ruppin

concept by suggesting that the environment may produce, by natural or social selection, new stocks of humans that differ from the original ones, and the greater environmental variation is, the longer it lasts. "By exterminating the people who are less adapted, the environment affects the development of the properties that are important for existence under the given conditions, in one group this way, in another group the other way. What nowadays we call human races are nothing but stocks produced by way of hybridization and selection" (Bloom 2005, quoting Ruppin, 1940, p. 16). Ruppin admitted that he never studied anthropology in a systematic way, but only heard some lectures by Felix von Luschan; his knowledge of biology was also rather limited.[12] Notwithstanding, Ruppin became one of the central demographers of the Jews and their history (see Ruppin, 1930a and b on "The Sociology of the Jews").

Ever since Ruppin's first encounter with "Jewish nationalism" and "Zionist aspirations" in February 1892, he maintained an interest in Zionism, although his attitude to these ideas remained ambivalent. A change took place in 1902–1903, at about the time that his interest in the Darwinian impact on social process increased. At that time, Ruppin went on an educational tour of Jewish life in Eastern and Western Europe. He turned to Zionism only in 1904, when he came in contact with the circle of what he called "practical Zionists," namely Martin Buber and his colleagues in Prague. On the face of it, there could be no greater dissonance between Buber, the philosopher and Ruppin the practical Zionist, but one has to remember that at that time Buber embraced the idea of "nation" and its application in the mystique of Jewish history. The young Ruppin was interested in the deliberations of the group of intellectuals in Prague, who at the beginning of the century were trying to reconcile humanism with nationalism, and "dreamt about Jewish settlement in Palestine" as a practical activity.

In 1904, Ruppin summarized his impressions from a trip in eastern and western Europe in his book, *Die Juden der Gegenwart* [The Jews of Today]. A second edition appeared in 1911 and the book was translated into several languages the following decade (Ruppin 1911, 1913). Shortly thereafter, Ruppin was named editor of a journal for demography and statistics of the Jews. In 1907, he made a private visit to Palestine and on Ajpril 3, 1908, he once more went to Jaffa and settled in Palestine, where he inaugurated the Palestine Office as the representative of the Zionist Organization. Ruppin eventually became the person responsible for land purchase and construction of all possible settlement forms in Palestine. He founded, among other institutions, the *Hachsharat haYeshuv* namely, the Palestine Land Development Company that bought land and primed it for settlement.

---

[12] In Ruppin's (1903) extensive review of the achievements of the life sciences, heredity included, no mention is made of the Mendelian theory of inheritance, even though it contains some references from 1902, including the works of Hugo de Vries. Although Ruppin's work is a thesis on the social sciences, which examined arguments in light of developments in natural science, Ruppin was treated as an anthropologist, even as a geneticist. In 1922, on the hundredth birthday of Gregor Mendel, the founder of modern genetics, a statue of him was erected in Brno, Czechoslovakia. Arthur Ruppin of Jerusalem is mentioned among the leading geneticists who sponsored the statue.

> I joined the Zionist organization under the motto "against the political Zionism [namely, against Herzl's idea for a Charter], and for practical work in Palestine." I wished to establish the right of the Jews to come to Palestine, not on the basis of some "political" agreement or concession, but rather on the historical and racial link to the Land of Israel, and I wished that they would acquire this right in the country by their toil. (Ruppin in Bein 1971, p. 207)

Ruppin emphasized the significance of the variability of the human species. He was concerned with maintaining the uniqueness of most individuals and communities. Therefore, it was in the interest of the human species to preserve the biological and the cultural uniqueness of the Jews, rather than to promote their assimilation:

> Have the Jews the right to a separate existence? The very question is an insult to the Jewish people, since no other people is required to defend by argument its right to survive [...]. The Jews might justly claim that a history extending over 3000 years is sufficient justification of their continued existence. [...] But these, after all, are only sentimental arguments; [...] we must first answer the following question: "Can the Jews do more for humanity by remaining a separable nationality than by becoming absorbed in other nations?" A people can be of use to humanity in two ways, firstly through its race-value, *i.e.*, through the spiritual and mental powers incorporated in it, and secondly through its culture. Whoever defends the right of the Jews to a separate existence must do so either in view of their racial or their cultural value. (Ruppin 1913, pp. 212–213)

As Alex Bein, Ruppin's biographer noted, all his books published after *The Jews of Today* were merely permutations of it. In his major opus, *The Sociology of the Jews*, published 26 years later, Ruppin repeatedly emphasized what he believed to be well-established facts on the race and the racial properties of the Jews.

Ruppin's Zionist world view was a function of his universal, humanist notions more than a direct deduction from the history of the Jews. From an historical perspective, many authors deny altogether any unity of race of the Jews. Ruppin was aware that in reality, the Jews, in the course of their 3000-year history, have assimilated foreign elements, though by and large, they represent a well-characterized race.

> But this very likeness to the Asiatic people, from whom they have been separated for 2000 years shows that the Jews have remained unchanged, and that in the Jews of today we may say we have the same people who fought victoriously under King David, who repented their misdeeds under Ezra and Nehemiah, died fighting for freedom under Bar-Kochba, were the great carriers of trade between Europe and the Orient in the early Middle Ages, and finally were excluded from culture in the isolation and misery of the Ghettos from the end of the Middle Ages onward for 500 years. (Ruppin 1913, p. 216)

The Jews have not only preserved their natural racial gifts, but through a long process of selection these gifts have become strengthened. Consequently, we have in the Jew of today what is in some respects a particularly valuable human type. Other nations may be superior in other areas, but with respect to intellect, the Jews can scarcely be surpassed by any nation. On the basis of this fact alone, the Jews may well claim their right to a separate existence and resist any attempt to absorb them.

> Just as it would be absurd to destroy specific kinds of fruit in order to produce one general kind, so it is equally absurd to wish to wipe out national differences. Mankind today aims not at uniformity, but at making use of individuality of every nation for the common good. (Ruppin 1913, p. 217)

4.5 Arthur Ruppin

Ruppin was convinced that in spite of the great variability between their communities, the Jews are distinguished by their special biological uniqueness, rather than by merely being a cultural-religious congregation, and emphasized the biological universal dimensions of this uniqueness. Their settlement in Palestine was necessary in order to maintain this uniqueness for the benefit of humanity at large.

* * *

The notion of political Zionism is to a large extent the reaction of the Jews to the romantic ideas of Western Europe: Emancipation and anti-Semitism, the "spirit of the nation" and the fatherland. The notion of "blood and soil" of the romantic German nationalists was adopted by Zionist intellectuals as a response to the biologization of Jew hatred in the form of anti-Semitism at the end of the nineteenth century and the beginning of the twentieth century. A straight line may be drawn from the insight of Moses Hess to that of Vladimir Jabotinsky, and further on to that of Arthur Ruppin and that of Martin Buber, each of whom assimilated the notion of "blood and soil" of the nineteenth century European philosophy in his own way and conceived of the formation of a Jewish nation in its homeland as a guarantee for the continued contribution of the Jews to universal culture, no less than to the solution of the plight of the Jews.[13] It must, however, be admitted that the Zionist idea acquired status among only a small minority of the Jewish Diaspora in the first half of the twentieth century and among those, only very few were ready to overcome the difficulties of carrying it out.

In retrospect, it is difficult to avoid the conclusion that the idea of national resurrection as an ideal of universal progress, an ideal that was thwarted by "world villains" in Buber's words, was what led to the Zionist response and the survival of a remnant.

---

[13] Etan Bloom (2008, I, p. 69) cites a quotation from young Ruppin's *Tagebuch* [Diary], "Only a Volk engaged in agriculture can be healthy […]. England and other states (whose agricultural populations are steadily declining) will always present only the aggregate of individual people who have been haphazardly thrown together." In response, Bloom (sadly) comments that it "may very well affirm what Hans Kohn and Hannah Arendt stressed decades ago, that Zionist nationalism was shaped by the German model which rejected 'Western civic ideals' and the democratic, universalistic models of the US and French revolutions."

Chapter 5
A Jewish Race Notwithstanding?

> If there are no races, how can Jews be a "race"? (Kaplan 2003)

In his book, *The Jews: A Study of Race and Society*, published in 1911, the Jewish anthropologist and demographer Maurice Fishberg (1872–1934) rejected the claim that Jews had a common biological denominator. The fact is that the variability of the Jews' physical characteristics parallels the variability of the people in their countries of origin. Fishberg claimed that what were characteristic-Jewish in their facial features were not the physical properties in the narrow sense of the word, but rather an expression of a Jewish soul. To support his argument rejecting a common Jewish biological denominator, he quoted the impressions of the Jewish British author, Israel Zangwill (1864–1926), who described the facial features of the attendants of the First Zionist Congress in Basel:

> A strange phantasmagoria of faces. A small, sallow Pole, with high cheek-bones; a blond Hungarian, with a flaxen mustache; a brown hatchet-faced Roumanian; a fresh-coloured Frenchman with eye-glasses; a dark Marrano-descended Dutchman; a chubby German; a fiery-eyed Russian, tugging at [his] own hair with excitement, perhaps in prescience of the prison awaiting his return; a dusky Egyptian, with the close-cropped, curly black hair, and all but the nose of a negro; a yellow-bearded Swede; a courtly Viennese lawyer; a German student, first fighter in the University, with a coloured band across his shirt-front; a dandy, smelling of the best St. Petersburg circles; and one solitary caftan Jew, with ear-locks and skull-cap, wafting into the nineteenth century the cabalistic mysticism of the Carpathian Messiah. (Fishberg 1911, p. 99)

Fishberg promoted assimilation, and he was optimistic that the socio-political atmosphere of the early twentieth-century culture in Western countries, such as the United States, England, and Germany, was ripe for its fulfillment. To his mind, if there was such a thing as the 'Jewish type', it was a product of the social and political environment. Fishberg's anthropological and demographic systematic studies were based on data that proved that the Jews and the non-Jews among whom they lived were biologically similar. According to him, the Zionist call for national identification would only be an obstacle to the social assimilation of Jews.

As noted in Chap. 3, although most researchers agreed that Jewish anthropological types did exist, not all anthropologists of the early twentieth century, Jewish or non-Jewish, accepted the claim that Jews comprised a race. The questions of what exactly a race is and what its characteristics might be remained open (see Andreasen 2004).[1] Furthermore, those researchers who did claim that the Jews had been a race in the past wondered what kind of race it had been and what had been its fate over the generations: Is there an anthropological connection between the Jews of the present and an ancient race? Were Jews originally the progeny of one ancient type, or were Jews from long ago an entity founded on a mix of two or three "races"? To what extent has the biological "blood connection" of Jews been maintained over generations? And, of course, was variability among Jewish communities a hereditary consequence of the traits that had been acquired as they adapted to environmental conditions, or were those traits an indication that "foreign blood" had been introduced into the original Jewish race?

Opinions are divided about the extent of the reproductive isolation of the Jews and there are different explanations for its causes. Some claim that Jews were isolated not because of outside persecution, but rather because of religious norms pertaining to sexual life and dietary customs; however, there is no agreement about the impact of these isolating factors. Others attribute reproductive isolation to "Jewish eugenic" thinking; they claim that this enabled strong natural selection and breeding for characteristics specific to Jewish communities – such as an emphasis on intellectual activities rather than muscular development. Still others suppose that it was acceptance and support of intermarriage in Jewish law that prevented further degeneration and invigorated them ('hybrid vigour' in the terminology of animal and plant breeders), bringing evidence like the story of Ruth, the Moabite convert to Judaism, who established the dynasty of the House of King David from the sperm of Boaz (see Biale 1992, pp. 13–16).

There are indications that Jews had a positive attitude toward proselytizing in ancient times. For example, in 538 C.E., the city council of Orleans considered it necessary to enact a law against proselytizing slaves in Jewish households. Still the voluntary Judaization of Christians continued thereafter, although the phenomenon became rare in medieval times.[2] A major example of proselytizing was the conversion

---

[1] Andreasen (2004) noted that genetic and independent variation arguments have been widely accepted as showing that races are biologically unreal. Yet the author argues that these arguments do not work against the *cladistic race* concept. "According to the cladistic race concept, races are ancestor-descendant sequences of breeding-populations that share a common origin. […] A 'breeding population' is a set of populations that are reproductively connected to one another and is reasonably reproductively isolated from other such sets." Many evolutionists agree that it is possible to "accurately represent human evolution as a branching pattern. As long as this is possible, it is possible to define races cladistically." Notwithstanding, since human history is one of intensive geographic as well as social migration (and the history of the Jewish people excels in these), human evolution shows a *trellis* pattern rather than a *branching* pattern (see e.g., Templeton 1998, 2008). Thus, imposing a cladistic model (as is the inbuilt assumption of most phylogenetic computer programs) may be grossly misleading.

[2] Corcos (2005, pp. 95–109) notes the abundance of proselytizing in Jewish history. The major theme of Sand's *The Invention of the Jewish People* (2009) is that the Jewish nation of today is a

5 A Jewish Race Notwithstanding?

of the Khazar tribe. As we shall see further on in Chap. 6, this event and its effects on the biology of the Jews are still vehemently disputed today.

Regarding the biological essence of the ancient Jews who lived on their homeland until the Babylonian exile, the question of whether and to what extent they maintained their biological identity during that exile has significant implications. Obviously, the more tribes were involved in the establishment of the original Jewish nation (and such founding tribes were clearly also not homogeneous with respect to their biological inheritance), the more difficult it is to identify defining characteristics that single them out as Jews. The greater the variability of Jews at their origin, the more difficult it is to identify the extent of their intermingling with the nations they lived among during their exile. Even a claim that it would be possible to recognize a 'Mediterranean nation' living among Europeans is meaningless, since many who argue for the multiracial origins of the Jews include 'Europeans' among the contributors to the ancient stock, whether these were assumed to be 'Galileans' or 'Philistines' (see Chap. 6).

As expected, many investigators settled on a combination of historical data – reliefs in ancient temples or descriptions in ancient texts – to establish the racial identity of the Jews, rather than on anthropological and demographic data of present-day Jews. By examining such ancient data, researchers believed that they could identify the multiracial origins of Jews (obviously always assuming that those Jews whom they chose to study represented 'real' Jews). The not-so-hidden intentions of the authors often became obvious when they maintained that Jews stem from two races: one from the Galilee, the 'Galilean' fair-haired Aryan; the other from Judea, the dark-haired Semite. Still, most physical anthropologists who had examined their contemporaneous Jewish communities speculated that Jews originated from three races. Thus, according to von Luschan, Jews originated from a mix of at least three different races: Semites, Amorites-Aryans, and Armenoids or the progeny of the Hittites. His conclusions were based on his travels through the Middle East and Asia Minor (Turkey).

> From these three varieties came all the different types of modern mankind, generally by local isolation. A very interesting example […] is found in the earliest known inhabitants of Western Asia. This is the land of those extremely narrow and high-arched noses we generally call Jewish or even Semitic. These remarkable noses, however, do not belong to the Semite invaders, of whom Abraham is the eponymic hero, but to the pre-Semitic population which might be called Hittite or Armenoid, as the modern Armenians are their direct descendants. (Von Luschan as translated in Kautsky 1926, p. 73)

Such conclusions, however, were drawn from the populations that the researchers happened, or managed to examine. Arthur Ruppin, who accepted in principle the idea of the multiracial origins of the Jews, noted that if von Luschan had traveled through Italy instead of Asia Minor, he would have suggested another classification

---

heterogeneous conglomerate of populations that over history joined or became Jewish, rather than being essentially the direct descendents of a historically integrated entity that had been exiled from its homeland.

of the races of origin of the Jews. I will come back later to this point and to the positions presented by different Zionist leaders.

As noted, not everyone accepts the notion that there is a scientific biological basis for the claim that the Jews are a race, or even the notion of any biologically significant racial differences between human populations. The very issue of racial biological differences is closely linked to the worldview of the persons involved. A humanistic view of the essence of a free man, on the one hand, and social conceptions of the responsibility of the individual to society, on the other hand, raise doubts about the biological meaning of ancient divisions of humans into racial categories. Zionists who endeavored to impose a humanistic and universal belief on their concept of race had to face not only non-Zionists and assimilationists among their own people, but also socially conscious thinkers, Marxists and others, who considered the very idea of a revival of the national notion a threat.

The Jewish physician and anthropologist, Samuel Weissenberg (1867–1928), born in the Ukraine, is one of the prominent researchers at the turn of the twentieth century who considered the Jews to be a race whose origins he tried to trace.[3] Weissenberg completed his medical studies in Germany in 1890 and returned to his home township where he spent the rest of his life. He often made anthropological expeditions throughout Russia, mainly in the south, and also visited the Near East and North Africa. It is difficult to define Weissenberg as a Zionist, yet he attended the Fifth Zionist Congress in 1905.

Weissenberg devoted a great deal of effort to comparative anthropologic research, trying to determine the extent to which environmental conditions affected what he considered to be characteristics of the Jewish type in Eastern Europe. He showed that the growth rate of babies is greatly affected by circumstantial factors, such as nutrition, economic situation, and more. Thus he rejected the claims that it was heredity rather than the environment that was responsible for the physical condition and characteristics of the Jews in the Pale.

Not race, but rather the environment – lack of proper nutrition and clean air – was responsible for the poor development of breast circumference of the Jews of central Russia and Galicia.

At the same time, he adored the traditional Jewish life style and was proud of the family purity laws, the low infant mortality, and the minuscule rate of venereal disease among the Jews of the Pale. He stressed, however, that these too were not inherited characteristics of the Jewish race. This could be proven by the fact that these advantages were soon lost among Western Jews when they gave up the Jewish life style. Weissenberg concluded that if Jews had advantages that resulted from their unique biology, then these did not come from selection of some physical racial characteristics, but rather from intellectual selection of a life style that preserved Jewry as a cultural entity. Thus, contrary to the Zionist perspective that conceived of life in exile as degenerative, Weissenberg saw Jewish life in exile as a stimulant for intellectual achievement, although he sharply criticized the poor social conditions in which the

---

[3] For a detailed survey of Jewish physicians who researched the origin of the Jews as a race at the end of the nineteenth century and the beginning of the twentieth century, see Efron (1994).

Jews lived. He attributed the poor physical condition of the Jews to the Jewish education system and its curriculum, and to their social habits (Efron 1994, pp. 91–122).

Although Weissenberg insisted that the characteristics of the Jews of the Pale were the consequence of environmental conditions and had nothing to do with the specific hereditary traits of these Jews, he jealously defended the unique hereditary origin that unified all Jews. He wholeheartedly believed that Jews originated from one prototype that could be traced anthropologically. According to him, this racial prototype (*Urjude*) is the one who eventually went into exile. On his visit to Palestine at the beginning of the twentieth century, Weissenberg believed to identify in the Jews of Peki'in, Safad, Shfram, and other settlements in the Galilee – who, according to him, were living there since the destruction of the Second Temple – as the true representatives of the Jewish prototype, *Judaeus primigenius*.

5.1 The Zionist Claim

It seems that the first person who defended the ideas of political Zionism with explicit anthropological arguments was the physician Aron Sandler (1879–1954). Sandler was among the pioneers of Zionism in Germany who participated in Herzl's first Zionist Congresses. Prior to World War I he founded an organization to fight the spread of malaria in Palestine. In the 1920s, he was one of the founders of the "Pasteur Institute" in Jerusalem (together with the 'Nathan Strauss Institue') (see Kolat in Katz and Head 1997, p. 44). Later on, he was among the founders of the *Jüdische Volkpartei* [Jewish National Party] that opposed the Zionist focus on migration to Palestine as the only solution to the Jewish problem (Reinharz 1985, p. 273). In 1934, Sandler eventually immigrated to Palestine and served as the physician at The Hebrew University of Jerusalem until 1948.

At the beginning of the century, in a discourse *Anthropologie und Zionismus*, Sandler endeavored to defend Zionism against attacks from assimilationist circles in Germany (Sandler 1904) by raising the anthropological question of whether Jews were a race. In response to the contention that the Jews were a distinct race – as most researchers at the time agreed – the assimilationists argued that the Jews were never a pure type, but rather a mixed group, and that even if that mixture was considered a race or stock, this "race" disappeared through the introduction of 'foreign blood' over the generations, in spite of the fact that Jews had maintained a hard core of inbreeding [*Inzucht*]. Although Sandler agreed with experts who claimed that it was difficult to determine the validity of the argument concerning the existence of races, since it had been widely accepted that humans are divided into races and since even the "insight of every gutter dwelling lad" [*Weisheit eines Gassenbuben*] could spot racial differences, 'race' had to be accepted as an *empirical fact*. Consequently, Sandler concluded that de facto "a large group of people that differentiates itself from another race, is itself a 'race'" irrespective of whether the origins of the group were from one ancient prototype or from many prototypes, and irrespective of whether they had intermixed to a certain degree over the generations.

The important issue became whether some markers exist that (presumably) allow distinguishing the races; these would be traits that according to the prevailing knowledge, were *acquired* by adaptation to living conditions that over the generations were transmitted as *inherited* traits. Once such premises are accepted, argued Sandler, it follows that the Jews are a race with unique characteristics and identifying markers. One of these identifying markers is the wonderful capacity of the Jews to adapt or assimilate.

This adaptive capacity, under varied circumstances and peculiar cases of exile, only further emphasized the uniqueness of the Jew and the Jewish life style that induced a high rate of inbreeding and further strengthened their specific traits. Sandler thus concluded that de facto Jews were anthropologically unique. According to him, the characteristic traits of Jews, whether acquired voluntarily or under compulsion, were inherited, and ultimately the Jews became a "race," whatever the historic-biological justification for racial uniqueness may be.

According to Sandler, Zionism does not depend on the issue of race: "there is no need to emphasize repeatedly that the anthropological foundation of the Zionist idea is not the only one on which Zionism is established, and that in the narrow sense of race as noted above, it is not needed at all for the theoretical foundation of Zionism" (Sandler 1904, p. 28). Sandler believed that the national identification of the individual depends on hidden instincts and impulses, and that these impulses are essentially cultural.

Although, according to Sandler's perception of acquired characteristics, if such acquired impulses persisted for enough generations as "racial" markers, it makes sense that these characteristics eventually became inherited. The impulses that are liable to raise national feelings or sentiments, whether acquired or inherited, varied. Among Zionists, for example, both humanitarian sentiment and historical identification could provoke impulses that stimulate a national consciousness as a free and voluntary act. In other words, the expression of national impulses depends upon the cultural-social environment of an individual. Following this logic, anyone who identifies in himself Jewish characteristics will naturally oppose assimilation, and sooner or later his inborn impulses will stimulate his identification with Zionism (Sandler 1904, p. 34).

One argument against Zionism was that it induced negative biological processes. Whether Jews formed a separate race or not, their culture and life style would create reproductive patterns unique to them (presumably fostering inbreeding and selection) that would be disrupted when the Zionists materialized their goal. Thus, Zionism harmed the causes of the eugenics notion, the movement for the preservation and improvement of the human species, which – as mentioned earlier – was popular at the beginning of the twentieth century. At first glance, such a claim by the anti-Zionists is puzzling, since the Zionists, and especially the immigrants in Palestine, emphasized the eugenic aspect of their project (see Chap. 7). Yet, here, once again, we encounter the problematic approach of Zionism: while claiming the continuity of the ancient inhabitants of Palestine of thousands of years ago to their direct descendents, present-day Jews, Zionism also endeavored to change (the Diaspora-type of) those Jews. The intensive use by all parties of neo-Lamarckian arguments concerning the inheritance of acquired traits illustrates how scientific ideas may flexibly serve to support, or to oppose, any particular conception.

5.1 The Zionist Claim

An accusations leveled against the Zionists was that "Zionism sanctions consanguineous matings, while consanguineous mating leads to degeneration" (Sandler 1904, p. 35). Indeed, anthropologists generally agreed that consanguineous mating played a crucial role in the development of humanity throughout history, and more specifically in the degeneration of cultures.[4] Sandler accepted as a proven fact that all nations in which consanguineous mating were common died out. But there are exceptions.

In Sandler's opinion the very survival of the Jews, in spite of consanguinity that is common among them, indicates that insofar as signs of degeneration appear among Jews, these are not the consequences of consanguinity. Rather the poor hygienic conditions of their ancestors over the centuries, and of their occupation in learning and in vocations suppressed the physical development of the body. Furthermore, Sandler insisted that consanguinity may also have positive consequences: This was the fastest and most efficient route to select positive traits and to maintain them against the "dilution" entailed in racial mixing. If someone raised misgivings concerning the negative effects of Jewish consanguinity, it was precisely Zionism that offered a solution: Contrary to the situation in the past when Jews from each community were inclined to marry among themselves, Zionism and the Ingathering of the diverse Exiles in "the Land of Israel" would inevitably lead to many marriages among Jews from different countries, and obviously the rate of consanguinity would decline considerably.

Upon characterizing the races, physical anthropologists, led by Virchow and von Luschan, insisted that languages, traditions and cultural traits should not be confused with physical characteristics, such as facial features, skull dimensions, and the like, which are inherited. Yet, the puzzle remained: to what extent can explicitly traditional and cultural characteristics become hereditary?

Sandler repeatedly struggled with the problem of the material-biological basis of the intellectual characteristics of the Jews, and the extent to which these are inherited. In his paper he responded to a publication by Albert Reibmayr, "On the effect of consanguinity and mixture on the political nature of a population."[5] Such publications that endeavored to interpret in biological terms social and political arguments were quite common at the time. Reibmayr asserted that the conservative or the liberal tendencies of an individual were primarily hereditary.

To Reibmayr's mind, there was a correlation between consanguinity and hereditary conservatism. According to him, the Jews at the time of Ezra, the ancient Egyptians, and the Brahmins of today were nations that encouraged consanguinity as well as conservatism. "The Pharisees formed a conservative party and led a lifestyle founded on strict consanguinity, whereas the Sadducees, among whom intermarriage was common, were the liberal party."

Reibmayr drew parallel conclusions with reference to his era: The Jewish people were divided into two camps – the Zionists and their opponents; the Zionists repre-

---

[4] Notice Nordau's *Degeneration* (1895) and the discussions it instigated (see Chap. 4).

[5] In 1897, Dr. Albert Reibmayr published a book, *Inzucht und Vermischung beim Menschen*, with an appendix of a chronology of Jew-persecution since the birth of Christ. It was supposed to be a "cultural history" of man, and to provide evidence that inbreeding and miscegenation played a significant role.

sented the nationalistic, strongly conservative party while their opponents, the assimilationists, were more liberal. Such a subjective subjugation of "historical facts" as evidence for "hereditary Zionism" was quite prevalent at the beginning of the twentieth century: At this age of "Social Darwinism," Darwin's principles of the evolution of species were extended to a wide range of social phenomena in human society and attributed to the struggle for survival among individuals or groups of various cultural and social characteristics (see Chap. 3). Sandler too accepted the ideas of Social Darwinism and, thus, agreed with many of the ideas of Reibmayr, although he refrained from associating the Pharisees with contemporary Zionists, given the fact that many Zionists professed extremely liberal ideas.

Sandler, like many thinkers of his generation, believed that characteristics *change* over generations, while at the same time he believed that the essence of biological entities was inherited, i.e., *constant*. Likewise, the Zionist neurologist-physician Dr. Isidor Isaac Sadgers, wrote in 1897 that "with reference to certain traits that cannot be changed the Jew was compelled to adapt himself to the climate of his country of residence and to the social atmosphere in which he lived. Both the culture and the rate of progress of his land of birth affected the change that occurred in him, and inheritance further fixated these changes." And Dr. Ignatz Zollschan, at the time a courageous resolute campaigner for Zionism and against racism, wrote about the effect of environmental conditions on the unique character of the race: "races are something that emerges" (Doron 1980, pp. 406–407).

Another decade passed before the science of genetics, founded at the turn of the twentieth century, provided the tools that would eventually allow researchers to unravel, conceptually and empirically, the issue of *nature* versus that of *nurture*, and to distinguish between the traits and the factors for these traits (Johannsen's 1909 distinction between the *phenotype* of a trait and its *genotype*). Traits are not inherited; traits are the consequence of hereditary and environmental factors, and the interactions between those factors are complex and can change from one situation to another. This is especially true for mental and behavioral characteristics (see Chap. 3).

If the character of these anthropological arguments of the beginning of the twentieth century seem odd to contemporary readers, it must be noted that such disputes on the relative importance of hereditary factors versus those of the environment in determining behavioral and mental characteristics continued to engage scientists and politicians not only among Zionists and their opponents, but also among those in the socio-political arenas in Europe and America at large. As a matter of fact, the dispute still continues in the present age of genomics.

One of the great achievements of population genetics in the 1930s was the analysis-of-variance (ANOVA), which by statistical methods parsed the variance of properties into the proportion contributed by heredity and the proportion contributed by the environment, (and the residuum that was due to the interaction between genes and environments). This, however, inadvertently gave new impetus to segregating *nature* versus *nurture*. The established lines of thought of the ancient social and political notions of the disputants have not vanished and are still raised even today by scientists. No wonder they also have been adopted by the popular conception of heritability (see, for example, Nelkin and Lindee 1995. See also Evelyn Fox

Keller 2010). The current daily news about complex characteristics such as homosexuality, shyness, or alcoholism, being reduced to isolated "genes for" do not vary in principle from Albert Reibmayr's proposition of the hereditary basis of conservatism and liberalism proffered at the beginning of the previous century. One may add here the continuous dispute over the inheritance of intelligence (see, for example, Murray and Harnstein's book *The Bell Curve,* 1994). The use of concepts from the natural sciences for the social and political needs of various interested parties has not been exclusive to thinkers and researchers of the past.

5.2 People of the Middle-East?

Dr. Elias Auerbach (1882–1971) immigrated to Palestine in 1905 and lived in Haifa for the rest of his life. Auerbach was among the first who argued that the Jews were not only a race, but a "pure" race. In 1907 he published an article claiming the racial purity of the Jews and tried to explicate the historic conditions that, according to him, were active in shaping the physical features of the Jewish race.

Like many, Auerbach began by making the traditional claim that contemporary Jews had their origins in the ancient population of Palestine. According to Auerbach, Jews who were dispersed to all corners of the world preserved their biological uniqueness, rather than merely their national and religious identity. Auerbach tended to adopt this story uncritically and interpreted it often freely, without even considering the historical evidence at his disposal. According to Auerbach, since the days of Titus, the instinct for the self-preservation of the Jewish race confined them to marry exclusively among themselves: Jews were encouraged to settle in border zones and areas of fast development and, thus, advanced the economy and the middle-class in these areas until competition engendered their persecution and expulsion. Whereas at the beginning of their settlement in a distant zone, their economic and cultural superiority led to their social isolation, later jealousy and hatred caused their segregation.

Either way, the rate of intermarriage was, in Auerbach's opinion, irrelevant both in ancient times and during other historical periods. Furthermore, those intermarriages that did take place did not significantly change the Jewish gene pool, because the progeny of such marriages lived as a rule outside the Jewish community (Efron 1994, pp. 127–141; see also Weiss 2002, pp. 151–152; and Doron 1980, p. 407).

Arthur Ruppin took a very different position in his 1913 book *The Jews of To-Day*. Ruppin claimed to be a person who "considers the facts objectively to the extent that this is possible." As discussed in the previous chapter, Ruppin deduced from his anthropological analysis that the Jews of the present were a national entity based on racial-biological, historical-cultural, as well as religious foundations. However, when he followed the wanderings of the Jews from their homeland and considered their mixing with the nations of the lands they traversed throughout their exile, he concluded that their biological uniqueness was that of the *nations of the Middle East,* rather than specifically *Jewish.*

According to Ruppin, when the main body of the nation was forced out of the Land of Israel, the emigrants moved along two major arteries: one westward, the other to the north. The westward artery moved through Egypt to North Africa and then split and spread in Greece, Italy, and Spain. The northbound artery passed through Syria to Babylon, Persia, Asia Minor, and the shores of the Black Sea.

Ruppin denied the racial purity of any community and noted that "both migration arteries absorbed into them over the two thousand years of separation a flood of external elements that were assimilated in them" (Ruppin 1930b, p. 26).

> The old dispute of whether the Jews in Exile maintained their "purity" must be settled as follows: from the moment that they left the Land of Israel the Jews absorbed the blood of many non-Jewish nations. However, from the racial perspective, many of these nations were primarily of the same three racial components of which the Jews of Palestine were shaped at the first instance. This is the reason that Jews of later generations resemble in their *racial* structure their ancient ancestors in Palestine, even though only very few individuals' *ancestors* reached as far as the Jews of their fatherland. (Ruppin 1930b, p. 30)

As a matter of fact, Ruppin suggested and developed the notion that the history of contemporary Jews fits the model of a trellis or grid, rather than the accepted model of repeated branching from a single stock. However, Ruppin contested Felix von Luschan's proposition that there are Hittite and Armenoid elements in the origin of the Jews. Eventually, Ruppin gathered a large collection of photographs of Jewish facial patterns categorized according to their communities and countries of origin (see Fig. 6.2) in an effort to support his main conclusion, namely, that whatever the exact composition of the original Jews and whatever changes occurred in their composition during their exile, racially they "belonged" to the Near East (Hart 1995, pp. 168–169). As for the biological aspects of the processes of the racial mix of the Jews, processes that have acquired new dimensions in recent generations, "[a]ny highly cultivated race deteriorates rapidly when its members mate with a less cultivated race." Yet he minimized the significance of claims concerning physical degeneration by marriages between Jews and their Gentile neighbors, because the relevant populations were related and, thus, very similar in their anthropological properties. "The racial difference between Jews and Europeans is not great enough to warrant an unfavourable prognostic as to the fruits of a mixed marriage" (Ruppin 1913, p. 227).

Can a race be altered by external conditions? The answer, according to Ruppin, depends on the point of departure. Politicians with limited horizons will undoubtedly answer in the negative, whereas a biologist, who is less interested in this or that change than in the *potential* for such change, will probably answer in the affirmative.

> Absurd as it would be to expect a European standard of culture from the Negroes in one or two generations of altered social and economic conditions, it would be equally presumptuous to declare against their being able to produce a high state of culture hundreds of thousands of years hence. […] It is highly probable too that an ancient culture affects and moulds the character of a race. […] The rationalism of the Jew thus becomes a result of his ancient culture, and the peoples of Northern Europe will at some future time arrive at the stage at which the Jew stands today. (Ruppin 1930b, pp. 218–219)

5.2 People of the Middle-East?

Ruppin was convinced that the socio-cultural institutions that Jews maintained over generations were what preserved and possibly even fostered their significant biological traits. He repeatedly mentioned the traditional preference of Jews to choose "learned pupils" as grooms for their daughters, and the effect of this preference on the breeding of an improved race. Moreover, intermarriage leads to the loss of the racial characteristics. "It follows that it is necessary to try and prevent it and to preserve Jewish separatism. The only possible way to succeed is to put a stop not only to intermarriage, but to the whole process of assimilation, which begins in denationalisation and ends in intermarriage" (Ruppin 1930b, pp. 227–228).

Thus, humanistic universal arguments for preserving the well-being of the human species were the main arguments for Ruppin's need to preserve the racial characteristics of the Jews. Toward this end, Jews should live together in communities protected from assimilation: "Just as an army in hostile territory is much more easily destroyed when it is divided into small groups than when it is concentrated in a mass, so Jews could best withstand assimilation by concentration in great numbers in one area. The defensive value of local segregation is not its only recommendation; it has the added positive value of creating a center for the production of an individual civilization" (Ruppin 1930b, p. 265).

Eitan Bloom (2008, p. 13) insists that Ruppin thought of himself as a *culture planner*, and as such was a loyal supporter of the theories of the cultural superiority of the nations of Central Europe and of the significance of eugenic measures for the preservation of their achievements. Since the 'original' Semitic race was considered akin to the inferior Bedouin type, Ruppin segregated the Ashkenazi Jews from the main Semitic stock. Referring to von Luschan's analyses of the multiracial origins of the ancient Jews, Ruppin endeavored to distance Semitism from the image of the 'original jew' and to bring him closer to the Indo-Geanirmc races.

The bio-historic assumption of Ruppin was that the deterioration of Jews – such as the acquisition of the "commercial instinct"– took place prior to the destruction of the First Temple, once intermarriage with the "Bedouin Type" [*Beduinentypus*] began. Thus, Ruppin explicitly followed the interpretations of the school of Chamberlain and Gobineau, who attributed the fall of Rome to uncontrolled racial mixing.

Bloom contends that Ruppin's universal humanism was directed at the Ashkenazim, whom Ruppin identified as the definitive Jewish type in modern times. "As far as he was concerned, the original and healthy Jews, who are responsible for the virtues of Jewish culture, belong racially mainly to the Indo-Germanics." Furthermore, Bloom claims that according to Ruppin "modern race research proved that the Semitic element in the Jewish race is degenerating, and the Zionist process of national resurrection [...], being eugenic in nature, gradually dismisses the Semites racial and cultural elements" (Bloom 2008, pp. 104–109).[6]

---

[6] Bloom goes even further in his conclusions: Ruppin's "analysis of Judaism [...] and, more important, its practical implementation, exposes the roots of Palestinian Zionism's discrimination against the 'Middle-Eastern Jews,' and clearly demonstrates the presence of internal Jewish racism and the anti-Semitic aspect of Modern Hebrew culture" (Bloom 2008[II], p. 436).

According to Ruppin, there were three alternatives for establishing massive common Jewish life: the first was to concentrate Jews in Eastern Europe and organize them there as a nation; the second was Israel Zangwill's proposal to settle the Jews in some territory in Africa or America that had not yet been settled by Europeans; the third solution, the Zionist one. Considering the urgency and terrible distress of the Jews of Eastern Europe, only the Zionist solution seemed realistic. Would it succeed? Ruppin was not sure, but the chances were good: the climate was healthy for the settlement of Europeans; the country was not too developed for emigrants from the countries of culture to be enticed to assimilate with the indigenous population; the country was agricultural and would remain so for many years to come, therefore, it was unlikely that the immigrants would give up cultivating the earth and flock back to traditional commercial professions. Furthermore, there was already a sound base of some hundred thousand immigrants in Eretz-Israel.

Unlike the researchers who concentrated on the anthropology of the Jews, the British physician and biologist, Redcliffe Nathan Salaman (1874–1955), early on took the position of a geneticist, or rather that of a eugenicist. Salaman was close to William Bateson (1861–1926), one of the founders of modern genetics. Being a specialist in microbiology, he discovered the genetic basis of the resistance of potatoes to a viral infestation in 1908. In 1926 Salaman founded the British Institute for Research of Viral Diseases in Plants, which focused on the study of potatoes. In 1935 he was elected Fellow of the Royal Society (FRS). Salaman was neither an anthropologist nor a scholar of Jewish studies; his interest in Zionism and the biology of the Jews stemmed from his involvement in the fate of his people and his identification with them, which led to his desire to apply his scientific insights to serve them. Thus, Salaman wished to align Zionism with eugenics. He claimed that the Jewish nation must be rehabilitated in its own country to preserve its uniqueness and the quality of its biological attributes.[7]

Salaman's family, which arrived in England from Rumania three generations before Redcliffe Nathan was born, was well integrated in British society. Brenda Salaman, his sister, was an anthropologist in her own right and was married to the well-known Oxford anthropologist, Charles Seligman (1873–1940), the author of the book *Races of Africa*. Redcliffe's first wife, Nina, the daughter of a successful engineer and Bible scholar, was an outstanding scholar of Hebrew of the Golden Age in Spain. Of their six children, one was a pathologist, one a physician, and a third an architect; another was an artist whose daughter was a singer. Redcliffe Salaman's biographer testified that he was "a man of culture and of wide interests" (Smith 1955, p. 242). He was intensively involved in the life of his community, especially the Jewish community. He was also involved in the affairs of The Hebrew University of Jerusalem and contributed a great deal as a member of the university's Board of Trustees. In the early 1930s he was a member of the Hartung Committee that examined the administrative and scientific organization of the young university in Jerusalem. He apparently played a crucial role in the reorganization of research

---

[7] See Stone (2004) and Endelman (2004). Both authors make the point that Salaman's interest in eugenics provided a link between his scientific interests and his Judaism.

5.2 People of the Middle-East?

and teaching of the life sciences at the university (personal conversations with the late Jacob Wahrman. See also Smith 1955, p. 243). An active member of the eugenics movement, he recommended that the researchers of the university take advantage of the unique and fortuitous opportunity to study the anthropological sources of the various communities of Jews in Eretz-Israel.

During World War I, Salaman became the medical officer of the Hebrew Regiments, and in this capacity, more than 5000 men "passed under his hands." In April 1918 he arrived in Egypt and later joined the forces that conquered Palestine (Salaman 1920).

Unlike most anthropologists of his time, Salaman's professional training and his family background prepared him to use the modern methods of genetic research. It was no coincidence that Salaman was the first to try to base the biology of the Jews on the foundations of the new science of genetics. Already in 1911 his paper "Heredity and the Jews" was published in the first volume of the professional English *Journal of Genetics*. In this paper, Salaman tried to apply the new tools of Mendelian inheritance, which provided the basis for modern hereditary theory, to anthropological research:

> The object of this paper is to lay before Anthropologists some results in the domain of Ethnology which, though arrived at by methods as yet foreign to anthropological research, promise a rich harvest in every direction. Mendelian methods [...] have for the last decade been the all-powerful weapons of the modern student of heredity (Salaman 1911a and 1911b–1912, p. 273).

Salaman emphasized the fact that the Jews comprised a coherent biological entity. He pointed out that "Ethnologists may be said to agree that the Jew is not racially pure, but on the other hand [...] the Jews constitute a definable people in something more than a political sense, and that they possess though not a uniform, still a distinguishable type" (Salaman 1911a and 1911b–1912, p. 278). Jews vary with respect to colour, cephalic index, and stature, like any other population; "Jews cannot be defined according to any of these standards. There is, however, one characteristic which rarely escapes attention, and that is the Jewish facial expression" (Salaman 1911a and 1911b–1912, p. 190). A Jew, according to Salaman, may be recognized by his facial features.

With the help of 'unbiased judges' Salaman classified the progeny of 136 families of intermarriage between Jews and Gentiles. The progeny of these families were classified as 328 'Gentiles,' 26 'Jews,' and 8 'Intermediates.' The 'Intermediates' together with the 'Gentiles,' were almost all (91 per cent) non-Jewish-type progeny. Among the progeny of 13 families of intermarriage of a 'Hybrid' and a Jew or a Jewess ("backcrosses" in the genetic terminology), there were 15 'Jews' and 17 'Gentiles,' i.e., a good approximation to a 1:1 ratio. This, Salaman claimed, suggested that Jewishness is inherited and may be reduced to a single Mendelian factor, whereby the Jewish allele is recessive to the Gentile one (Salaman 1911a and 1911b–1912, pp. 281–285).[8]

---

[8] A gene may have several modes or *alleles*. In sexual reproduction, the progeny receive one copy of each gene from the mother and one from the father. When the two copies of the gene are simi-

In other words, the Jewish type has a solid biological basis, according to the most advanced scientific achievements of the time. This approach led Salaman on a parallel path to that of the Zionist political movement: One may assert that biology served as a rationale for Salaman's Zionist outlook.

Salaman's research applied the modern methods of genetics to the theoretical claims of physical anthropology, which is typical of the first two decades of the twentieth century. Eugenicists conducted genetic research specifically in order to prove that modern social life distorted the processes of natural selection that in the past had formed the characteristics of (European) human culture. Thus, these eugenicists identified the relatively simple means required to stop that deterioration. For example, the American researcher Charles Davenport (1866–1944) claimed that the skin colour of Blacks, or their curled hair structure, differed from that of Whites in the alleles of not more than one or two genes, and the same was true for eye colour and many other differentiating traits.

Such claims of simple, straightforward inheritance of known characteristics supposedly indicated that the solutions were also simple and unequivocal. Yet, it must be pointed out that in spite of Salaman's intensive involvement in the life of his community and the Zionist project, he repeatedly stressed that his interest in Zionism was purely professional. The Zionist experience gave him, as he claimed, a unique opportunity to examine universal principles of population genetics and Darwinian evolution in human beings.

> But to me the interest of the whole Zionist movement is, I think, much more scientific than idealistic. What will evolve from an unchained Jew on a land and in an atmosphere of his own? We shall hardly see it ourselves unless it be a failure, but if not a failure, then the fine fruits cannot be expected till we have the third and fourth generation on the land. (Salaman 1920, p. 192)

As many before him, Salaman accepted the assumption that Jews were originally a mix of several tribes and races. However, contrary to his predecessors – once Mendelian genetics was established – he rejected the claim of the inheritance of acquired characteristics. For him, the Zionist project of settling farmers provided a golden opportunity to follow natural selection in action in the Jewish gene pool. On that occasion, another interesting, subsidiary anthropological detail regarding Jews was revealed:

> It is with no small amount of hesitation that one attempts to deal with a problem, the subject matter of which has hitherto been the legitimate terrain of the Historian and Archaeologist. The origin of the Philistines, or rather their cultural and social relationship to other nations and peoples, has been the work of the Egyptologist and the students of the dawn of European History. Their history as a people is only known to us from the Bible where they played the unenviable part of a feared and dangerous foe and rival to the Israelites. [...] But it is as a student of Genetics, that youngest offshoot of Biology, that I venture to attack the problem and attempt to show not only who the Philistine was, but where he disappeared and where he may be found. (Salaman 1925, pp. 1–2)

---

lar alleles, the progeny is *homozygous* for that gene. When the two copies of a gene are of different alleles the progeny is *heterozygous* for that gene. Often only the input of one of the alleles is expressed in heterozygous individuals: This is the *dominant* allele, and the other, non-expressed one, is the *recessive* allele.

5.2 People of the Middle-East?

To achieve this, Salaman took us first on a tour of the streets of London of his day:

> Among the Jews of purest Jewish descent it is common to find that there occur individuals who are usually considered to be quite non-Jewish looking. By non-Jewish [facial patterns] it is found that what is meant is that the facial appearance is totally unlike the Hittite type and very different from the Semitic. […] The general cast of features is predominantly Western European in character. […] individuals of this type may be quite fair. […] In Cairo and Palestine the same type was found amongst Sephardim as well as the Ashkenazim. (Salaman 1925, pp. 3–4)[9]

These features in the Jewish community, together with such features as "smallness and refinement," led some authors like Fishberg to conclude that they indicate a mix of Teutonic and Slavic blood with Jewish blood. In Salaman's opinion, such a conclusion is wrong. He categorically rejected the claim that the "variability of the Jewish types parallels that of their surrounding non-Jewish environment," presumably as a consequence of intermarriages with their neighbors.

Salaman agreed that in some locations, like the Caucasus and Yemen, facts support the legend that non-Jewish local blood had infiltrated the Jewish race; also, "in the Mediterranean basin where the Jewish communities are darker in colour than those of the rest of Europe, this is due to the fact that such communities are made up of Sephardic Jews, who as Marranos [Christianized Jews of Medieval Spain] brought with them into exile no small amount of Moorish and Iberian mixture" (Salaman 1925, p. 5). However, according to Salaman, intermarriage between Jews and Gentiles was very rare among Ashkenazi Jews in the rest of Europe until fifty years before his time (see also Chap. 6).Salaman also denied the possibility that the adaptation to environmental conditions produced "Gentile features" among Jews. Thus, Salaman concluded that the so-called Gentile characteristics among Jews were nothing but the emergence of the ancient Philistine genes. In his opinion, there were good reasons to believe that the disappearance of the Philistines as a nation and as a racial entity coincided with their complete assimilation in the main body of the Jewish nation.

To be honest, this was not a new idea. It was suggested by von Luschan and others already in 1892, and was evident in Emoritic figurines of red hair and beard, found on Egyptian tombstones. As mentioned, Fishberg emphatically rejected such an explanation, claiming that blond Jews indicated the non-purity of the Jewish "race." But the claim came up again later: In 1934, Tschurtakover, a Jewish physician in Lvov [Lemberg], examined this concept in a paper published in the [Hebrew] journal, *Harefuah,* and finally rejected it (Tschurtakover 1935). Also Salaman (1925) studied the historic sources and the archeological data of the tribes that were called Philistines, or similar names, in the civilizations of Israel, Egypt, and Babylon.

> The outstanding character of all, [...] is that in appearance they form a group who are unquestionably European in countenance. […] It is moreover a markedly dolicocephalic [long skull] race with which we are dealing. […] These two differential features of dress, the helmet and the shield, give the clue as to the more immediate origin of these

---

[9] See also note 11 in Chap. 2: Adolphe Bloch (1913) on the origin and evolution of the blond Europeans.

proto-Philistines as we may call them. Both the head-dress and the shield are typical of Caria, which was a Cretan settlement in the South-West corner of Asia Minor. (Salaman 1925, p. 11)

On the basis of the Biblical text, Salaman concluded: "The Philistine was gradually absorbed into the Israelite nation – he was never lost. His presence with us today is proved by the existence of the Pseudo-Gentile type in our midst" (Salaman 1925, p. 16).

With a battery of arguments such as these, Salaman upheld his claim that "the racial purity" of the Jews (meaning predominantly Ashkenazi Jews. See Chap. 6) was maintained even though they lived among non-Jews for centuries. He "proved," so to speak, that the racial variability of the Jews, which others said was the result of the infiltration of European characteristics into Jewry over the generations of exile, was not Teutonic, but rather indicated their multiracial origins in ancient times. Salaman went even further: It was not the Jewish race that was diluted by the injection of foreign blood over two thousand years.

On the contrary, the Hellenic race is the one that was injected with elements of Jewish as well as non-Jewish blood! "So it is with the Jews, [...] the recessive Pseudo-Gentile type asserts itself as the laws of Mendel would lead us to expect, and gives us today Jews who are physically indistinguishable from their Philistine enemy of old" (Salaman 1925, p. 16). Salaman apparently forgot that he had "proven" that the Gentile-type is dominant, whereas the Jewish type is recessive.

5.3 A Political-Social Perspective

Social-political thinkers, such as Karl Kautsky (1854–1938), a Marxist and one of the leaders of the German Social-Democratic Party, opposed Zionism and its reliance on the hereditary and environmental determination of the characteristics of the Jews. Kautsky fought strenuously against violence and for a gradual progression towards social order. In the second edition of his book, which was published in 1914 and translated into English in 1926, *Are the Jews a Race?* Kautsky probed the social-political meanings of the wide use of the concept of race. Contrary to Sandler, who accepted the existence of race as a given fact that every "gutter dwelling lad" could identify, Kautsky examined the concept of race in light of the arguments of the biologists. He concluded that not only was there no biological basis for this concept but, moreover, it was meant to serve as tools in the hands of one group of people to exploit another group of people and deprive them of their basic rights. Zionism, by adopting the claim that the Jews are a race, dragged the Jews into a negative social process of exploiting nationalism instead of honoring the rights of assemblies of persons and increasing understanding among them.

Kautsky noticed that together with the industrial revolution and the establishment of the new capitalistic theories of production, scientific reasoning had replaced religious reasoning. The bourgeois intelligentsia accordingly anticipated that its social expectations would correspond to advances made in the life sciences. Thus,

5.3 A Political-Social Perspective

in order to justify colonial policy as a natural necessity, bourgeois theoreticians advanced the concept of race as the foundation of the argument that it was Nature that created master races that oppressed other races. Just as in the past, religion served to deny and suppress any expression of secular interests, now life sciences supported exploitative interests as being natural and eternal (Kautsky 1926, pp. 16–17). The claim to the absolute truth of the life sciences became an instrument in the hands of the politicians of his time, just as the uncontestable truth of religion had done in the past. This also served as the foundation of modern anti-Semitism. Contrary to naïve anti-Semitism of previous generations, the anti-Semites of his day proudly presented the scientific approach as justification for their liberation from past religious and social prejudice. Kautsky thus asked: what were the social targets of those who conceived of Jews as a distinct race, and why did the Zionists consider it appropriate to support such an argument for a distinct Jewish race?

Kautsky accepted the Darwinian argument that humans had undergone a process of evolution, just like any other living organism. However, whereas most of the evolution in other creatures was expressed in their *adaptation to the environment*, Kautsky claimed that, in man and the animals that he bred, *the environment was adapted* to the creature.[10] Therefore, argued Kautsky, human evolution favored the development of a special trait, namely, the capacity to adapt creatures to the environment that humans controlled – in other words, the evolution of mental and cognitive capabilities.

At the same time, Kautsky emphatically rejected the proposition of the constancy of hereditary factors. In his opinion, acquired characteristics are inherited. As evidence, he cited Maurice Fishberg's data on the similarity of the traits of New York Jews to those of non-Jews, proving that typical Jewish traits may change. He found further support for this claim in the work of Franz Boas (1858–1942), the German-Jewish physical anthropologist who immigrated to the United States in 1887.

Boas demonstrated that the classical parameters of anthropologists, which serve to associate individual attributes to their "typical" race, particularly the dimensions of the skull, were culture dependent, and were found to change among the immigrants from Europe to the United States the longer their sojourn in the New World.[11] Although Boas took care to stress that he was not discussing the stability of inherited traits and that his comments related only to the methods by which contemporary anthropologists determined stability according to their prejudiced conceptions, nevertheless, his findings were widely interpreted as proof of the inheritance of acquired traits.

Kautsky rejected that there was any *scientific* basis for racial classification. He found support for this in his interpretation of the process of evolution based on the inheritance of acquired characters, and in the inconsistency among the natural scientists themselves when referring to the concept of race. With respect to the

---

[10] This is an important insight. Today we understand that this is not an issue related only to man. There is no process of adaptation of a living being to its environment without a corresponding adaptation of the environment to the organism. See, e.g., Futuyma and Slatkin (1983).

[11] Kautsky did not discriminate between the inheritance of traits or properties and the inheritance of the factors (genes) involved in such traits and properties.

inconsistency of scientists, he noted that even Darwin had the impression that different scientists classified humans into races in various ways – as many as 63! In desperation, many anthropologists classified humans according to semantics. Kautsky quoted von Luschan at the first congress on race in London, 1911:

> Coloured people are often described as savage races, but it is comparatively rare to find any attempts to give a proper definition of *coloured* and *savage*. [... M]any books have been written on the differences between races of men, and serious scientists have tried in vain to draw up an exact definition of what really constitutes the difference between savage and civilised races. (Kautsky 1926, pp. 68–69)

Von Luschan, however, did not deny the *existence* of races; he only denied that races descend from a common origin and that some races were *inferior* to others. The question of the number of human races has become more a subject of philosophical speculation than of scientific research. The constantly changing environment in Europe, Western Asia, and Africa, of constant inventions, discoveries, and acquisitions, of incessant trade and traffic, have made us what we are, contrary to the Australians who remained isolated for fifty or hundred thousand years.

> We have thus three chief varieties of mankind – the old Indo-European, the African, and the East-Asiatic, all branching off from the same primitive stock, diverging from each other for thousands, perhaps hundreds of thousands, of years, but all these forming *a complete unity, intermarrying in all directions without the slightest decrease in fertility*. (Kautsky 1926, pp. 71–73)

Kautsky, as mentioned, saw the division into races as a social-economic rather than biological differentiation. Therefore, according to him, the future belonged not to the segregation of races into exploiters and exploited, but rather to the amalgamation of the races into one human race. The first stage would be intellectual amalgamation and economic equality, which would cause a "reduction or weakening of the differences." His attitude to the Jew and hence to Zionism must be understood as deriving from such expectations.

Kautsky denied Zionist arguments that Jew hatred was nothing but one aspect of the attitude toward race. We find such hatred among those most related to the Jews, just as we find it among Europeans in general. The racial argument for anti-Semitism is exploited by interested parties, whether local or international, for their own purposes.

> It may therefore be assumed in advance, in the case of a group of humans that have marched for tens of centuries in the front rank of the process of economic evolution, that have undergone the most extensive migrations, economic and political revolutions, that there is no possibility that such a race may be a unit or a pure race.
>
> But we are told this statement does not apply to the Jews. It is claimed, again and again, that the Jewish race has maintained itself in its purity since time immemorial [...]
>
> This view [of the continuity of the Jewish race] is widely accepted to this day as an irrefutable and unquestionable fact, a fact which is so irrefutable and unquestionable that its advocate forgets to state what are the appallingly constant and immutable traits of the Jewish race. The race theorists usually hand over this scientific task to the cartoonists and the comic papers. (Kautsky 1926, pp. 90–91)

5.3 A Political-Social Perspective

Most competent scholars found that the main identifying feature of the Jews was the shape of their nose. Fishberg, who measured the noses of 2836 Jewish males and 1284 Jewish females in New York, found that only 12–13 percent of them had "a Jewish nose," but the aquiline nose is under no circumstances limited to Jews. The findings were similar for hair colour or eye color, as well as for the form of the skull. The Jews of his time, claimed Kautsky, were not a pure race by any geographic or chronological criteria. He accepted the argument that as long as the Jews dwelled as a nation in Palestine, environmental conditions necessarily predisposed them to form a uniform geographic "race." However, this unifying environmental factor did not exist while in exile. Kautsky concluded that the Jews gradually were transformed from a nation to an international brotherhood, and the unifying connection that continues to exist among them is the relic of their national life, namely their religion (Kautsky 1926, p. 113). He conceded that it was only natural that the success of the anti-Semites in grounding their hatred on pseudo-rational arguments of racial differences had instigated a counter-reaction among the Jews that was expressed by a desire to take advantage of race theory, as the Zionists did: "If this theory permits Christian-Teutonic patriots to declare themselves demigods, why should Zionist patriots not use it in order to stamp the people chosen by God as a race of nature, a noble race that must be carefully guarded from any deterioration and contamination by foreign elements?" (Kautsky 1926, p. 17). In other words, the exploitation of the racial argument by the Zionists is as faulty as that by the anti-Semites, and for the very same reasons. Furthermore, even though it might sound like a paradox, the fact is that many Jews were worried about the penetration of emancipation in Eastern Europe, because assimilation would consequently increase there too. The Jews would be assimilated in their environment and disappear if treated as equal and free humans. His dispute with the Zionists was that they saw the preservation of traditional *Jewry* as a more important objective than the elevation of the status of the *Jewish* person as an individual.[12]

Kautsky did not belittle the persecution of the Jews in Eastern Europe that eventually led to mass migration to Western Europe and the United States. In his words, a Jew tired of persecution who still had the energy to act – would undoubtedly emigrate. He was aware, however, that wherever the Jew arrived as an immigrant, he would be an unwelcome stranger. The reactionary powers in America, who rejected the Japanese, Chinese, and Blacks, opposed the immigration of the Jews with the same passion. Thus, he was sympathetic of the Jew who believed that he might be safe from suppression only in a country where he would not be a stranger, in a country of his nationhood. Only in a real Jewish state will the emancipation of the Jews be possible.

---

[12] See also Jabotinsky's dispute with Kautsky (Bilski Ben-Hur 1993, pp. 119ff): "Jabotinsky developed his first theory of Jewish nationhood in response to critics of Zionism (primarily Kaotsky)" [sic!].

> This is the guiding thought of Zionism. Even among the circles of Western European Judaism, this idea has in recent years been replacing the idea of assimilation, of equality of rights within the existing states, [...]
>
> Zionism meets anti-Semitism halfway in this effort, as well as in the fact that its goal is the removal of all Jews from the existing states. (Kautsky 1926, p. 183)

He had no doubt that the Zionists would find a sympathetic ear upon taking the first step to lead the Jews back into society. But he doubted that the Zionists would also successfully accomplish the second step, namely settling the Jews in a country of their own: In the "world of culture," all areas were occupied. A Jewish state would be feasible only outside the world of culture under the patronage of a non-Jewish power. For some time, the Zionists considered settling in an East African country, although by the time Kautsky wrote his book, the Zionists were focused on Palestine as a Jewish state. Kautsy did not doubt the powers of the Jews and the ideology in their hearts, but he doubted that a Zionist project in Palestine could succeed because it was ridiculous to expect the Jewish race to be capable of adapting to agricultural life, and so where would they find farmers in the new Zion? The problem was not that the Jews were *Jews,* but that they were *city residents*. It was absurd to try to suggest to an urban community anywhere in the world to become farmers! And how would it be possible to develop a meaningful industry in Palestine? The local market was not big enough, and the need to export would encounter heavy competition. Also, with respect to natural resources, Palestine had not yet proven itself. Kautsy quoted Ruppin's estimate that out of 2,000,000 Jews who emigrated from Eastern Europe in the years 1881–1908, 1,600,000 went to America, 300,000 to Western Europe, and only 26,000 to the "Land of Israel." Enormous amounts of money had been invested in the settlement project, but with only minor success. Of those who had settled in Palestine, many later emigrated, mainly the young persons.

Kautsky initial pessimism about the Zionist project in Palestine apparently changed to some extent in the second edition of his book, which was published after World War I. A few years after the war, the situation in the country actually improved – roads were built as well as irrigation systems; agriculture developed, and even cultural institutes were established. Furthermore, the immigrants had changed considerably overall from being mere beggars to productive workers and intellectuals with a pioneering spirit. The great problem remained as to how it might be possible to continue at that rate. In this respect, Kautsky was less optimistic. One could already witness the Jews flocking into towns and returning to the typical Jewish professions.

> At best, it might bring about the following partial accomplishment: the number of Jews in Palestine may increase more rapidly than the number of non-Jews in the country (the Arabs) and the new Jewish state, although it will never embrace the great mass of the world's Jewish population, may nevertheless be predominantly Jewish in tone.
>
> But even this prospect is not likely to be fulfilled.
>
> To be sure, the length of time that would be required by Jewish colonisation in order to impress a Jewish stamp upon Palestine would be no argument against such colonisation, [...] if the conditions for the realisation of Zionism were progressively improving in the course of the economic and political evolution. But these conditions do not apply in the case of Zionism [...]

5.3 A Political-Social Perspective

> It is a delusion to imagine that the Jews arriving from Europe and America will ever succeed in convincing the Arabs that a Jewish rule in this country will ever redound to the advantage of the Arabs themselves.
>
> In the early days of Zionism, people were blind to this difficulty. Little more attention was paid to the Arabs than was paid to the Indians in North America. Only occasionally is it remembered that Palestine is already an occupied country. It is then simply assumed that its former inhabitants will be pushed aside in order to make room for the incoming Jews. (Kautsky 1926, pp. 206–209)

Kautsky's analysis concerning the *biology* of the Jews, while not even valid according to the life sciences of that era, was quite characteristic of the many social scientists that endeavored to apply the principles of the life sciences to their concerns at that time. On the other hand, the *socio-political* analysis, which may have seemed to be a utopian ideology of a socialist revolutionary, presented a very sober view of the conditions and forces that shaped the Zionist project. Of course, Kautsky did not envision the takeover of the Nazis and the Holocaust that followed. Yet, he did not conceal his intentions in analyzing the situation of the Jews and the chances of the Zionist movement. His entire book is directed at the idea of the socialist revolution and the important role that the Jewish workers in Eastern Europe were to play in its realization. Therefore, according to him, it was not in Palestine that one should find the solution to the plight of the Jews, but in Eastern Europe, where their greatest number was concentrated. Their fate was intimately linked to the revolution in their country of residence. As far as he was concerned, Zionism was not a progressive movement, but rather a regressive one. The aims of Zionism did not follow the obligatory road of evolution, but rather threw sticks in the wheels of progress.[13]

* * *

The appearance of the Zionist movement as a political movement for the return of the Jews to their homeland did not *raise* the issue of the biology of the Jews, and as pointed out by Sandler, it did not even depend on it; yet the issue of the essence of the "Jewish race" was inherently bound to it from the start. The claim may be made that Zionism merely addressed these issues, and that the ongoing concurrent discussions of the biology of the Jews served as an effective tool in the hands of its followers, as well as of its opponents. Although Kautsky presented a sympathetic position – assigning to Jews, as a cultural-social entity, an important universal task – from his perspective, Zionist national aspirations were an obstacle.

At the onset of the twenty-first century, writes Steven Kaplan, of The Hebrew University of Jerusalem, in a paper cited as an epigraph of this chapter:

> While there may be no clear consensus as to what it means to be Jewish, in recent years authors have been virtually unanimous in rejecting the idea that the Jews are a race […]. However, almost without exception, authors who have argued against Jewishness as a racial classification have based their arguments on the premise, that, while races do exist, Jews do not fit into such a category. Unlike "real" racial groups, they claim, Jews are or have been only *mistakenly* identified as a race. (Kaplan 2003, p. 79)

---

[13] See also the discussion of Laqueur (1972), pp. 416–421.

Repeated attempts to identify Zionism with racism were countered by vast scientific literature that challenged all attempts to sketch the Jews as a racial group. Thus, Kaplan wonders: what is the meaning of the statement that Jews are not a race? According to him races exist as categories of *social structures*. Jews are often described by others, and also often perceive themselves as a "race" in the sociohistoric sense (Kaplan 2003). The Ethiopian immigrants' community in Israel helps support Kaplan's claim: their being "black" as well as Jewish challenges the prevailing conceptions in the dominating society with reference to racial classification and, thus, exposes the essence of race as a social structure: "Funny, they don't look Jewish!" Physical, genetic, and historic characteristics that distinguish the Jews of Ethiopian origin from other Jewish communities recently received special attention (see Chaps. 7 and 8), mainly because they apparently reveal the "racial" conceptions that are very much at the root of the question, "who is a Jew," thereby bringing to light their nature. It appears that there is no good reason to assume that the first decades of the twenty-first century would be significantly different from the first decades of the previous century.

Chapter 6
Eidoth

> From the differentiation between Ashkenazi "idealist workers" and Yemenite "natural laborers" during the Second *Aliya*, through Prof. S. N. Eisenstadt's discernment between Ashkenazi "*olim*" and Oriental "immigrants" during the 1940s, and up to the branding of "outstanding" Ashkenazi and "deprived" Orientals in the 1960s – the casting out of Orientals to the social margins of Israeli life has always been justified by the presumption that the Ashkenazim were considered to be culturally superior in two ways: they were more modern and more ideologically committed to Zionism. (Peled 1999, p. 325. TRF).

As the Zionist ideal materialized, it intensified and sharpened the issue of the historic-biological relationship of Jewish ethnic groups, or *eidoth*.[1] Tensions due to geographic origins and class differences between *eidoth* in the Diaspora are well. Knowing many of them presumably centered on differences in the details of religious traditions, such as prayer styles.[2] But as the Zionist project of settling the Land of Israel progressed, and the religious definition of the Jew seemed to recede, the issue of the biology of the Jewish *eidoth*, their historic origins and ethnic identity, surfaced. As we saw in previous chapters, there was no agreement on the (genetic) nature of Jewish ancestry. Whether or not Jews stemmed from one proto-

---

[1] There is no good English translation for the Hebrew word, *eidah* (pl. *eidoth*). See note 4 in Chap. 1. Whereas communities are usually "locally" defined, *eidoth* are rather "ethnically" defined. Ethnicity is a kind of racism. Max Weber defined ethnicity as "the belief in group affinity, regardless of whether it has any objective foundation" (Gilman 2010, p. 6). Most would agree that the Ashkenazi and Sephardic Jews are two distinct *eidoth*. They are presumably of different geographical-historic origins and have distinct versions of the religious rites. But are Yemenites, or Mugrabs, distinct *eidoth* of similar status, or merely Sephardic (or Middle Eastern) communities? Likewise, how are the Bnei-Israel from India, the Ethiopian community, and others defined? Muhsam (1964) considered it to be a confirmed observation that "in any society where the Jewish minority has lived for several generations, the Jewish group to which we shall refer in the following as *eidah* [...] resembles to a certain degree the non-Jewish majority." He seems to identify any Jewish ethnic group living for an unspecified number of generations next to an accepted non-Jewish ethnic group as a genine *eidah*.

[2] There are different versions of the story of a lonely Jewish inhabitant of a deserted island who constructed two synagogues. When asked why, he answered: "one for me and one for them."

type or more, many years of living in distant communities in the Diaspora certainly generated ethnic differences. Intermarriage with the local population along with the impact of diverse environmental conditions and other random processes that affected mainly small isolated ethnic groups were all factors that widened the biological gaps between distinct *eidoth*. On the other hand, there were continual cultural-religious as well as economic-commercial ties between various communities that often led to the formation of family connections. Traveling Jewish merchants, delegates officially dispatched to foreign communities (often for philanthropic purposes), rabbinic leaders hired by remote communities[3] – all these occurrences, together with the forced expulsions of whole populations,[4] contributed not only to cultural inter-relationships between communities, but also to biological intermingling of ethnic groups and a blurring of lines between distinct but related *eidoth*.

Thus, even if all *eidoth* were linked by a common Jewish tie, obviously mainly secondary, cultural considerations guided their alliances and disaffiliations, and these have not necessarily corresponded to biologically meaningful differences: Why is there one overwhelming Ashkenazi Jewish *eidah* spread all over Europe while the rest of the Jewish populations are segregated into Sephardic, Middle Eastern, Yemenite, and other *eidoth*?[5] Was mainly the basic division between the Ashkenazi and the Sephardic a socio-cultural construct that was devised by historians and researchers to segregate Ashkenazi from non-Ashkenazi, rather than to

---

[3] One of my students pointed out to me that the family name, Ben-Harush (literally: The son of Harush), which is common among North African Sephardic families, is of Ashkenazi origin: Rabbi Asher Ben Yechiel (1250–1327), a great authority on the Talmud, whose acronym was *Harosh* (Hebrew: the head), was born in Germany and active in Rothenberg and Worms. In, 1303, he was forced to flee the threat of persecutions and settled as a spiritual leader in Spain (Toledo). His progeny later moved to North Africa, and the 'Son (= Ben) of Ha-Rosh' became eventually Ben-Harush.

[4] The influx of Sephardic communities to Amsterdam and Hamburg after 1492 is well known. Although legends mainly emphasize the segregation of the Sephardic and the Ashkenazi within the communities, such as that of London's East-End, Romeo and Juliet affairs were not confined to the families of Montague and Capulet.

[5] Defining "Sephardic" exposes most clearly the problem of defining an ethnic group. Presumably, the Sephardim are the progeny of the Jews exiled from Spain and Portugal in the 1490s, who settled mainly in North Africa and in the Ottoman Empire, including Palestine. Formally, they are identified mainly by cultural indicators, such as their distinct prayer style. Were the conditions of centuries of Jewish life in Spain so different from those of other European Jewish communities that a different ethnic group was formed with significant cultural-religious, as well as genetic differences? An interesting attempt to explore the "racial" identity of Portuguese Crypto-Jews, the progeny of Jews forcefully baptized by the Portuguese Inquisition, at the historic as well as the biological (Y-chromosome and mitochondrial DNA) levels, was attempted by Nogueiro et al. (2015). It became clear that the phenomenon of the Crypto Jews is more complex than the hiding of a persecuted identity and maintaining a group identity of Judaism. As further emphasized by Marcus et al. (2015), methodological ambiguities with respect to the inference of Jewish ancestry, such as those of the molecular clock, preclude their usage as "reliable Jewish ancestor predictors."

indicate the variability of Jewish communities?[6] When we consider the history of the *eidoth*, a major question arises about the sources of the biological commonality of Jews: is it due to their ancient origins in the land of Israel, or to the secondary histories of the *eidoth* in the Diaspora? The improvement of biological diagnostic tools at the age of genomics to the level of detailed DNA-sequencing of individuals and populations, compels us to continually question to what extent do the empirical results *dictate* the narrative of ethnic groups, or how far should the detected biological variability be *interpreted* and *guided* by history?

Throughout the more than hundred years of Zionism the division of the Jewish people into *eidoth* was a prominent reality and a large number of anthropological papers were published that claimed to uncover the original Jewish prototype and the origins of the different Jewish ethnic groups, including the explicit identification of "Jewish leftovers" among overt non-Jewish people.

One of the major indicators for racial identification was (as has already been mentioned) the shape of the skull: elongated-skulls (dolichocephalics) versus broad-skulls (brachicephalics). Among the Jewish ethnic groups, the Sephardic were said to be dolichocephalic, whereas the Ashkenazi were classified as brachicephalics. With the establishment of the science of genetics, Salaman and others attempted to identify specific genes, such as those for Jewish facial features, although most scientists agreed that facial features cannot be reduced to a few simple variables.[7] Anthropologists were constantly struggling to invent up-to-date research methods for justifying their science by basing it on time-honored physiognomy and other facial image techniques. I mentioned Galton's (1878) most ingenious method of superimposing photographic negatives to extract basic facial patterns (see Chap. 2). Another technique, also using photographs, was a catalog of faces and facial parts (noses, ears, eye-brows) organized by Cesare Lombroso. These were actually attempts to revive Lavater's physiognomy by replacing subjective drawings with modern, presumably objective, photography (Pick 1989; Sekula 1989). Another official forensic device, finger prints, was introduced in the 1870s.

Systems of classification were used not only by anthropologists, but also by behavioral scientists who classified humans into types by linking body traits with behavioral patterns. Emil Kraeplin (1856–1926) and Ernst Kretschmer (1888–1964) were among the most renowned German researchers of mental diseases. They showed that mental illness was often linked to severe physical deformity and deduced accordingly that both were caused by disturbances in hormonal balance (hormones were discovered at the beginning of the twentieth century). Their conclusions were extended to healthy people by reducing them to three or four types based

---

[6] Nurit Kirsch (2003, pp. 643–644), calls attention to the methods used by several Israeli researchers who divided their subjects into meta-categories of Ashkenazi and non-Ashkenazi, without any explication (see also Chap. 6).

[7] See, for example, the outspoken attack of Harold Laski (1893–1950), a British professor of political science, of Jewish descent who was an avid fighter against racism and colonialism. As a young man, Laski was associated with Karl Pearson, in whose journal *Biometrika,* he published his attack on Salaman (Laski 1912).

Fig. 6.1 Classification of human types by three criteria (Sheldon 1954)

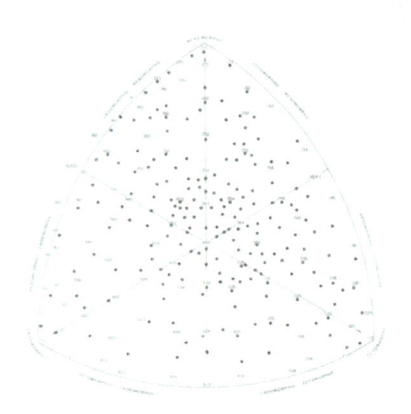

on physical patterns and character traits that accorded with hormonal activity (see Kretschmer 1921 and 1936/1970). Such systems for diagnosing behavioral types and personal character, correlated to body structure, were in use for many years. In the 1940s, the American William Sheldon (1898–1977) suggested a continuum of seven grades for each of three soma types (ectomorph, mesomorph, and endomorph), resulting in a total of 343 types (Fig. 6.1). Sheldon's (1954) classification was quite commonly accepted at the time in universities and colleges, though I am not aware of its applications for classifying races and communities.

Arthur Ruppin joined the researchers who mobilized the photographic technique to classify Jews and the search for the biological prototype common to all Jewish communities. He collected a large number of Jewish types from different *eidoth* and from different countries of origin, as well as from ancient sculptures and reliefs in which Jews were depicted. These were organized as an appendix to his *Soziologie der Juden*, so that it would be possible to use them to draw conclusions about a biological relationship that crosses cultural and linguistic barriers, in line with the ancient tradition of physiognomy. Thus, for example, the photograph of Albert Einstein – the German Ashkenazi prototype – was placed next to Lord Reading's and Sir Herbert Samuel's, together with that of a woman born in Jerusalem, whose father was Sephardic and mother Ashkenazic, obviously insinuating common facial patterns (Fig. 6.2). The legend of the photographs emphasizes that the facial types presented persons from different backgrounds and climates, and that although the lives of these persons followed different paths, they were related by racial ancestry (Ruppin 1930a, picture-appendix; Hart 1995, pp., 168–170). Such classifications often reflected the (not so) hidden aims of their authors: Kretschmer, Lombroso, and Sheldon needed evidence to anchor their classifications in distinct biological, hereditary entities. Ruppin, on the other hand, accepted that there were biological differences between Jews and their Semitic relatives, but claimed that given proper circumstances, the intra-racial (inter-ethnic group) differences were not so much biological as cultural and environmental.

Fig. 6.2 From Ruppin's (1930a) gallery of Jewish types: Albert Einstein ("Mediterranean impact"); Lord Reading and Sir Herbert Samuel (England); and a Jewess, the daughter of a Sephardic father and an Ashkenazi mother, born in Jerusalem

An increasingly important indicator for ethnic differences became "ethnic-specific" diseases. Rare mutations that occurred in reproductive isolates became "flags" for group association. "Of all the age-old categories through which 'disease' is comprehended certainly the one that is most discredited is 'race'" (Gilman 2010, p. 6). Obviously, such ailments may be related to specific environmental variables, like climate and nutrition, as much as to local "racial" genetic factors.

In retrospect, studies of Jewish ethnic groups may be divided into two categories based on their interpretations: the "biology-as-history" approach regards the variability of Jewish communities and ethnic groups as a biological source for understanding the history of the Jewish people, whereas the "biology-as-variability" approach regards this variability as an historic source to learn about the sociocultural, as well as the biological, relationships between variables.

6.1 The Middle Eastern Jew: The Jewish Prototype?

The segregation of Jews into *eidoth* was not initiated by the Zionist movement during the colonization of Palestine. "The Jews do not form one exact anthropological type, but are composed of several types, which are not everywhere the same," commented the anthropologist of the Jews, Samuel Weissenberg (Efron 1994, p. 98), who based his classification on the form of the skull, although he vehemently opposed the current opinion among physical anthropologists at the time that Jews were easily discerned by this criterion. As noted earlier, although Weissenberg insisted on the impact of living conditions and styles on appearances, he objected to the impressions of "travelers" that were not based on systematic research but on superficial markers, such as the Jews' unique garments or their side-locks (*vide infra*). The seven facial types that he identified among Jews were proof for him that the Jews were able to *adapt* to a wide range of environmental circumstances – a capacity that was essential for their survival. Notwithstanding, Weissenberg concluded that the great variety of types of Jews of, say, South Russia was evidence that they were not a pure race. Thus, the obvious question for him was, in what way do contemporary Jews reflect their *Urtypus* (Efron 1994, p. 105). As noted, in his search for the Jewish prototype Weissenberg traveled to South Russia, Asia Minor, and Palestine. During his visit to Jerusalem and Jaffa, he conducted anthropological measurements on some fifty Yemenite Jews. Although it was accepted among anthropologists that the Yemenites were a separate Arab race who had converted to Judaism in the distant past, Weissenberg supported those who claimed that they were related to the (dolichocephalic) Jewish prototype. He examined two communites of Sephardic Jews, one in Constantinople and one in Jerusalem, and found them to be very similar. He thus concluded that the dolichocephalic Sephardis maintained more of the Semitic purity than do the Jews of Eastern Europe. He later examined the Jews of North Africa and concluded that they too mainly resembled the Jews of Palestine and Yemen, thus it was the Sephardic and Middle Eastern Jews who best represented the original Jewish prototype (Efron 1994, pp. 105–119). The "average Jewish type," according to Weissenberg, namely the Ashkenazi Jew, who comprised the majority of the Jews of his day, the Jews of the Russian Pale of Settlement, was very different from the Semitic dolichocephalic type. But, contrary to the common procedure of interpreting the anthropological data according to the historical givens, Weissenberg, the scientist, inverted the procedure and reformulated the history according to anthropological data. At the time, the common version, based on accepted historical claims, was that Jews stemmed from a mixture of dolichocephalic and brachicephalic races, and that the Jews of Eastern Europe came from the emigration of small groups of Western European (brachicephalic) Jews who escaped the eleventh century persecutions. However, on the basis of anthropological evidence, Weissenberg concluded that the origin of Eastern European Jews could be traced back to (dolichocephalic) Jews who lived in Russia in ancient periods. He believed that the socio-political impact of those Jews was so strong that it was a major factor in the conversion of the (brachicephalic?) Khazars to Judaism in the eighth century C.E. and in their absorption into the original Jewry.

The notion that the Sephardic Jews were the more authentic representatives of the original Jews, both with respect to their culture and their origins, prevailed among the *Wissenschaft des Judentums* at the end of the nineteenth century, and many Zionists at the beginning of the twentieth century (Efron 1994, p. 106). John Efron, who analyzed this trend, noted that a group that defined itself as a distinct race would usually emphasize its superiority versus the deficiencies of the others, yet Ashkenazi Jews were those who placed the Sephardim at the top of the hierarchy. According to Efron, the self-image of the Ashkenazi Jews reflected the figure of the miserable, bleak residents of the Russian Pale, which anti-Semites had attached to Jews for generations. Thus, these Jews conceived of the suntanned brown persons of southern Europe and the Maghreb as the ideal racial figure that they could not ascribe to themselves. There was also something of the respect for the 'wild noble' that the Europeans ascribed to the natives of unknown lands, who supposedly developed their own thriving culture (Efron 1993). Many Zionists thus saw in the Sephardic Jew the romantic figure of an ancient forefather. In his 1923 book in Hebrew, *The Knowledge of Our Nation: Demography and Nationology*, Jacob Robinson spelled out the distinctive marks of the Sephardim and their manners: "A thread of graciousness (*grandeza*), nobility of generations and ancestral merit is manifest on the faces of the Sephardim" (Robinson 1923, p. 30). This attitude apparently induced Ashkenazi Zionists to adopt the Sephardic accent as the authentic pronunciation in the renewal of the Hebrew language, instead of the Ashkenazi accent that they had brought with them from the Diaspora. Also, in response to the "assimilationists," the immense contribution of Sephardic Jewry in the past to the culture of their countries of residence provided a kind of promise for the potential of the awakening of Jewish culture within European culture, if only the Jews would be allowed to integrate into society, as they did during the Golden Age of Spain's Jewry. For some non-Jewish researchers, on the other hand, this segregation justified their hostility toward the (Ashkenazi) Jews among them, presumably without being accused of anti-Judaism, *per se*, because the Sephardim are also Jews. In Efron's words, this modern form of racism adopted the ancient anti-Semitic tradition of segregating Jews into good ancient ancestors and bad contemporaries: "In this way, the modern Jew of Central and Eastern Europe, a figure long vilified, was juxtaposed with, by being separated from, the more praiseworthy ancient Semite (Sephardic Jew)" (Efron 1993, p. 81).

This distinction, which identified the Middle Eastern and Sephardic Jews rather than the Ashkenazi Jews as the authentic Jews presented a problem when bearing in mind that Herzelian, political Zionism emanated from the culture and politics of Western Europe. The demand to recognize the Jewish people as a nation returning to its homeland was rooted in European nationalism, and the state envisioned to assimilate Middle Easterners, whether Arabs or Jews was to be "Western." Such a notion of cultural superiority appears self-evident in Herzl's *Altneuland*, and this was the keynote for most of the Zionist leaders that settled in Palestine. How does one put down one's biological roots in the East and one's cultural roots in the West?[8] Ruppin, the great *culture planner* and settler, unequivocally adopted Western standards in the absorption and settling of new immigrants, whereas Jabotinsky, who had devoted much thought to the theory of race and the Jewish *Volk*, encouraged tolerance for the human being, while maintaining a sense of superiority over the culture of "the East." Jabotinsky distinguished the objective conception, which is the racial constitution of the individual, from the subjective conception, which is the

---

[8] Many Western writers and artists made efforts to assimilate elements of the Eastern culture into their European culture. Suffice it to mention writers like S. Tchernichovski, painters like A. M. Lilian and Abel Pan, and composers like Mark Lavri and Paul Ben-Haim.

consciousness of the individual. For him "East" was not a geographical concept, but rather a notion that expresses an early stage in the development of nations. Consequently he vehemently attacked the trends in Zionist thought that ascribed to the Jewish nation "Eastern" origins and character:

> We Jews have nothing in common with what is called "the East," and thank God for that. The uneducated masses must be weaned from their antiquated traditions and laws which are reminiscent of the East […]. We are going to the Land of Israel, first, for our national convenience, and second, as Nordau said, "to expand the boundaries of Europe up to the Euphrates River." In other words, to sweep the Land of Israel clean of all traces of the "Eastern spirit." Concerning the Arabs who live there, that is their own business; however, if we could do them a favor, this would be the one: to help them free themselves from the "East." (Jabotinsky in Bilski Ben-Hur 1993, p. 132).

Even today, when most of the population in Israel is of Middle Eastern origin and much of its culture is increasingly a Middle Eastern culture, Israel is considered by most Israelis to be a 'Western country.'

6.2 On Khazars and Ashkenazim

The history of the Khazars is of special interest to the study of the biology of Jewish ethnicity. The Khazars, one of the 'Turkish' tribes that arrived in Europe in the fifth century C.E. probably from internal Asia, settled in the region between the Caspian Sea and the Black Sea. In the 7th and fifth centuries, some of these tribes acted as barriers to the invasions of the Muslim Caliphs to North Eastern Europe, as well as to the invasion of more distant Asian tribes from the east. Likewise, they prevented the Norman and Viking tribes from invading the kingdom of Byzantium from the north.

The Khazar king and part of his court allegedly adopted the Jewish religion around 740 C.E. The truth of such a conversion and its extent has been the subject of many discussions, and the topic of vehement disagreements in our age of genomic DNA analyses. It may be that the act of adopting Judaism was a way to express the native peoples' objection to pressures from both the south to turn to Islam and the west to turn to Christianity; also, it is likely that quite a few Jewish refugees of the persecutions, primarily in Byzantium but also in the Islamic world, found refuge and settled in the land of the Khazars.

The extent to which the Khazars contributed to the Jewish gene-pool, and more specifically to the Ashkenazi ethnic-group(s), has become a charged issue among expert scientists as well as nonprofessionals. National and ethnic prejudices play a central role in the controversy. Already in the early nineteenth century, "there was lively interest in the lost Jewish kingdom, especially among the Jewish Russian scholars," and interest in Khazaria intensified in the second half of the century (Sand 2009, pp. 230–231). But, apparently, fear of compromising Russian nationalism on the one hand, and Jewish Ashkenazi ethnic group identity on the other hand, combined to suppress researching such claims both in the Soviet Union and among

6.2 On Khazars and Ashkenazim

Jews.[9] In recent years, the historical-archeological research on the Khazars has been revived, hopefully in a less emotionally charged atmosphere. Yet, opinions appear to be divided today as they were in the 1950s when the studies of Ab. N. Poliak, *Khazaria: The History of a Jewish Kingdom in Europe* (Poliak 1951), of Douglas Dunlop, *The History of the Jewish Khazars* (Dunlop 1954), and the book of Arthur Koestler, *The Thirteenth Tribe* (Koestler 1976), were published and bitterly criticized.[10]

According to Poliak,

> The extended existence of a large Jewish kingdom [...] compels us to revisit the truth of the concept, common among us, of the situation of the Jew during the Middle-Ages. We are used to conceiving of the people of Israel in those days as wanderers who occasionally succeeded in constructing, with the grace of foreign rulers, a physical and spiritual center in some country, only to be rejected after some time, then moving on to erect for themselves another temporary center. This notion has been fixated in the modern Jewish historiography from its initiation, under the influence of the Christian religious notion of the Jew who led an eternal life of wandering and subjugation since the time they took upon themselves the responsibility for Jesus' blood. This notion affected Christian researchers [...] to relate to the history of Israel after the crucifixion of Jesus merely as a proof and foundation for that notion, rejecting a priori any fact refuting it. (Poliak 1951, pp. 11–12).

It makes sense that some Khazars gradually adopted a Jewish way of life (rather than formal conversion), and probably many increasingly adopted a form of basic Judaism. There is evidence that in the Crimean peninsula, which was known by the nickname "Khazaria Minor," there existed a kind of Karaite community during the reign of King Bula (the presumably first Jewish king at about 740) and Ovadia (circa 800). Chasdai Ibn Shaprut, the chief minister at the court of the Caliph of Cordova, apparently corresponded with King Joseph of the Khazars somewhere between the years 954 and 961. In these letters, King Joseph attested that he was not

---

[9] In a 1950 paper, allegedly authored by Stalin under a pseudonym, the author conceived of the Khazar story as offending Russian nationalism. Also in Israel, emotions are still high when it comes to the history of the Khazars, as I witnessed in a symposium on the issue at the Israeli Academy of Sciences in Jerusalem (May 24, 2011). Whereas Prof. Shaul Stampfer believed that the story of the Khazars' converison to Judaism was a collection of stories or legends that have no historic foundation, (and insisted that the Ashkenazi of Eastern Europe of today stem from Jews in Central Europe who emigrated eastwards) (cf. Stampfer 2013), Prof. Dan Shapiro believed that the conversion of the Khazars to Judaism was part of the history of Russia at the tume it established itself as a kingdom between the pressures of the Moslem rulers on one side and the Christian Byzantines on the other. Prof. Amitai Reuveni suggested that the Khazars were not really "converted," but simply turned to Judaism without meticulously following the religious laws, while Prof. Israel Bartal suggested that at the Age of the *Haskala* and onwards, the modern pamphlets against the Khazars were the activity of Sephardic organization opposed to the "Khazaro-Ashkenazim." On the other hand, Arthur Koestler's (1976) story of the conversion of the Khazars to Judaism was interpreted as a means to fight anti-Semitism.

[10] Even in recent years, leaflets were distributed in the streets of Jerusalem with the caption: "The Ashkenazim (East European Jewry) are Khazars," with quotations from Poliak's and Koestler's books, and ending as follows: "This message is intended to arouse the public to a renewed concern concerning the processes that shape the Jewish-Israeli society and the Jewish Diaspora of today." Needless to say, the leaflets were not signed.

of Semitic origin, but rather descended from Japheth and Togarma, the forefathers of all Turkish tribes.

However, the Khazars were gradually worn out by their struggles for power with both Byzantine and Russo-Viking tribes. Their impact declined in the tenth century and they completely vanished by the twelfth and thirteenth centuries. It seems probable that they were the victims of the Mongol invasion of Genghis Kahn. Thus, in the thirteenth century the Khazars are mentioned only in Russian folklore as "Jewish heroes" in a "land of the Jews" (*Zemelya Jidovskaya*). Even today in the Crimea, there is a people in Karaite villages, probably of Khazar origin, who speak Turkish. There is no mention of the Khazar in the Medieval Jewish literature in the West. Yet many testimonies indicate that Khazar extended to Slovenian countries, and many of these also mention relations with Jews and Judaism; however, much of the evidence is indirect and circumstantial, such as Jewish names derived from Khazarian names throughout Russia, the Carpathian Mountains, Poland, and Lithuania. The novelist Arthur Koestler, in *The Thirteenth Tribe,* accepts this evidence to support his theory concerning the foundations of large Jewish centers in Eastern Europe. However, it must be remembered that Koestler was not applying scientific methodologies and that his novels are often biased by his preconceived notions. Thus,

> Ethnically, the Semitic tribes on the waters of the Jordan and the Turko-Khazar tribes on the Volga were of course 'miles apart,' but they had at least two important formative factors in common. Each lived at a focal junction where the great trade routes connecting east and west, north and south intersect, a circumstance which predisposed them to become nations of traders, of enterprising travelers, or 'rootless cosmopolitans' – as hostile propaganda has unaffectionately labelled them. But at the same time their exclusive religion fostered a tendency to keep to themselves and stick together, to establish their own communities with their own places of worship, schools, residential quarters and ghettoes (originally self-imposed) in whatever town or country they settled. This rare combination of *wanderlust* and ghetto-mentality, reinforced the Messianic hopes and chosen-race pride, both ancient Israelites and mediaeval Khazars shared – even though the latter traced their descent not to Shem but to Japheth. (Koestler 1976, pp. 125–126)

It may be reasonable to believe that the expanding kingdoms of Poland and Lithuania in the tenth to fourteenth centuries needed the immigrants who came mainly from Germany in the west, but also from the countries in the south and the east, including ex-Khazar elements. It is not necessary to accept Koestler's speculation that the majority of the Jewry of Poland and Lithuania stems from the remnants of the Khazars who were later joined by Jews from the west. Koestler, however, identified many customs of Eastern European Jews (such as the architecture of the synagogues and their ornamentation), their professions (wagon/cart owners), and even their attire (the Yarmulke, the Streimel), as relics of the culture of eastern tribes, to which the Khazars belonged. He goes as far as to suggest that the traditional Jewish *gefilte fish* is a relic of the days their forefathers lived on the coast of the Caspian Sea (Koestler 1976, p. 129).

Stories of the Khazar origins of the Jews were also prevalent in common traditions. The author Joseph Roth (1894–1939) mentioned in a novel he wrote in the early 1930s that in Austria there were many red-haired Jews, "kind of a joke of nature, perhaps an expression of a mysterious law of nature referring to the unknown origin from the legendary Khazar tribe" (Roth 1950; see also Sand 2009, p. 236). This subject is mentioned also in other contexts, but it is difficult to see how evidence at the biological level could be found.

Zvi Ankori of the Tel Aviv University Department of Jewish History criticized Koestler for taking advantage of his literary talents to uncritically sell Poliak's claims, which had been rejected by most historians as unsubstantiated speculation (Ankori 1979). Ankori asserts that Koestler exceeded his professional competence in his interpretations of the findings and writings. Koestler's understanding of genetics is facile. His use of linguistic relationships or physiognomic characteristics is also invalid. Nevertheless, there are others, among them historians such as Prof. Shlomo Sand, who offer evidence that supports the significant contribution of the Khazars to the modern Jewish gene pool. In recent years, molecular biologists like Doron Behar and his colleagues have examined the distribution of the inherited Y-chromosome's DNA markers in the Jewish paternally-inherited priestly castes of Cohanim and Levites (and the remaining Israelites) (see Chap. 8), and they have found that some Ashkenazi Levites carry a unique haplogroup marker not found in other Jewish castes and communities. After struggling with different explanations, they reluctantly admitted that "[a]n alternative explanation, therefore, would be a founder(s) of non-Jewish European ancestry, whose descendents were able to assume Levite status. [...] One attractive source would be the Khazarian Kingdom" (Behar et al. 2003). According to Sand, the "silent lapse in the Jewish Israeli memory" of the Khazars' contribution is due to the "anxiety about the legitimacy of the Zionist project, should it become widely known that the settling Jewish masses were not the direct descendants of the 'Children of Israel'" (Sand 2009, p. 236).[11]

> Although claims of the Khazari contribution to Ashkenazi Jews are old, and are nearly always fiercely rejected by opponents, Sand is not alone. This opposition attests to the common conceptions of the Jewish historians, especially the Zionists among them, and their fear of political consequences. We have already mentioned Weissenberg's claims at the beginning of the twentieth century to the possibility of the Khazar element among the Ashkenazim. Jacob Robinson published in the early 1920s a book intended to serve as a textbook for school children in Palestine. He praised the characteristics of the Sephardic Jews and claimed that the discussion concerning the origin of the Russian Jews was not over. He went so far as to state that "some believe that they came not from the west but from the south, namely that they are the progeny of the Khazars" (Robinson 1923, p. 30). Not surprisingly, the immediate response of Ben-Zion Robstein was a devastating review calling such statements "exaggerations, inaccuracies, drawing of non-scientific conclusion" (Robstein 1924).[12]

---

[11] See also http://www.khazaria.com and http://www.zionism-israel.com/ezine/Jewish_Origins.htm. The language used by several speakers at the symposium in the Israeli Academy with reference to Prof. S. Sand is seldom heard in academic circles. See note 9.

[12] See also modern Palestinian claims, such as Ashkenazim being "clearly closer to Turkic/Slavic than either is to Sephardim or Arab populations" (Chap. 9).

Nevertheless, considering the later ethnographic-linguistic studies of Paul Wexler of Tel Aviv University on the origins of Yiddish, there are convincing indications that Central European Jews came from the Balkan region of Serbia-Kosovo, and there is also some support for an explicit Turkish-Khazar element in Jewish culture, though not of a biological element.[13]

Eran Elhaik reexamined the Y-chromosome DNA haplotype sequences of Jewish, especially Cohanim, descendents, and a wide array of Caucasian and South Russian populations (Elhaik 2013). His study suggests links between the Caucasus populations and Eastern European Jews, supporting the hypothesis that Khazars contributed to the contemporary Jewish gene pool. This is at odds with the narrative that views modern European Jews as being immediate descendents of an assortment of Israelite-Canaanite tribes of Semitic origin. Recently Das et al. (2016) went further and localized most of the origins of Ashkenazi Jews along major primeval trade routes in northeastern Turkey along the Silk Roads.[14]

Not surprisingly this evidence was countered by Doron Behar and coworkers, who suggested that the wrong populations were sampled: More specifically, the people of the South Caucasus who, according to Elhaik, have Khazar blood, were claimed by Behar et al. to actually be the progeny of Mediterranean people who emigrated northward from Mesopotamia and Iran, rather than those of Khazar origins who inhabited South Russia (Behar et al. 2003). Thus, the circular argument recurs and rather than the historians and anthropologists probing genetic-molecular data to provide clinching evidence, it seems that the geneticists seek historical and anthropological evidence to support their molecular data.

6.3 The Merger of *Eidoth*: Assimilation or Amalgamation?

Whereas Salaman's interest and involvement in the prospective contribution of immigrants were unequivocally the consequence of his identification with his people (or with the Ashkenazim) as a British citizen, Ruppin's attitude was primarily the result of his Zionist-humanist notions. His scientific approach was that of an anthropologist who attempts to make use developments in genetics.

As a universal culture planner, Ruppin believed in the major role of the Jewish people, and he feared that the collapse of a religious framework might threaten the cohesiveness of the Jewish nation. Thus, he conceived of the Zionist settlement project as one of the universal, political, socio-economic *devices* in the realization of his goal of preventing the disappearance-by-assimilation of the Jews. "All the higher cultures," wrote Ruppin, "degenerate quickly when their members start to mate with members of inferior races. In most cases the mixing of distant races brings negative results" (Ruppin, in Bloom 2005). Once Ruppin conceived of the

---

[13] Talk by Prof. Paul Wexler in a conference on "Genomic Views on Jewish History." *Ma'ale-Ha'Hamisha*, May 31, 1999. Also Wexler (1993).

[14] See The Trail of Y-Chromosome Haplotypes in Chap. 9.

urgent need for both a physical and conceptual transformation of the Jewish people, he practically adopted the argument of the German economist Werner Sombart (1863–1941) in his 1911 book *Die Juden und das Wirtschaftsleben* [The Jews and Economics]. Sombart ascribed modern capitalism to the Jews who were, according to him, "endowed with a mercantile instinct." The Jews were "planned" biologically, intellectually, and morally for capitalism. Ruppin accepted Sombart's claim that the "ruthless Jewish capitalist behavior" stems from the fact that Jews had an "over-developed commercial instinct," the bio-historic roots of which were implanted in them even prior to inhabiting Europe. This biological characteristic was common among them and among all other Semitic nations. Thus, eugenic means were needed to curb this biological characteristic. Bloom believes that the eugenics program that Ruppin advocated in *The Jews of To-Day* and elsewhere was a "Haeckelian-Lamarckian" attempt to collectively subjugate the commercial instinct that was a component of a system of defects that Ruppin wished to repair in the body of the nation and the Jewish race (Bloom 2005, pp. 98f).

Ruppin represents, to my mind, an extreme example of the conflict inherent in the political-Zionist movement, the roots of which are in the humanist-colonial culture of Western Europe at the junction of the nineteenth and twentieth centuries. On the one hand, he conceived of the *settlement of Palestine* as only a partial solution for the Jewish people, being also a contribution to the human species as an expression of Western culture. On the other hand, he believed that the specific needs of the Jewish people was the *assimilation* of the various Jewish communities (those that were not Western) in their non-Jewish environment, rather than to amalgamate all of them and satisfy their cultural needs by the colonization in Palestine.

As noted in previous chapters, one aspect of Ruppin's conception was his attempt to treat the present hereditary pool of the Jewish people and their ethnic groups in relation to that of the hereditary pool of the people of the Middle East as one extensive web. Rather conventionally stating *a priori* that they all stem from a common stock that had diversified in various countries of the Diaspora. According to Ruppin, the separation of Jews into ethnic groups was primarily a result of their Exile; the Ashkenazim, the Sephardim, and the Middle Easterners all absorbed a considerable amount of foreign blood, each ethnic group according to its specific routes of wandering. None of the *eidoth* is a better representative of the prototype or prototypes of the ancient inhabitants of the country. The inter-ethnic group gap is secondary to the common origin, and according to him, natural selection played a great role in increasing the gap brought about by the Diaspora. In the ghettoes, for example, mental acuity was actively selected by wealthy Jews who preferred to marry their daughters to scholars. Ruppin did not suppress, however, his biased prejudice for the superiority of Western European culture: "It is perhaps owing to this severe process of selection that the Ashkenazim are today superior in activity, intelligence, and scientific capacity to the Sephardim and Arabian Jews, in spite of their common ancestry" (Ruppin 1913, p. 217 footnote).

Ruppin was very conscious of the socio-cultural barriers that were erected during generations of separation between ethnic groups. Already in the *Jews of To-Day* he noted the deep gulfs between the Israeli *eidoth*: "Even the slight religious difference which exists between the Ashkenazi and Sephardic Jews of Amsterdam was sufficient to prevent marriage between the two until the middle of the nineteenth century" (Ruppin 1913, p., 160). These Jews experienced profound demographic change; yet Ruppin was afraid it would be difficult to overcome the inter-*eidoth* barriers in attempting to regroup the Jews of the Diaspora along non-ethnic lines without constructing a critical cultural-linguistic core: "The contrast between Sephardic and Ashkenazi Jews in Argentina, Brazil, and Egypt is founded more on linguistic factors than on economic ones. In Palestine, where the day to day language has become Hebrew, animosity among ethnic-groups has indeed decreased" (Ruppin 1930b, Vol. II, p., 113).

Already at the beginning of his Zionist path, Ruppin saw the breakdown of the barriers between ethnic groups as one of his primary tasks. In 1911 he wrote: "In Palestine, marriage between Sephardim and Ashkenazim is still quite exceptional. Here, certainly, there is the additional hindrance of difference of language and culture" (Ruppin 1913, p. 160). However, the success of the revival of the Hebrew language in Palestine "has its effect on the population, and makes for a *rapprochement* between [east and west as well as between] the Sephardim and Ashkenazim. The coolness which still exists in the East between Sephardim and Ashkenazim has tended to disappear in Palestine, and this is greatly due to the common language" (Ruppin 1913, p. 264).

> It is Zionism, again, which has re-established the bond of unity between the Western and the Eastern Jew. Before its advent the Western Jew remembered his brother in Eastern Europe only when his sympathy was aroused by bloody persecutions in Russia. Apart from these catastrophes there was no connection between East and West. The relation between the Western and Eastern Jew was not greatly different to that between the Sephardim and Ashkenazim in the eighteenth century, when the Sephardim in London forbade marriage with the Ashkenazim, and actually induced the town of Bordeaux to expel the Ashkenazim Jews. The Western Jew did not know the Eastern, and did not wish to know him. He was ashamed of him as of a poor uneducated relative, whom one pities and supports in private but denies in public. [...] The Western Jew had no idea of the wealth of idealism, of the undiscovered spiritual treasures of the Russian Jews.
>
> Zionism has changed all that. The gulf between East and West is not yet filled in, but it has been bridged, and vast possibilities have been opened up. Even the Sephardim of the East (in Palestine and elsewhere) have been touched by the breath of Zionism. (Ruppin 1913, p. 281)

Ruppin did much to relieve the distress of the Jewish communities, including those under the Nazi regime in the 1930s. Though he did not foresee, nor believe, the physical annihilation of European Jewry in the 1930s and 1940s. The more he became acquainted with his Jewish brethren and their ethnic groups, the more he was convinced that except for all the communities of Middle Eastern racial origins, Jewishness *per se,* was biologically meaningless. The common denominator uniting Jews, as well as splitting them into ethnic groups, was essentially cultural-educational. Out of respect for his Jewish brethren, Ruppin aspired to establish a socio-cultural system that would assimilate them all; such frameworks would be, of course, those of Western culture, the creators of which were Western Jews, rather than those of the "failing East." At the same time, from early on, he was aware that

6.3 The Merger of *Eidoth*: Assimilation or Amalgamation?

there was also a non-Jewish population of more than 500,000 persons in Palestine. Concurrent with the Jewish national revival in their homeland, an Arab national movement was taking shape, which became an increasingly painful issue. "It is clear that these will not leave the country to make room for the Jews. This even the Zionists would not desire; Zionism does not wish to have Palestine exclusively for the Jews; it only seeks to create, by steady immigration, a large, coherent, united population of Jews which will be protected from the dangers of assimilation. [Although t]he backward state of the culture of the native population nullifies the danger at the outset" (Ruppin 1913, pp. 290–291). In other words, the socio-economic gap between Jews and Arabs would, according to Ruppin, at least at the beginning, overcome the threat of the assimilation of the Jews among the Arabs. It seems that Ruppin was so convinced of the power of Western culture (and of the readiness of the Jewish ethnic groups to adapt to it) that it would act as an automatic barrier between the peoples. It did not occur to him that the Middle Eastern and Sephardic *eidoth* minority could provide a bridge to the Palestinian Arabs who might even assimilate the Western ethnic groups and their culture.

Ruppin believed, even at the decline of the colonial era, that the economic and social utility of the Zionist settlement would convince the local Arabic population of the country to trade their national and religious zeal for socio-economic advancement, and that at the end of the long-term process, the Arabic and Jewish populations would fuse, since they were blood-related. Following such a trellis model, even if there was ground for apprehension about interbreeding, this did not apply to the peoples of the Middle East. And even if there was some ground for the claims of eugenicists of the detriment of racial mixing, this did not apply to Jews and Arabs. Notwithstanding his incisive economic-social point of view, Ruppin is torn between his hopes and his apprehensions:

> But the time may come when the Jews, by introducing into Palestine large industries and modern agricultural methods, may become, not merely buyers and consumers, but very dangerous rivals. It may well be that they will buy the land at prices higher than the primitive Arab fellah can afford, and thus deprive the Arab farmer of the chance of extending his property. At present the danger of this is not imminent, as hardly one-half of the land is cultivated [...]. But when it comes to corn-growing, the increasing immigration of the Jews is likely to cause friction. This might be mitigated somewhat if the Arabs are clever enough to imitate the superior agricultural methods of the Jews. They would then have nothing to fear from the competition of Jewish producer, while the change from *ex*tensive to *in*tensive agriculture would necessitate their using a fraction of their present agricultural area. In this way the needs of the cereal grower could be satisfied, and need not necessarily cause the Arab to be expatriated.
>
> If this economic difficulty could be satisfactorily overcome, there is not much to fear from the national jealousy of the Arabs. (Ruppin 1913, pp. 291–292)

Eitan Bloom attempted to address the simplistic eugenic conception of Ruppin: The eternal life [*Unsterblichkeit*] of a race is contained in the biological material rather than in the human spirit. The Jew is a Jew because he has the *biological structure* of a Jew, thus a change in his biology will also bring a change in his mentality. What did Ruppin mean when he wrote: "We may use the verse from Ezekiel, 'The fathers have eaten sour grapes, and the children's teeth are set on edge'"? (See

Ruppin 1930b, and Bloom 2005, pp. 15–16. The citation is from Ezekiel, 18:2). What is "sour" in Ruppin's biological interpretation? And why were the children's teeth set on edge? According to Bloom, the answer may be formulated as follows: The reason for the deterioration of the original Jews (the *Urjuden*) was the introduction of the racial Semitic element among the Jewish people, primarily the Bedouin or Middle Eastern type. For Ruppin, the original Jews, those who were farmers and lived prior to the destruction of the First Temple, were non-Semitic tribes. Thus, when they started to mix with the Semites, the principle of racial preservation was disturbed – these are the "sour grapes." The Semitic element in the Jewish race, which gradually became dominant, severed their contact with nature, with their land and agriculture, and intensified their uncontrollable "commercial instinct" – these are the "teeth set on edge" of the sons.

According to Ruppin, the Zionist settlement in Palestine was a "natural eugenic development of the Jews." Consistent with this approach, he rejected the immigration of any tribes that he deemed were not biologically related to the Jewish tribes. When in 1934 Dr. Yakob Feitlowitz asked to bring the Jews of Ethiopia to Palestine, Ruppin argued that the Ethiopians are "negroes, who were turned to Judaism by force of the sword in the 6th century B.C.E. They had no blood relations to Jews […] Therefore there was no reason to increase their number in Palestine" (Bloom 2005, quoted from a protocol of the Jewish Agency [14/10/34] register no. 20210, Ben-Gurion Archive).

Ruppin, as the head of the Palestine Office, thus led the practical Zionist settlement project in Palestine with the explicit intention to resurrect the Israeli *nation* as one distinct, well-defined entity for the future evolution of humanity. From the moment he arrived in Palestine in 1908, he tried to identify the group of Jews who would help to construct the healthy national body [*Volkskörper*] – the Eastern European immigrants – and simultaneously strengthen the non-Semitic element through the cultural and biological assimilation of the members of the Middle Eastern *eidoth* with the Jews of the West.

6.4 Jewish Diseases

> There is a new type of human, who relates to health as if it were nothing but a disease.
> (Franz Kafka)

The differential distribution of diseases has always been an important indicative variable of the biological kinship between Jews and Gentiles as well as between the members of different Jewish communities and *eidoth*. Many diseases are not randomly distributed but rather clustered among people in certain locations or among members of certain ethnic groups, specific age groups, or among people consuming certain foods or using specific medicines and drugs. Epidemiologists try to discover the causes of such clustering – whether they are the consequences of environmental

6.4 Jewish Diseases

circumstances, specific cultural habits, or sensitivities related to some inherent biological causes (Scriver 1992).

Specific (and peculiar) cultural, behavioral and health characteristics, were always ascribed to Jews. To what extent are these characteristics due to heredity or to the special living conditions that were ordered by Jewish tradition, by persecutions, or by social and geographic isolation? Obviously, conditions such as population density and oppressive living conditions in the ghettos and in the Pale gave rise to the outbreak of diseases and drew attention to the so called 'Jewish diseases.' However, to the extent that Jews were physically and culturally isolated from non-Jews, it is also possible that morbidity differences between Jews and non-Jews reflect their specific biology, i.e., the composition of their gene pool. Max Nordau claimed that the very life in exile as a distinct and persecuted congregation brought the Jews to a state of biological degeneration of both body and soul. Others adopted a diametrically opposite position, namely that the Jews had developed immunity to diseases that afflicted the modern world, especially Europe. Theodor Lessing (1872–1933), who invented the term "Jewish self-hatred" [*Der jüdische Selbsthass*], compared the Jewish people to an organism that successfully survived a plague through the acquisition of antibodies to the plague's specific ailments (Almog 1991). For Lessing, it was not the Jewish people who were sick, but rather the non-Jewish world. The symptoms that characterized Jews were the antibodies developed against the plague's diseases. Finally, rare mutations that have occurred haphazardly may be mainly limited to isolated or semi-isolated communities, such as Diaspora Jews living among foreign Gentile communities. To what extent was the distribution of diseases among these Jews in comparison to their neighboring non-Jews the consequence of their living conditions or of their specific biology?

As pointed out repeatedly, properties are neither 'inherited' nor 'environmental.' Any clear-cut distinction between *nature* and *nurture* is superficial and misleading. Each and every property is the product of both; it might be preferable to say that hereditary factors – genes – are only one component among all 'environmental components' involved in the development of a common characteristic.[15] An instructive example is the disposition toward diabetes and heart failure among the immigrants to Israel from Yemen and Kurdistan. Genetic factors that affect the probability of being afflicted with these diseases have been well known. Following immigration, in the early years, the frequency of these diseases was conspicuously low among all age groups of immigrants to Israel from Kurdistan and Yemen, compared with other ethnic groups. The frequencies, however, increased dramatically the longer these persons lived in Israel, to the extent that the frequencies among the veterans of these communities did not differ from those of members of other communities who emigrated from Western countries, or from those born in Israel (Cohen 1963). The determining factor of the difference between the communities that seemed to be genetic turned out to be largely a difference in diet.

---

[15] In 2010, Evelyn Fox Keller further elaborated on this theme in her book, *The Mirage of a Space between Nurture and Nature*.

In his article entitled "Nervous Diseases and Eugenics in Jews," published in 1918 in Warsaw in the Hebrew periodical *Hatekufah,* Dr. Shneor (Zygmunt) Bychowski (1865–1934) attempted to analyze the effects of environmental variables versus hereditary factors of a disease traditionally considered to be "Jewish," the biases involved in the collection of data, and their interpretation (Bychowski 1918. See also Falk 2003–2004). Bychowski was born in Koritz (Wollin), Poland and studied medicine in Vienna. He was active in Zionist organizations and attended the First Zionist Congress. Although he did visit Palestine, he pursued his work as a physician and public activist in Warsaw until his death in 1934. Analyzing the contribution of environmental and hereditary factors to the appearance of diseases and the biases involved in their interpretation, Bychowski concluded that there was nothing special in the *biology* of the Jews: "It is agreed among experts of neuropathology that Jews are especially prone to nervous diseases. Nevertheless, this opinion has no solid foundation: pursuing it would reveal that it deserves re-examination and must be further contemplated and debated" (Bychowski 1918, p. 289).

Bychowski responded to the well-known neurological researcher, Jean-Martin Charcot, who defined a special disease, "The Wandering Jew" (*Le juif errant*).[16] Besides Charcot's wealthy and refined patients, he was approached by poor and miserable Polish and Lithuanian Jews, seeking treatment. Clinically, he could find nothing of interest in their disease. However, with no knowledge of the French language, or for that matter, of any European language, these patients had dragged themselves all the way to Paris and came explicitly "To Charcot in person." They had already consulted all the famous doctors of Europe on their way to Paris, were already acquainted with all medical means and knew by heart all the contrivances, although they despaired of their utility.

> Although their disease extended over many years, they never used any medicine in a regular or systematic way. Rather they leaped from one expedient to another. When it was suggested to them to stay in a clinic for an extended period of time, they rejected this and went on wandering. Each of them had in his pocket plenty of prescriptions and doctors' orders. They kept these carefully in their pockets, but never followed them up. (Bychowski 1918, pp. 290–291, in Falk 2003–2004)

Bychowski called attention to the statistical deviation that such a phenomenon could cause; for example, each of these patients could appear several times in the records according the number of physicians he consulted.

> There is no point in analyzing in detail the statistics published with regard to the distribution of nervous diseases among Jews. None of the statistics that had been carried out is reliable. Researchers ignore the fact that they were dealing with living persons who migrate and change places of residence. […] I even found in American professional journals pictures of my very patients from Warsaw! (Bychowski 1918, pp. 293–294, in Falk 2003–2004)

Bychowski, however, was not only critical of researchers' method of collecting data, but also doubted the scientific relationship between "endogenic diseases that

---

[16] See Goldstein 1985, concerning Charcot's and his students' attitude towards Jews and their suffering.

6.4 Jewish Diseases

originate in the body itself" – a term related to inherited properties – and "exogenic diseases, that are imposed from the outside" (Bychowski 1918, p. 295). At the time, emphasis was shifted to the external environmental factors as major causes of nervous diseases. What truth was there in the claims of the degeneration of the Hebrew people and of the excess of neuropathies among them?

Bychowski rejected as "nonsense" suggestions that the Jewish people had acquired immunity against drunkenness and syphilis over thousands of years. However, he suggested that besides the explicit exogenic and endogenic factors contributing to mental illnesses, there were also *perigenic* causes, factors dependent on the immediate environment, such as the education of a child during the first years, and the input of the parental home. Bychowski concluded that such "circumstantial factors" that were the products of the socio-cultural background of individuals and the direct consequences of persecution and humiliation were the perigenic causes of 'Jewish' nervous diseases like that of the 'Wandering Jew':

> This means that the causes are not dependent on the nervous system itself, but rather are due to environmental factors. It was simply the life of the Jews of Russia that was ridden with so many conflicts and full of anomalies and sickness, all of which must have caused loss of the corporal capacities, which were aggravated by the Jew's hard work, his grievous life. [...] We do not find the usual kind of struggle for existence encountered all over Europe among the Jews of Russia and Poland. Their lives were a specific 'Jewish' struggle for each piece of bread, for a sip of water to drink, and for some air to breath. This was a struggle for the privilege of spending the night outside a freight-truck, for the right to enroll in school, and even the right to be treated by a doctor. (Bychowski 1918, pp. 303–304, in Falk 2003–2004)

Bychowski claimed that the causes of their neuropathies were the persecution of the Jews, the abuse, and the imposed poverty. By observing the sons and daughters of the Russian Jews who immigrated to western countries and America, and who were extricated from the mental yoke and its hardships, Bychowski noted that one generation of relief of the suffering would be enough for these diseases to disappear.

All the same, there were diseases that were relatively frequent among Jews that Bychowski accepted as hereditary, such as Tay-Sachs disease. Interestingly, Bychowski, who was keen to follow eugenic procedures to maintain the health of the Jewish population, did not consider the need for such measures in the case of Tay-Sachs disease, because "anyhow, these children did not reach the age of puberty" (Bychowski 1918, pp. 298–299). Obviously, Bychowski's knowledge of genetics was limited: Tay-Sachs is a recessive disease that is transmitted by two healthy *carrier*-parents, so that eugenic means to prevent it may well be (and are nowadays) taken.[17]

---

[17] Recessive: see note 8 in Chap. 5. Sir Archibald Garrod published his book, *Inborn Errors of Metabolism*, in 1909. In it he showed that human diseases and other "deviant" properties may be conceived as due to changes at a specific Mendelian factor. Garrod identified alkaptonuria, cystinuria, albinism, and pentosuria as due to such errors. Tay Sachs disease was only later identified as another inborn error of metabolism.

Developments in the medical sciences and hygiene have since completely revolutionized life in Western societies and brought a significant reduction in peripheral factors that may cause disease, whether by the development of therapeutic means or by introducing efficient hygienic methods. Also, developments in genetic research contributed to diverting significant attention from environmental factors to inherited factors. Although less than 2 percent of known diseases may be linked directly to a major single genetic factor (so called *monogenic* diseases), much attention is given to such diseases as modern biochemical and molecular methods allow the detailed identification of the specific metabolic defect, as well as of the mutation involved, down to the level of a specific enzyme and even to a single nucleotide in the DNA sequence. Even though we are inclined to call such diseases 'genetic diseases,' it must be kept in mind that it is *not the diseases that are inherited*, but it is rather the inherited genetic factor(s) that play a role in the predisposition to the diseases.

Clearly, the prevalence of a 'hereditary disease' in one population and its rarity in another may be due to differences in the frequencies of alleles[18] of some genes or to environmental circumstances (climate, nutrition, sanitary, and medical conditions) that affect the conditions of the appearance of the phenotype among the individuals, even among individuals of identical genotypes. Furthermore, when both parents contribute to the chance of their descendents being affected (as in the case of recessive diseases), then factors such as mating habits in the population (consanguinity), and the population's effective size, are significant: the smaller the population and the more consanguineous the mating, the higher the probability that affected children will be born.

There is no doubt that several hereditary diseases of various degrees of severity are differentially distributed among Jewish ethnic groups and even among specific communities such as Ashkenazi families from Lithuenia, whereas anemias, such as β-thalassemia, are found in Jews of Kurdistan origins, and α-thalassemia in Yemenites. The anemia due to a deficiency of the enzyme G6PD (Glucose-6-Phospho-Dehydrogenase) is common among Middle Eastern Jews, and Familial Mediterranean Fever (FMF) is practically restricted to the Libyan Jewish community. Three inherited founder mutations in the tumor suppressors BRCA1 and BRCA2 genes mutations in which predispose to a high risk of breast and ovarian cancer comprise 11 percent of breast cancer and 40 percent of ovarian cancer in the Ashkenazi Jewish population of Israel (Gabai-Kapara et al. 2014). What brought about these differences and, especially, what caused the relative prevalence of disease-bound alleles of the genes involved? Are these diseases indicators of the long history of communities, and what kind of links may be inferred from the presence of any of these diseases also in the neighboring Gentile communities? Do they indicate the prevalence of intermarriage or are they the effects of common environmental variables? As we shall see (Chap. 9) the introduction of DNA-sequencing analyses allowed new dimensions of such analyses.

---

[18] Allele: one of the alternative states of a gene. See note 8 in Chap. 5.

For medical doctors and others who diagnose peoples' diseases, the ethnic-biased distributions of diseases are certainly important indicators. However, since various genetic and non-genetic biases could cause what appears to be "the same" disease – and no less significant, mutations in different nucleotides of the same gene's DNA-sequence that can and do occur – special care must be taken in relating people to ethnic communities by their disease, or in diagnosing a person's suffering by virtue of the community to which he belongs (see, e.g., Ross et al. 2015).

6.5 Immigrants and Natives

Discussions about the link of contemporary Jews to the ancient Israelites extend far beyond the contested contribution of the Khazars to today's Ashkenazi Jews. For example, as noted earlier, researchers of the *Wissenschaft des Judentums* believed that the Sephardic Jews best represented the descendents of the original Jews. Redcliffe Nathan Salaman, who adopted the eugenic perspective and insisted that the Ashkenazi Jews were racially purer than the Sephardim, took a completely different position. He claimed that the relatively better circumstances of the Jews in Spain and the Middle Eastern Diaspora resulted in more intermarriages with Gentiles in these communities, compared to the low rate of intermarriages found among the Ashkenazim, where persecutions and boycotts by their neighbors kept them effectively reproductively isolated, thus preserving their biological uniqueness. "[D]uring the last 1800 years, there is no doubt that the Ashkenazim can show a far cleaner bill than the Sephardim who are known to have absorbed in no small quantity both Moorish and Iberian blood" (Salaman 1911a, p. 276).[19] Salaman thus concluded that "the Ashkenazim are racially identical with the Jews of Ezra's time" (Salaman 1920, p. 227). He also brought empiric evidence for his claim that the facial features of progeny of intermarriages between Ashkenazim and Sephardim are dominated by Sephardic lines. This, according to Salaman, is so because the Sephardim carry non-Jewish genes, which he previously claimed to be dominant over the Jewish alleles. Salaman's assertion that the Ashkenazim are racially identical to the Jews from the era of Ezra – and his derogatory view of the non-Ashkenazi – is accompanied by statements quite common at that time concerning the "other," namely, the Middle Easterners and the Sephardim. Yet one may have expected otherwise of a person like him; whose first wife was a researcher of Hebrew literature in Spain and his brother-in-law, the anthropologist, Charles Seligman, was on his mother's side of the family a descendent of Emanuel Mendes da Costa, the second Jew who in 1747 was elected Fellow of the Royal Society of Sciences. Towards the

---

[19] "Then one is reminded that in the eighth century the kingdom of the Kozars in South Russia was converted to Judaism. […] [A]ccording to Joseph Jacobs after the destruction of the Kozar Empire it was the Jews of that district who formed the Karaite sect, and this sect has remained absolutely distinct from the rest of the European Jews" (Salaman 1911a, b, pp. 276–277).

end of World War I, when Salaman served as the medical officer of the Jewish Battalions in the Near East, his impressions of the Yemenite Jews whom he encountered were most blatantly disparaging:

> The Yemenites are for the most part undersized and rather poor spirited *natives*. They are *not* racially Jews. They are black, long headed, hybrid Arabs. [...] The real Jew is the European Ashkenazi, and I back him against all-comers. [...] [Notwithstanding,] the Yemenites display a really passionate love for Judaism and have withstood centuries of bitter persecution. (Salaman 1920, pp. 28–29, emphasis in original)

Such an extreme statement, though not rare in encounters with Europeans and "natives," is still exceptional. The Hon. W. Ormsby Gore, British MP, who wrote the introduction to Salaman's memoirs, commented: "Some of Dr. Salaman's statements in his appendix on the 'Bonds of Jewish Unity' [...] are controversial. [...] For instance, many will dispute his statement that Yemenites are not racially Jewish" (Salaman 1920, p. xii).

As noted by Todd Endelman, Salaman overturned the Western Zionist belief that the Sephardic Jew was "the Jew who could be authentically linked to both an ancient and glorious past, and by extension, could serve as a model for a future rejuvenated Jewry." It is not clear to Endelman why Salaman "reversed the hoary myth." It may appear as if Salaman merely followed the pattern of praising his own kind by defaming the others (Endelman 1987, 2004). I believe, however, that these statements reveal much of his political plight: Salaman, the eugenicist was eager to provide a 'scientific' argument in favor of the immigration of East European Jews to Britain (and probably also to Palestine): He repeatedly suggested that such an immigration not only would not damage the gene pool of the population of the British higher classes, but on the contrary, would improve and increase their level because the immigrants of Ashkenazi stock carried a superior gene pool.

Endelman doubted such an interpretation, since "by the 1920s the immigration question was no longer the issue that it had been two decades earlier" (Endelman 2004, p. 73). It is important to call attention to the fact that in the first years after World War I, the immigration of Jews from Eastern Europe to England increased considerably and with it the voices calling for the introduction of limitations on this immigration. Scientists were among those who made a case against this immigration, pointing out the deleterious effects on the gene pool of the native population. The main expression of opposition was probably the paper by Karl Pearson and Margaret Moul entitled "The problem of Alien Immigration into Great Britain, Illustrated by an Examination of Russian and Polish Jewish Children," published in 1925. Pearson (1857–1936), who was one of the leaders of modern statistical demographics, explained that he could "sympathise with a man who has suffered hard treatment, but that in itself is not an adequate eugenic reason for granting him citizenship in a crowded country" like Britain. In order to grant that citizenship "we demand physical and mental fitness; we need the possibility of an ultimate blending" and we need "full sympathy [of the immigrant] with our national habits and ideals." To evaluate the presence of such properties among the immigrant there is no better way than a cool measured statistical test, since "we [scientists] have no axes to grind, we have no governing body to propitiate by well-advertised discoveries; we are paid by nobody to reach results of a given bias" (Pearson and Moul 1925, p. 8). Based on this declaration of the built-in objectivity of scientific evidence, Pearson claimed that he had proven in statistical detail the inferiority of the immigrants. According to Pearson, the children of Jewish immigrants deviated from the locals not only in hair and eye color, but also in height, weight, the amount of hemoglobin in their blood, cleanliness of hair and clothes, frequency of various diseases, and in the level of their intelligence. It did not occur to Pearson that these data may be attributed to the terrible living conditions of the immigrants in the degenerate slums of London (and in their countries of origin). He avidly rejected the principles of Mendelian genetics, and consequently the parsing of (phenotypic) variations of properties into hereditary (genotypic) and environmental components, as proposed by Johannsen (see Chap. 3). Instead, Pearson supported the notion of the fixation of characters, including those acquired by a slow and gradual process of Darwinian selection.[20]

In 1926, in response to Pearson's paper, the Jewish Health Organization of Great Britain called for a study of the achievements of Jewish school children in Britain that reached diametrically opposite conclusions (Davies and Hughes 1927; see also Endelman 2004, p. 73). Bearing in mind this atmosphere, Salaman apparently considered it of upmost importance to convince upper classes and the authorities that the Jews who immigrate to England from Eastern Europe (the Ashkenazim) were not "natives" (like the Sephardim and Middle Easterners); rather than bringing foreign biological components to the British nation, they could even improve it. He apparently expressed similar considerations in reference to the Jews of Middle Eastern countries concerning the Zionist immigration to Palestine. He endeavored to convince the British authorities in Palestine that the high standards of the Jewish settlers – meaning the "Europeans" – would contribute positively to the British Empire. A couple of years prior to Pearson's paper, Salaman published a note in an effort to prove the intellectual superiority of Jewish children, in spite of the debasing conditions to which their fate had subjected them (Salaman 1923).

One of the most important opponents of Salaman was the anatomist Sir Arthur Keith (1866–1959), an avid Social-Darwinist. He maintained the position that "racial sentiment was [essential] to understanding the development of nations, particularly through warfare." Although both believed that the Jews had "a legitimate claim to be regarded as racially different from the general population," Keith upheld the superiority of the "Western Caucasians" and advocated their "right to occupy territories currently inhabited by races that would die off in the evolutionary struggle." Keith argued as late as the 1930s, that "immigration into the lands of northwestern Europe was impermissible" (Stone 2004, pp. 232–234). Naturally, Salaman presented his genetic insight to advance the vested interest of his people in the face of the circumstances of Britain's colonialist outlook at the time.

---

[20] On the problems involved in Pearson's objective methods of compiling data, see Gould (1981), Chapter 22, "Science and Jewish Immigration," especially pp. 296–302.

A general view of European racial superiority was also a component of British sentiment and policy. The, 1903 proposal of the British Government to Herzl to settle Jews in Uganda stemmed precisely from its colonialist policy: White settlers in the "Colonies," or what is called today "the Third World," were essential for the developing regions, such as East Africa and the Near East, and consequently benefitted the "Natives." However, it became increasingly difficult to find enough British candidates who were willing to immigrate to these countries after the Boer War in South Africa in 1900–1902. Since there were not enough candidates in Britain proper, substitutes were needed:

> Policy on population was fundamental to the underlying approach of both the practitioners and the theorists of contemporary British colonialism. It derived in part from the then virtually universal, 'Darwinian' tendency to think in, and ascribe enormous importance to, ethnic categories. [...] 'The Negro,' wrote Sir Harry Johnston, a contemporary authority on Africa (who later joined in the ensuing public debate on Jewish settlement in East Africa), 'seems to require the intervention of some superior race before he can be roused to any definite advance from the low stage of human development in which he has contentedly remained for many thousand years. [...] We desire to make of the native a useful citizen and [...] we consider the best means of doing so is to induce him to work for a period of his life for the European. (Vital 1982, pp. 156–157)[21]

Colonial settlers were expected to serve the interests of all: Settlers from the "world of culture" will obtain a just share in economic progress, whereas the 'Natives' will gain gradual progress. The Boer War, however, made it clear that the lives of Europeans would be less comfortable than expected. To Joseph Chamberlain (1836–1914), the Secretary of State for the Colonies, who met Herzl on April 23, 1903, it was perfectly natural that it "occurred to him that the East European Jews might serve the overarching imperial purpose as well or better [by settling in Uganda rather than in Palestine]. [...] The Jews were bound to serve the economic interests of the territory well." No less important, "[b]ack in England, there was the other, undesirable stream of immigration and the prospect of having to cope shortly with the unpleasantness of a Bill to be introduced into Parliament to bring it to a halt" (Vital 1982, p. 158). It is no wonder that against this background, Salaman would respond by emphasizing the eugenic value of Ashkenazi immigrants to Britain.

It goes without saying that Salaman, in his position as a head of the social pyramid, adopted unhesitatingly the claim that "the tendency of social stratification is as natural as the sedimentation of the rocks" and that "as in nature so in human society an inversion of the strata can only be effected by vast and cataclysmic upheaval." Present-day Jewish communities were not constructed on such a basis: "There has never been an outstanding aristocracy of the ghettos" (Salaman 1923, p. 135); but the Jew differed from all people around him in respect to the subjects that eugenics endeavored to promote:

---

[21] The quotation of Sir Harry Johnston is taken from: "The Development of Tropical Africa under British Auspices," *Fortnightly Review* (November, 1890), p. 705. The last two sentences are drawn from Sir Henry Belfield's *Proceedings of the East Africa Protectorate Legislative Council* (1917).

6.5 Immigrants and Natives

> The Jewish scholar of the ghetto [...] had drunk deep of the wisdom of his forefathers, and their views were curiously enough extremely modern for they were essentially eugenic.
>
> And hence it comes about that the Jewish communities of the last thousand years have been steadily increasing their intelligence at the expense of their lower classes and have existed without conscious class segregation because without the means to make those distinctions visible. [...]
>
> The outstanding difference [...] is that the emigrant Jews by reason of the peculiar circumstances which have been already outlined are on the one hand of a higher intelligence than any other group of emigrants from European people and on the other reach their new home as it were in disguise. The external circumstances make them appear as members of the lower classes whilst in point of fact they are an unsegregated but highly gifted mass deficient in both the extremes common to a normal freely moving population – an aristocracy and a criminal class. (Salaman 1923, pp. 136–137)

Selection acted not only on the unique mental properties and on the social characteristics of the Jews, but also at the physical level:

> The lower death-rate [among the Jews] is so general and so considerable in amount that it cannot be a matter of chance. [...] The lower death-rate at all ages, especially after the first year, means that the Jew offers greater resistance throughout life to all the inimical influences of the environment, that he is on the whole a tougher and a more resilient specimen of humanity. (Salaman 1923, p. 148)

All this was intended to show unequivocally that nature, rather than nurture, is on the side of the Jewish babies. Jewish immigration will not damage the British tradition and its heritage, not even that of the upper classes.

Salaman vacillated between his alliance to his Jewish heritage and his identification with his English homeland in his attitude to the Jews of Eastern Europe who wished to immigrate to Britain. He agreed with Pearson that immigration was a eugenic issue of primary importance:

> There would appear to be no question more suitable for the consideration of eugenists than this [of the admittance of emigrants from Eastern Europe]. The whole problem is a relatively simple one: are these emigrant people of value to the state or not? Do they bring promise of greater gifts beneath their tattered garments than the jaundiced eye of a relieving officer can appreciate? (Salaman 1923, p. 152)

Furthermore, considering the perspective of their new homeland, was it preferable for the immigrants to assimilate in the general population or to maintain their uniqueness? It appears that Salaman presented arguments based on population-genetics to try to justify the existence of a distinct Jewish minority in Western countries, as well as their aspirations towards one large Jewish population in Palestine:

> Whether the state gains more by the fusion of a small and gifted minority in the general population than by enjoying the concentrated output of a highly self-conscious group, is a very difficult question. If we could assume that the specific and hereditary intelligence of the Jew were controlled by [discrete] Mendelian factors, it is highly probable that the decision should be against amalgamation when the minority is as small as it is in most countries. (Salaman 1923, p. 152)

According to Salaman, a minority which was almost equal to the population of a large city could be allowed only in Palestine, where an amalgamation of all Jewish ethnic groups would overcome, or might soon overcome, their current status of a minority. In Palestine, accordingly, it was fitting that amalgamation of communities should occur, whereas in other countries of immigration, where the Jews would remain a minority, it was advisable that ethnic groups maintain their separateness. "If these conclusions are correct, then there would appear but one answer to our question. The Jewish emigrant is a bearer of qualities, which are of essential value to any civilized state" (Salaman 1923, p. 153).

It is educational to observe how the means to preserve a minority and its values change with the advances in genetic research: We will discuss the case of the ultra-orthodox Jewish isolate (see Chap. 8). The Bedouin community in southern Israel is another traditional society where genetic diseases are prevalent. Carmi et al. (1998) have implemented a carefully designed educational program in order to apply the molecular age tools to "attend the needs of the Negev Bedouin community and be sensitive to its traditional values."

Chapter 7
Pioneers as Eugenic Agents

> Many who in the past died young because they were unable to withstand the struggle for existence, did not manage to leave children; who knows if the survival of weaklings might not bring about the degeneration of the whole human species. (Jacob Talmon [1916–1980], professor of History, HUJI. "The idea of the Hebrew University in the past and present." *Haaretz*, 28 September, 1966. TRF)

Zionism aspired to produce a new Jew according to the national conceptions of the *fin de siècle* and the beginning of twentieth century Europe. At the Zionist Congresses, in art, in professional photography, as well as in social discussions, the image of the new Jew as conceived in the eyes of Central Europeans of the era was presented as a sun-burned farmer, tilling his land with his bare hands, against the background of an idyllic Mediterranean landscape. This Jew, or perhaps better call him, this *Hebrew* pioneer (Almog 1993), was a figure that was expected to construct a new world of mental and physical activity, a figure in which fantasies and reality intermingled (Fig. 7.1).

In their brave new world, the Zionists often depicted these pioneers as living in a cultural vacuum, as if the local inhabitants, the Arabs, were not a population with its own values and habits. As a rule, it was taken for granted that the moral, cultural, and the educational level of the Jews in Palestine were higher than that of the local Arabs. Special attention was directed at the cultural revival established by the impact of Hebrew as the everyday language of the Zionist pioneers. Indeed, there is no doubt that the mental, professional and scientific activity conducted in the rejuvenated Hebrew language was a unique phenomenon. Furthermore, the revival of the Hebrew language by the Zionist movement immediately became the focus of a cultural revolution, not only in Palestine but among Jews all over the world. Professional periodicals, like *Harefuah* [Medicine] and *Hachinuch* [Education], or the periodicals of the authors' association *Moznaim* [Scales], as well as magazines printed overseas, like *Hatekufah* [The Period], published articles in Hebrew written

Fig. 7.1 Jewish National Fund poster, 1947

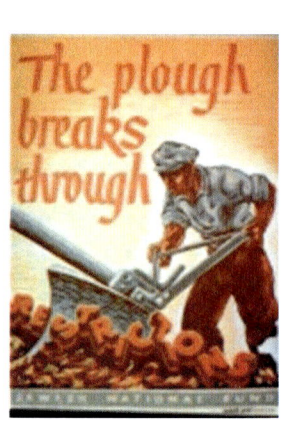

by the immigrants in Palestine, and also by professionals in Vilna, Paris and New-York.

Life in Palestine in the first decades of the twentieth century, especially in the days of the Second Aliya[1] (1904–1914) and the Third Aliya (1919–1923), was not easy even for mentally and physically healthy individuals. In spite of the image presented to the world of vigorous cultural life in the country, conditions demanded enormous daily effort and personal sacrifice. A great deal of dedication and belief in long-range goals, even fanaticism, were needed to withstand the hardships of pioneer life and carry on working towards the distant target. There can be little doubt that the greatest rebellion of the pioneers against the traditional Jewish scale of values was promoting physical work to the top of the cultural scale and unseating the life-of-study from its traditional high status of many generations (Almog 1993, p. 341).

Hardships notwithstanding, cultural life did not stop. The dominant mood in Palestine was pride in constructing a healthy, modern community upon strong European foundations. Emphasis was placed on the role of the new circumstances in changing the biological essence – rather than the spiritual-cultural essence – of the Jewish image. There was much of the romantic spirit of the *fin de siècle* that regarded urban life as being the source of degeneration of body and soul of human-beings, who had been destined to live in Nature. Urban living conditions were often compared to the conditions of animals in captivity, or to be more precise, human urbanization was compared to animal domestication.[2] And just as living conditions in the Diaspora harmed the biological pool of the Jews, living conditions in Palestine would ameliorate them. The notions of the Zionist pioneers in Palestine were thus completely consistent with the eugenic doctrines of those years. The pioneers

---

[1] Aliya: Hebrew for ascent. Immigration to the land of Palestine/Israel is considered an act of ascent. Zionist immigration to Palestine is divided into five "waves" of immigration.

[2] See footnote 10 of Chap. 4.

believed that they were saving the Jewish individual from the degeneration of many years in the Diaspora by acting according to the most modern scientific insights. When in 1911 Rachel Yanait (the wife of the future second president of Israel) spoke of devotion to working the soil as a way of liberating Jewry in Palestine from the harm of the "mental warts" that affected previous generations in the Diaspora, she did not discern between inherited "warts" and acquired ones. All would be healed by the change in living conditions. Yet, could a person become a farmer for the rest of his life just by will power? Must the turn from urban life to rural life involve mental restraint similar to that of the Jesuit Ignacio Loyola (Almog 1993, p. 333)? Or was Zionism a step toward a significant biological change, whether through the inheritance of acquired characters or by the selection of the proper types, as happened in the past to the "Philistine" genotype, according to Salaman (see Chap. 5)?

After rejecting claims that intermarriage and the inheritance of acquired characters affected the creation of the Jewish type (at least as far as the Ashkenazi Jews were concerned), Salaman went one step further and, based on his experience as the Medical Officer, claimed that certain facts became clear to him as his knowledge of the people became more intimate.

> In the first place the younger generation of Colonists are physically well developed and muscular. In sports they held their own against all teams of Gymnasts in the British Egyptian Expeditionary Force. Their average height was certainly greater than that of the Judeans soldiers recruited from Russia and America.
>
> It was, however, the facial type, of the younger generation that was most interesting. The outstanding fact was that the Palestinian youths presented a very considerably higher proportion of Pseudo-Gentile faces than did their foreign brethren of the other battalion. Indeed it would appear that some force was at work which was bringing into existence again the old Philistine type in the land of the Philistines. (Salaman 1925, p. 17)

Salaman's reasoning was that since the living conditions of the pioneers in Palestine and those of the Jews in the Diaspora were opposite, selection forces acted on them in a diametrically opposed manner. Under the living conditions of the Jewish pioneers in Palestine, natural selection acted in favor of the Philistine genetic elements of the Jewish gene pool, elements that had been gradually eliminated from the gene pool under Diaspora conditions. In the Diaspora different forces were active in favor of other genes that showed "correlation of mental and physical characters." Such characteristics, Salaman observed, upon "examination of a large collection of portraits of Anglo-Jewish worthies showed that those leaders who had been in their time outstanding philanthropists – and Jewish philanthropists are above all distinguished by that very loveable but formless type of charity which is so well known as 'Rahamonuth' – were almost all of an outspoken Hittite type of countenance" (Salaman 1925, p. 17).[3]

> It may therefore be forgiven the writer if, when looking at the young home-born Palestinian Jews as they were marshalled under their Zionist banner on the plain of Sharon, and noticing the prevalence of the Pseudo-Gentile type of face he fancied that here, too, perhaps was evidence of another correlation, a correlation between the spirit of adventure and the

---

[3] *Rahamonuth* – is a Yiddish word that means compassion.

> Pseudo-Gentile type of face, which would become active as a selection agent in respect to those immigrants who came to Palestine to found a new Judea. (Salaman 1925, p. 17)

Salaman expected that the frequency of Gentile-like facial patterns would increase with the establishment of the Jewish settlement in Eretz-Israel. Contrary to his predecessors, who claimed that circumstances may change the Jewish construct, once they turned to "normal" living conditions, Salaman believed that conditions in Palestine would select different components of the genetic pool of the Jewish race than those selected under conditions of exile. "And so it may be that, in the old home of the Philistine, there is being recreated that ancient race from the bowels of its one time enemy and victor, a race which, be its faults or its virtues what they may, was certainly dominated by that spirit of adventure and hardihood which made the Aegean of old the Viking of his day" (Salaman 1925, p. 17). Salaman obviously overstepped the alleged objectivity of the scientist and gave expression to his feelings. According to Todd Endleman (Endleman 2004), Salaman attempted to refute two anti-Semitic racial assumptions of his day by asserting, first, that the Jews were not just "Semites"; other people, such as the Hittites, had left their permanent imprint on the Jewish gene pool. Second, Salaman refuted the "Aryan" claims to Nordic or Teutonic exclusiveness of patterns like blue eyes or fair hair color. These features may be found also among Jews and Philistines, and are not exclusively Northern European features. Fritz Lenz (1887–1976) strongly disagreed with Salaman's comments. Lenz, the German researcher of human races, who together with Erwin Baur and Eugen Fischer authored the famous *Menschliche Erblichkeitslehre* [Human Heredity],[4] expressed eugenic ideas that were extreme even in those years when genetic discrimination between people and populations was acceptable. While Salaman interpreted the Philistine contingency as one of three components that had mingled with each other and eventually formed the Jewish race that went into exile, Lenz interpreted Salaman's conclusions to suit his own purposes, claiming that the Philistine element is the original Jewish race: "In the Zionist attempt to resettle Palestine with Jews, it has been interesting to find that few of the settlers are conspicuously Jewish in type; manifestly they are recruited for the most part out of the non-Jewish racial elements which have been incorporated among the Ashkenazi or eastern European Jews." Thus, Lenz drew his conclusion: "Taking them all in all, Jews constitute only 3.6% of the agricultural population of Palestine, and this percentage is declining. Owing to their deficient talent or inclination for the primary work of production it would seem that a State system consisting exclusively of Jews would be impossible" (Baur et al. 1931, p. 669).

As is well known, Fischer's and Lenz's expertise did not help to identify Jews by their biological characteristics, so the Nazis had to fall back on the "Yellow Patch" to identify Jews. As a matter of fact, following the Nazi interpretations with respect to race theories and eugenics Salaman altered to a great extent his views concerning the essence of the Jewish race (Endelman 2004, pp. 81–84).

---

[4] The book was first published in 1912. The English translation by Eden & Cedar Paul of the 3rd German edition (1927) is: Baur, Fischer and Lenz (1931). *Human Heredity*. New York: Macmillan.

Salaman was not alone in considering the project of Zionist settlement as a eugenic system for selecting various hereditary elements different from those prevalent in Jewish communities in the Diaspora.[5] The physician of "Herzelia," the first Jewish Gymnasium in Palestine in the 1920s, Dr. Aharon Benjamini, established a society for the preservation of the Jewish race, and was active in encouraging childbirth among the settlers. He used to measure and weigh the students and compared his data with those of various European studies. Excitedly he reported: "Our children exceed in height their age contemporaries in France and Germany [...]. They exceed in their development their age contemporaries in Kovno by approximately one to one-and-a-half years." The higher weight that the children in Eretz-Israel exhibited over those in Europe, explained Benjamini, was due to the excellent care given to the youngsters. Yet, he claimed that this did not explain everything. Excluding climatic and other environmental factors to explain their robustness, he asserted:

> The case in front of us is one of natural selection of people who inherited their physical robustness, and planned their march from the Diaspora to Eretz-Israel powered by their judgments and ideals [...]. Zionism was adopted, and was indeed competent to be adopted, only by individuals who were whole in their bodies, those who had physical strength, or those of potential strength, in whom the power of the muscles was hidden and concealed, so that it developed further among compatriot fellows. [...] As in any living organism, so also in our nation a process of selection ensued [...], biological natural selection. (Benjamini 1928, in Stoller-Liss 1998. p. 2. TRF)

Salaman was more concerned with genetics than most others who were carrying out research on the biology of the Jews, but like many in the early decades of the century, he was captivated by the ideas of eugenics. Like many of his Zionist colleagues, he believed that he could identify Jews by the phenotype of their physical statistics; fortunately they did not have an opportunity to put this alleged capability into practice. Starting in the 1920s, more geneticists grew critical of the eugenic doctrine, especially as several politicians and social reformers exploited it for their own purposes, although many continued to believe in the need to apply the findings and insights of genetic research for the advancement of human societies, in spite of the risks involved.[6] As already noted, attempts were always made to maintain a fragile balance between nationalism and humanism, to balance the obvious preference for one's own group against the understandable immanent rights of each

---

[5] For example, Ruppin argued that the Zionist enterprise of the settling of Palestine would strengthen the non-Semitic element of the Jewish nation.

[6] One of the major problems of the relationship between science and society was the conceptual reduction, which proved for many years to be a most effective *method* and *tool* of experimental science research. Salaman, as many of his colleagues, was an extreme reductionist in his scientific approach, as witnessed by his simplistic, single-gene-distinction of Jewishness, or that of the analysis of the outstanding properties of the pioneers. It was only half a century later that the successes of the reductionist methods, especially in molecular biology, convinced biologists of the need for *systems' perspectives,* not only in inter-disciplinary relationships. Obviously, the complicated social, as well as political, relationships could not be reduced to well-defined, discrete biological variables (see, e.g. Falk 2009, 2013).

human individual, which was also an integral part of Zionism. It is most difficult to evaluate in hindsight what should have been the preferred strategy.[7] It is a fact that this balance broke down repeatedly in Europe in the twentieth century. In the critical years on the way to establishing a state, Zionism managed – often at the very last moment – to give humanism priority over nationalism. In such a context, one may understand, for example, how Ruppin, who often viewed his Zionist activities from a humanistic perspective even more than Zionist ideology, maintained close connections with one of the worst racists and followers of the Nazi regime, Hans F. K. Günther. Ruppin and Günther, who were both eager collectors of photos of facial types, continued to correspond until the late 1930s. Salaman also continued to participate in eugenic meetings during the 1930s, in spite of the increasing espousal of social and political reforms that imposed simplified solutions of scientific methodology to a complex interactive human society. Future developments would not distract persons like Salaman from embracing humanistic principles. Thus, his support of Zionism was eventually based more on a cultural bond than on biological arguments (see, e.g., Salaman 1950).

7.1 Hebrew Work – An Insurmountable Challenge

Even a community committed to high moral values and ideology, such as that of the Zionist pioneers, required tangible means and solutions to practical matters. One of the major slogans of the Zionist settlement in Palestine was "Hebrew Work." The fact that Arabs worked as laborers in the Jewish colonies affected the very essence of the Zionist concept of Jews who abandoned the traditional professions of the Diaspora, returning to till their own land. In reality, this call also reflected to a large extent the duress of the unemployed pioneers who flocked into Palestine, notwithstanding the efforts of employers to obtain a cheap and efficient working force. Few paid attention to the threat that the crusade for Hebrew work would take away the means of livelihood from the local Arab population and, thus, would increase their sense of deprivation set off by the Zionist settlement. The presentation of this issue by Shmaryahu Levin (1867–1935) of Keren-Hayesod at the Eleventh Zionist Congress in 1913 reflects the typical views of the pioneers: "The Jewish laborer is used to cultural conditions, accordingly he expects better salaries and employment conditions than those given to the Arab competitor. Nonetheless, the higher salaries [paid to Hebrew worker] pay off because the quality of his work is higher." It soon turned out, however, that the "laborer used to cultural conditions" needed more than a pioneering spirit to compete with the Arab laborers and persist in doing the exhausting physical tasks. The solution, typical for Europeans, was and has

---

[7] For a thorough discussion of an *a posteriori* judgment, one should read Michael A. Bernstein's book, *Foregone Conclusions against Apocalyptic History* (1994). The book deals with the philosophical aspects of the research of history, and especially that of the Holocaust literature, with an emphasis on the writings of the Israeli author, Aharon Appelfeld.

7.1 Hebrew Work – An Insurmountable Challenge

remained, the recruitment from faraway of low cost efficient "natives" (or "Third World" persons, in modern day terms), who were used to exhausting labor under minimal working conditions. Thus, also the Zionist solution was to find "native" Jewish workers.

At the beginning of the Zionist settlement, several other, non-Zionist Jewish communities were living in Palestine: Spanioli speakers, who were probably the progeny of Jews expelled from Spain in 1492; Jews who emigrated from Eastern Europe in the eighteenth and nineteenth centuries for religious reasons; as well as other Ashkenazi Jews. During the years of Zionist settlement, many became professional workers or small merchants, serving the settlers and their agricultural companies. All these were no source for hired workers. Furthermore, since natives of the Ottoman Empire were not restricted by migration laws imposed on Jews from Central Europe, there were relatively small groups of Jews from Bukhara, Persia, Yemen to the south, Syria to the north, the Caucasus, Morocco, and others, who had settled in the country over the preceding two centuries (Ruppin 1930a, b). They migrated to Palestine for economic and religious reasons and did not provide laborers. European Jews often assumed that Middle Eastern Jews were not persecuted and thus had no ground to rebel against the conventionals of their forefathers (but notice, for example, Salim Fattal's [2003] memoirs from Baghdad), consequently they often found themselves in secondary positions in the society, in the economy, in the culture, and in the politics of the Jewish settlers (Eisenstadt 1948, 1954) and were not a potential source for labor: Although several of these Eastern residents were admired by the Zionists as romantic figures of noble fighters, riding horses and carrying a sword and a rifle, there was a rather clear-cut distinction between the European Zionists (the Ashkenazi), who strove to build a modern Western society in the country, and the Sephardic and Middle Eastern Jews. Each ignored the other community to the extent that they practically did not need each other (Berkowitz 1997, p. 144–162; Eisenstadt, 1954). Obviously, the (Ashkenazi) Zionists settlers did not turn to these communities for solutions to "Hebrew workers".

As the Head of the Palestine Office of *Hachsharat Hayishuv*, the society for purchasing land and preparing it for settlement, Ruppin wished to build the population of the Jews in Palestine rationally, on sound economic principles. His projections were that not more than two million Jews would be able to settle in Palestine. The number of Jews throughout the world was, however, 12 million. "Zionism is a cure for the moral distress of Judaism rather than for the economic misery of the Jews [...]. The great majority of Jews will not go to Palestine, and with the present scanty prospects of earning a livelihood there, it is undesirable that they should" (Ruppin, 1913, p. 296). Therefore, it would be preferable, even inevitable, to encourage selective immigration to Palestine. This idea of assigned-immigration of European Jews, which Ruppin had thought up prior to the Nazis takeover of power in Germany, which was vehemently rejected by the leaders of the community in Palestine, was however applied in an attempt to solve the problem of "Hebrew work-power." Indeed, Ruppin insisted on selective immigration of Jews on the basis of economic need rather than on biased community-based preferences, while his notions were essentially those of the Central-European colonial age.

To solve the problem of the need for a "natural" Jewish laborer in Palestine, it was suggested quite early on that the immigration of the Jews of Yemen should be encouraged: They were used to hard physical labor and very humble living conditions and appeared to be the ideal Zionist replacement for Arab workers. Jews from Yemen had been immigrating to Palestine for centuries, mainly for religious Messianic motives. Yet, since 1909 there was a conscious effort to promote the immigration of Yemenite Jews, who would be directed to agricultural work in the Jewish colonies.[8] In Hadera, Rehovot, Petach-Tikva, Kineret, and other colonies, special quarters in the outskirt were constructed to house these immigrants, who provided essential labor to farmers and citrus-growers.[9] Although no explicit eugenic arguments were mentioned in this context, these activities are considered the lowest points in the history of Zionist settlement in Palestine, and they can only be compared to the ethnic discrimination occurring in the United States during the very same years, which were explicitly justified by eugenic arguments. Various researchers pointed out the "mental callousness" and neglect of the basic needs of the Yemenite immigrants, which among other factors were reflected in their high mortality rates. From the beginning, they were discriminated against and viewed as merely cheap labor. Nevertheless, the number of Jewish laborers was relatively small. Towards the outbreak of the World War I, there were some 1,500 Jewish laborers in the colonies, compared to 6,000 Arabs employed laborers (Etinger 1972, p. 207).

The Zionist ideal of Hebrew work, as a symbol of the change of values in the renewal of Jewish society in Palestine and as an instrument to alter the biology of the Jews, utterly collapsed in the face of the economic-social reality in Palestine.

7.2 Education and Racial Hygiene

National education and its relationship to the values of Western culture were a source of dispute from the beginning of the Zionist settlement in Palestine. Already in the second volume of the journal *Hachinuch* for 1911–1912, a 'young Rabbi' wrote in defense of the emphasis on Hebrew national culture in the education system: "On the issue of religious education in schools…We believe that we have a great history, a history of men of science and intellect, who are equal to that of all the wise in the world, as well as heroic fighters who went through fire and water, spiritual heroes like none others" (*Rav Tzair* 1911–1912. TRF). Most of the fourth volume of *Hachinuch,* 1914, is dedicated to the Hebrew-Language-Warfare at the

---

[8] Immigrants from Yemen settled in Kfar-Hashiloah, near Jerusalem, as early as, 1885. In 1911–1912 the emissary of the Zionist organization, Shmuel Yavnieli, went to Yemen and organized what became known as the "Yavnieli Immigration." Much has been written about these events and their significance. See, e.g., Bloom (2007).

[9] There is ample literature dealing with the subject. The following are two references from the non-professional literature: Smilanski (1936), p. 75; Vilnai (1980), pp. 76–80.

7.2 Education and Racial Hygiene

"*Technikum*" (later: Technion) and the "*Re'aly* School" in Haifa. In a memorandum to the Board of Trustees of the Technion in Haifa, the Palestine Teachers' Association wrote:

> The power of the Jews and their cultural and political influence depend on uniting the various communities and ethnic groups, which comprised the Jewish settlement [...]. The educational system must consider living circumstances in Palestine and use the Hebrew language, in order to unify all the various groups into one whole body and into one public, united inwardly as well as outwardly. (*Hachinuch* 3(6–8), p. 38; 1914. TRF)

Starting in the 1920s, there was a surge in the impact of Western cultural discourse of original articles, worldwide literature reviews, and professional books in Zionist and related professional journals. Doctors and educators, who devoted much effort to learning about new theories of heredity and evolution, considered the Zionist effort a crucial step for the biological future of the Jewish people. Eugenic ideas overlapped precisely with their wish to cure Jews from the sicknesses of the Diaspora that had infected them. They repeatedly preached to the pioneers to take advantage of scientific developments, at least to prevent the degeneration of the Jewish race, whether it was caused by hereditary diseases or by a loss of mental capacity. The notion that the Jews were a genuine biological entity was always implicit in this endeavor.

The physician, M. Bruchow, published an article in 1922 in the Hebrew journal, *Hatkufah,* entitled "Mendelismus." In it, he pointed out that "the knowledge of heredity is needed not only in the abstract, for knowing the culture and history of people, but also in practice, for the survival of the individual and the nation, for politics and settling, education and public hygiene."

> Assuming this duty is the foundation of the power of researchers of public life, as well as ideologists, yet this movement has not properly penetrated all countries [...]. The spirit that stimulates this movement is the idea that the greatest offense that humans may perform to the God of Life is to beget sick children, imprinted by degeneration; the public concern to watch over the properties of the community and those of the individual requires taking rational means to prevent the birth of any helpless progeny. In this struggle of peoples, in this covert "cultural" war between one people and another, the one concerned with the improvement of the race, the improvement of the biological value of its progeny, triumphs. (Bruchow 1922. TRF)

In 1927, Dr. Zvi Rudi published an article on "The Biological Foundations of Education" in which he emphasized the great variability of the intellectual capacity of humans, which is based on the hereditary legacy [*Erbmasse*] of individuals. Yet, beyond these individual differences, he pointed at racial variation:

> The essence of the matter is that the property of the human species is far from being uniform. Even the large races are not uniform, as described by Gobineau in the mid-19th century, in his book on the inequality of nations. Now we know, through the work of Luschan, Günther and others, that the differences in the hereditary properties of humans are varied and so different that even in a population of one million, each individual possesses a unique hereditary combination. On these differing and varying combinations of hereditary predispositions are founded the roots of individuality. This, of course, does not glorify or promote the significance of races. Thus, members of one race may have more common hereditary-mass than do members belonging to other races. (Rudi 1927. TRF)

Rudi identified learning difficulties in school, "weakness of the senses" and the like, as problems due to hereditary-weight. "De facto, as shown by modern hereditary-theory, the cause of such abnormalities is hidden in the hereditary properties." Nevertheless, the author claimed that such hereditary differences correlate with "social strata," as well as with "geographic districts." Thus, "differences by race are also observed in the mental inclination of Jewish and Christian children in Eastern Europe." Notwithstanding, in spite of emphatically discerning social and racial hereditary differences in intellectual characteristics, he warns that one must be careful about drawing far reaching conclusions from the role of inheritance in the determination of traits, and "obviously, one should not make facts into rules."

The physician Dr. Jacob Sass, in his book *The Hygiene of the Body and Soul (A Guide to Parents, Teachers and Educators)*, devoted a whole chapter to "the hygiene of the race and inheritance." He insisted that "every race should eradicate the drawbacks, the bad inclinations, *which are inherited*, and nurture instead the good characteristics, which are inherited as well" (Sass 1929, p. 87). Furthermore, he suggested that "we need race hygiene more than other peoples, [because] we stand out as being excessively intellectual, with a disproportionately low number of craftsmen and farmers; on the other hand, we brought with us from the Diaspora many faults, physical and mental deficiencies – a hereditary load – as well as many talents. We must *strengthen* our talents and *eradicate* our deficiencies!" (Sass 1929, p. 92).

Sass also pointed out the advantages to the individual, as well as to the hygiene of the race and the nation, of maintaining a family archive that would record all health details over generations. He was aware of the burden of such a demand: "I hear the flood of words, such as: 'your demands are very tough, actually cruel, affecting the sense of love.'" There might be some truth in these claims, "but our demands are undoubtedly correct and just. At this occasion, when there abound dangerous venereal diseases and other problems that lurk everywhere, we must also be tough toward ourselves." Thus, he came up with what he believed to be racial hygiene requirements, the kind that had been introduced in several states in the United States. These included the isolation of very sick persons who may beget a sick generation, and surgery for the sterilization of the sex glands (Sass 1929).

In the 1930s scientific research activity among the settlers in Palestine intensified. Although in the world at large professional geneticists increasingly had reservations regarding the validity of eugenics as a science, among those interested in social issues and politicians, eugenics was very much in vogue. In Palestine, both physicians and educators were among its followers. Israel Rubin (1890–1954), an educator and literary critic, referred to the relationship between Zionism and eugenics in a commentary entitled, "The Ingathering of Exiles from a Eugenic Perspective," published in *Moznaim*, the Hebrew monthly of the Authors' Association in Palestine. Regarding inter-community marriages, he gave physicians and educators the following advice: "this coupling [...] not only 'looks beautiful' but is rather *inevitable*, when one comes to discuss the *improvement of the race*" (Rubin 1934, p. 89. TRF, italics in original). Rubin attached great importance not only to changes in life style, but also to inter-community marriages:

7.2 Education and Racial Hygiene

> Both are most important for *life in our homeland*, which is in its very essence, above all, a daring great national effort in the *eugenic* sense. Anyone who does not consider the return of the sons to the land of their ancestors a great *eugenic* upheaval in the life of the people fails by looking at single trees instead of at the whole "forest."
>
> "Economic relief to the needy and the persecuted in the Diaspora" – of course, "some political relief," [...] sure, but these are only details; the essence is the sum total: *the production of the restored and improved New Hebrew type.* Thus, a *psycho-biological approach to the issue of the settlement of Palestine* is an obligation for all of us!
>
> It is not merely the settlement of Eretz Israel, but all the attempts and efforts underway in the Diaspora to set our nation on a different life path than the one on which we marched, or more appropriately – limped – for many generations, must be conceived definitely as *eugenic* nuclei [...] The Lord of the Nation decreed: go and improve, go and strive toward a new Jewish type, improved, repaired [...]
>
> An elementary eugenic truth is, for example, the great eugenic value of *intermarriages* between peoples and races, [...] and thus the ingathering of the exiles in Eretz Israel allows "intermarriage," not between Jews and non-Jews, but rather between real Jews and other Jews, between Jews of different *eidoth* and different territorial origins. [...]
>
> [...] The eugenic imperative instructs us explicitly: An end to "*Landsmanschafts*" in education! An end to segregation by geography, that is, by the countries from where we came. (Rubin 1934, pp. 89–91. TRF)

Dr. Ab Matmon, a physician and the son of the educator Yehuda Leib Matmon-Cohn, one of the founders of the "Herzelia" Gymnasium in Tel Aviv and the "Hebrew Gymnasium" in Jerusalem, asserted in his booklet *The Improvement of the Race of the Human Species and Its Value to Our People*:

> [A]bove all, we must always remember the assumption that in order for the people not to degenerate, we must attend not only to their quantitative significance, but also to their quality. The best people are those who always go forward, pushing, or more appropriately, dragging the weak ones.
>
> A great many opinionated persons, especially religious men who usually do not rely on modern science, think that the fate of the nation, its future and strength, depend on the greatest number of marital links and the births that follow. They do not assign much value to the nature of the newborn [...] unlike those who base their opinions on science and especially the science of inheritance. They well know that this is not the right path. A people in which there are many individuals affected by inherited deficiencies is destined to degenerate [...] Let's take, for example, sight defects that are so prevalent among our people, and there is no doubt that we have, or had, a special predisposition for them; two thousand years in the dark Diaspora, in the ghetto, caused this. Or let's mention other weaknesses, such as fragile nerve, that many of us inherited. [...]
>
> Such a situation cannot go on, since every year the number of defective persons increases, because it is precisely those that are deficient who produce more children and bequeath to them their traits-properties. And let us not forget the many expenses that every culture-nation spends on these degenerate persons [...]. We face, therefore, the problem – how to withstand the flood. Shall we be satisfied that almshouses and shelter homes will be arranged for these miserable peoples, or will we let them go on unattended until they pass from the world? The latter is inhuman and does not provide any solution because, as mentioned, this section of the population bears the most offspring and we are not at all permitted to let chance or human impulses go their way. On the contrary, we must take the fate of these persons into our hands, give them the help and shelter they need, and at the same time affect their reproductive course, and direct it in the right way for the benefit of society at large. This is the new task of modern hygiene: to protect humanity from the flood of the defective and block their deficiencies from penetrating humanity by preventing them from bequeathing their defects to future generations. This is a new branch of hygiene, reproductive hygiene, eugenics or "*human breeding*." (Matmon 1933, pp. 3–6. TRF)

Indeed, starting in the late 1920s, intensive literature appeared that aimed at instructing and assisting in matters of health and family planning. Dr. Josef Meir (1890–1955), the chairman of the Workers Health Fund, later the first head of the Ministry of Health of the State of Israel, asked and answered the following question in a guide for parents issued by the Health Fund:

> Who is allowed to produce children? The search for a proper answer to this question involves eugenics, the science of the improvement of the race and its protection from degeneration. [Marriages of carriers of hereditary diseases] are not rare in all nations, but especially in the Hebrew people, who lived in exile for eight[een] centuries. And now that our nation is resurrected to life in nature in the land of our ancestors, is it not our duty to make sure that we shall have wholly healthy progeny, in body and soul? We value "eugenics," in general, and protection against the transmission of hereditary disease, in particular, even more than other people! [...] Physicians, sportsmen, national activists must spread the idea: Do not beget children when you are not sure that they will be healthy in body and soul. (J. Meir. *Ha'em Vehayeled*, 1934 [in Hebrew: Mother and Child]. TRF)

Nevertheless, it was not only the issue of the health of the people that engaged teachers and doctors. No less important were the intellectual capacities of the settlers, who were differentiated by *eidoth*. Many articles in the education literature published in Palestine refer to the biological disposition of the skills of the Jews and their ethnic groups. Of special interest is the article by the educator, Nissan Touroff,[10] in which he discussed the roles of heredity and of environment in the shaping of the individual and of society, and their significance to the settlement of Palestine.

> A public organism is far removed from being an extended replica of an individual organism, yet it is an organism of a kind [...], of course there are environmental influences, yet it is inconceivable that there are only environmental impacts. The people of Israel are the proof: There is no people in the world whose fate was to be exposed to so many different and diverse conditions (geographic, climatic, economic, cultural), yet there have remained, up to these days, after thousands of years of wandering, several physical and mental common traits, some manifest some hidden (that occasionally become exposed), in many, probably the majority of its members. (Touroff 1938, p. 275. TRF)

Tuoroff obviously supported Herder's concepts, conceiving the nation as an organism, a living entity. Accordingly, he accepted the reality of a national psychology, an approach popular mainly in Germany, where it attained monstrous dimensions at the very time that Tuoroff's article was penned. Thus, although many researchers had already distanced themselves in the 1930s from attaching material traits to social entities like nations, Touroff contests emphatically the opinion of anthropologists like Alexander Goldenweiser (1880–1940), who claimed that "national psychology" is founded on history, not on biology:

> The "history" of the people of Israel indicates in favor of biology [...]. Although there seems to be no absolute objectivity in Goldenweiser's words – their intentions are good: namely, to eradicate nationalistic arrogance; that incessant declaration of "thou hast chosen us," that looks down at other peoples.

---

[10] Dr. Nissan Touroff (1877–1952) immigrated to Palestine in 1907 and was a pioneer in pedagogic education in Palestine. In 1912 he founded the periodical *Hachinuch*. In 1919 he left for the U.S. In the 1930s he unsuccessfully attempted to return to The Hebrew University of Jerusalem.

7.2 Education and Racial Hygiene

> With reference to the people of Israel, the arguments run both ways. Some claim that we are the most pure blooded among peoples; others say the opposite, that our anthropological type is one of the most blurred [...] but what is important with respect to the present issue is that anyhow, there had been conserved in our people, in the external appearance of our people, in their general facial features, in the glitter of their eyes, in their movements, etc., something special and rather "typical," [... that] beyond a doubt points to *specific biological rudiments of common origins* – even if not to single origins – of a pedigree written by a family on one parchment, even though often with diluted ink or one of diverse colors. [...]
>
> Did centuries of persecution bring about a condition in which only the most clever ones would survive? [...] a probable assumption: Although it is questionable if, *a priori*, it was worthwhile to pay such a high price for this intellectual advantage, but in retrospect we should not be sorry about the unavoidable results. And, maybe, we should be allowed to ascribe with satisfaction to our special national psychology the fact that, at least until lately, we had the lowest number of heavy criminals compared to their number in other nations. *The tendency for bloodshed, for example, appears not to be an Israeli trait* [...].
>
> If indeed, with time, external conditions will change enough to guarantee us a free existence and the possibility of the true *self-expression* of our hereditary sources – we may be able again to participate proudly in the symphony of general human creation and our honor will be repaired.
>
> Is there any need to name the only corner on Earth where this dream may be realized? (Touroff 1938, pp. 291–292. Italics in origin. TRF)

Touroff stressed the decisive role of heredity in improving the intellectual capacities of individuals, which, according to his analysis, plays a crucial role in shaping our public image. During the years in the Diaspora, the distinctive hereditary pattern of Jews was maintained, although it was greatly affected by conditions and circumstances. Only the socio-cultural web of life in Palestine could lead to a new distinct shaping of the Jews in the spirit of the twentieth century. This, however, turned out to be more difficult to realize than anticipated.

In 1926, Dr. Brachyahu[11] reviewed the work of the Department of the School Hygiene in Palestine. He noted 13 *eidoth* among the pupils: Ashkenazi, Sephardic, Yemenite, Persian, Bukharin, Georgian, Urfal, Babylonian, Aleppoian, Baghdadi, Maghrebi, Kurd, and Ajami, and he claimed that "each one of these ethnic groups determines a specific type" (Brachyahu 1926, p., 188). One of the teachers quickly responded to this, asserting that people usually make fine and detailed distinctions in communities or groups of people that they themselves belong to rather than to those that are alien to them. Still, in Palestine, all European Jews, from East to West, were classified by authors as "Ashkenazi" – the group to whom the author himself belonged – whereas the Jews of the Orient were classified by them ti ethnic group that no contemporary person had ever heard of. Does this indicate a new kind of discrimination? (*Achad Hamorim* 1927 [Hebrew: One of the Teachers])

> The assertions of the physician, who has been scientifically examining thousands of children every year, would surely be instructive to many of his colleagues. Thus, it is sensible that he should enlighten us about the *racial* characteristics of each of the 13 *eidoth*. Of special interest is the difference between Baghdadis and Babylonians, and between Ajamis and Persians, since up to now we believed these to be synonymous or, to be more precise, the respective Arabic and Hebrew names for the same people. Up to now we had not heard

---

[11] Dr. Moshe Bruchow (*vide infra*) changed his surname to Brachyahu.

of differences in customs and prayer styles [among these people]. What is then the basis for the anthropological differentiation? And, just as this new differentiation must be explained, so it should be explained why all those who speak the German-Jewish dialect were included in just one *eidah* – are not differences within that group similar to those between Aleppoians, Urfals, and Baghdadis whose dialect is the same? Let the honorable doctor instruct us, and we will be grateful. (*Achad Hamorim* 1927. TRF)

There is no doubt that "One of the teachers" was sensitive to ethnic group discrimination, even if the source of this discrimination was ignorance. As far as I know, the query remained unanswered. But it is apparent that Brachyahu pointed out an accepted and undisputed claim of the great variability that existed among Jews from different regions in the Orient (compared to the presumably less conspicuous territorial differences among the more mobile European Jews). Neither Brachyahu nor most other researchers explicated what methods they used to classify people into categories. Most of the differences were, of course, based on external appearances, like dress and customs, although long relative isolation could have also brought about significant differences in gene pool frequencies.

The encounters of people from utterly distant cultures and living styles must have been a great shock to everyone involved, and empathy demanded a more profound search for a common denominator. John Glad quotes a pediatrician who asserted a year after he emigrated from Germany to Palestine: "The difference between Ashkenazi and Sephardic Jews is so great that biologically we can speak of two races" (Glad 2011, p. 208). Yet the physician, who could communicate with his patients only in Hebrew, a language new to both patient and doctor, was astonished when he found out that his patient had come from Aram-Naharaim, the Biblical home of the patriarch Abraham, thus indicating a cultural link to common forefathers, whether or not common genes linked the physician with his patient across the insurmountable cleft (personal communication).

The author and educator, Jehuda Burla (1886–1969), was born in Jerusalem to a Sephardic family that had immigrated to the Holy Land in the eighteenth century. His work, which centered on the life of the Sephardic Jew, exposed the plight of "the child of a Middle-Eastern *eida*, who finds himself in the minority among Ashkenazi children [in kindergarten] and under the supervision of an Ashkenazi teacher," deprived of the familiarity of his home-environment. Some asserted that in order to conserve the folkloristic essence of the Middle Eastern ethnic groups, it would be proper and justified to have special educational institutions for children of Middle Eastern extraction. But Burla consistently rejected such proposals: "If our concept of the Hebrew revival, of the return to our homeland, and of our Zionist enterprise is comprehensive, then obviously the merging of all our activities into one great amalgamation, may accomplish in due time the ideal of generations to become one people" (Burla 1927).

Still, researchers of European origins were concerned about the differences between "Ashkenazim" and "Sephardim," and although most stressed the difference in living styles that were typical of the ethnic groups, there were also utterances of immanent Western superiority. Dr. A. Ornstein performed in July 1930, at the end of the school year, comparative psychological tests between Ashkenazi, Sephardi, Yemenite, and other children:

> Let us turn to ethnic-group differences. [...] The achievements of the children of Middle-Eastern ethnic-group lag much behind those of their Ashkenazi brethren [...]. What is the cause of the difference in the intellectual development of members of our *eidoth*? One factor, probably the dominant one, is the social status [...]. However, beyond this external factor, there are also mental factors of two kinds: (1) Differences in inheritance and education, which determine the level of intelligence, (2) differences in the rate of development. It appears that the development of Middle Eastern children is retarded. (Ornstein 1938. TRF)

The researcher sums up his findings, stating that "it may be assumed that the skill of appraisal is developed in our children not less than in children abroad, and in Ashkenazi children probably even more."

Dr. Karl Frankenstein was more explicit in emphasizing the differences and the need to check Middle Eastern child birth. In his article "Delinquency and Neglect of Jerusalem Youth," he stated that "the contrast between the primitiveness of the Middle Eastern Jews and the civilization of the New Palestine" is that the great majority of felons are of Middle Eastern ethnic groups, i.e., of poor civilization circles (Frankenstein 1938). According to this author, the character patterns of persons of the Ashkenazi ethnic group "indicate a strong tendency to individualist positions and deeds [...]. Ashkenazim, as well as Yemenites, rarely participate in group felonies. The position of Ashkenazi to Middle Easterners is like that of an adult to a youngster, or like that of a mature person to a primitive person" (Frankenstein 1938, p. 31).

In 1937, A. Z. Eshkoli published *The Human Race: Race Research and Race Theory*, which presents the history of racial theories that were common at different times. He noted: "Also many Jews were attracted by race theories: Some by accepting the false claims of their enemies, some getting involved in apologetic and unfounded claims, such as proving the priority of the Jewish 'race' [...]. Others who treated the subject just formally, 'scientifically', were unable to choose between the positions of the Gentiles and the opposite ones of the Jews" (Eshkoli 1937, p. 14).

Nevertheless, articles on racial differences between Jews and non-Jews did not disappear even from the Hebrew literature. Ten years later, in 1948, Dr. Noah Nardi measured the brains of Jewish children in general public schools in Palestine and the Jewish schools in the United States. "Nearly all studies indicate a clear superiority of Jewish children. By this, the author did not intend to prove any racial superiority, rather an emphasis on the importance of specific factors that may explain the higher IQ ratios of the Jewish children, is underlined" (Nardi 1948).

7.3 Jewish Intelligence (and Disease)

For many years, repeated speculation about the role played by hereditary factors in the intelligence of Jews, primarily in relation to Ashkenazi Jews. Claims of prejudiced mating traditions such as marrying the "Wise Students" to the daughters of the community "Chieftains" i.e., marriages of persons with presumably high intellectual capacities with persons of promising socio-economic capacities, were comon (See e.g., Motulsky 1980, and MacDonald 1994). Many researchers tried to bring circumstantial coincidental events to support hypotheses of Jewish superior intelignce, such as:

- Ashkenazi Jews have the highest IQ, together with an exceptional cognitive profile, compared to all known ethnic groups. There is no evidence of a high grade of intelligence among Jews in ancient eras, neither among Jews of Middle Eastern or Sephardic origins.
- Among Ashkenazi Jews, there was only a negligible influx of genes from outside, thus optimal conditions for natural selection were secured.
- Extreme selective pressures were exerted on Ashkenazi Jews, especially in the ninth to eleventh centuries C.E., which demanded high intelligence to survive. Jews were pushed into professions of finance and management, which demanded high skills to handle complex situations and success among the populations around them who were mainly farmers and craftsmen. The Ashkenazi Jews were thus placed in situations that promoted economic success as well as reproductive success. Such explicit selection for professional preferences of the Jews, it was argued, was absent in the countries of Islam.
- Among Jews, there happened to be elevated frequencies of function-related hereditary factors, revealed by those individuals with diseases that affect neural activity, primarily diseases involved in fat metabolism (sphingolipoids), and those involved in DNA damage repair, and other groups of hereditary diseases the biochemistry of which is still less known. This is in contrast to the absence of such factors correlating hereditary intelligence and disease in the neighboring non-Jewish populations.

As a rule there has been no support to claims that physiological factors, such as climate, nutrition, or hygiene were responsible for this difference. Thus, it was only natural that all these claims elicited many responses both pro and con.[12]

Obviously, the reliability of the different arguments discussed above is the subject of intense controversy. Many authors, most notably Henry Harpending and Gregory Cochran (2002) – as many before them – ignored the problematic of defining the populations they studied, whether Jews or not, and of establishing the criteria to determine whether the Jews were Ashkenazi, Sephardic, or Middle Eastern. They also largely ignored the definition of intelligence, and their classifications, consciously or unconsciously, often remind us of the racist definitions of the

---

[12] See, for example, the British journal, *The Economist,* June 2005 that published a front page story with the picture of Sigmund Freud, Albert Einstein, and Gustav Mahler, declaring "The high intelligence of the Jews may be the result of their persecution." For a critique, see Howard Metzenberg. To his mind the article reflects the view of "The Merchant of Venice," and all talk about the high intelligence of the Jews is nothing but a product of culture. (see http://www.gnxp.com/MT2?archieves/004983.html)

Dark Age.[13] It is difficult to overcome the impression that preliminary assumptions of authors about the cognitive superiority of the Ashkenazi Jews were what directed the research and consequently the conclusions. To counter the argument that the high frequency of, say, diseases is a consequence of the populations being small and relatively isolated, the authors claimed that the founder effect or drift could explain the high allele frequency of one gene or another, but not of many genes (see e.g., the discussion of Behar et al. 2004a, b), especially not of a group of genes of physiologically related diseases. This may become particularly significant in other isolated populations, like those in Finland and the province of Quebec in Canada, where there is a high frequency of certain hereditary diseases, probably due to the founder effect and drift, but there is no clustering of physiologically related diseases. Harpending and colleagues accepted that Jews have a strong common East Mediterranean origin and if there was only a minor leakage of 0.5–1.0% of foreign genes per generation into the Ashkenazi gene pool over the 80 or so generations of the Diaspora, still roughly half of the gene pool remained Middle Eastern. Indeed, there are alleles of other genes among Ashkenazim that indicate their relatedness to the Middle East, like those for Factor-XI deficiency, BRCA1, and APC [*Adenomatous polyposis coli*], prevalent among both Jews of other ethnic groups and non-Jews of the Middle East. But they claim that it makes sense to assume that the products of some of the other genes are involved in providing the conditions for high intelligence, and that these were inadvertently selected for among Ashkenazim, who maintained a specific mating tradition. The genes for the lipid-storage diseases, they claim, provide such a cluster, since it is well known that these lipids are involved in the construction of nerve envelopes. They found further indirect support in patients with Gaucher's disease, a disturbing but not fatal disease that belongs to this lipid metabolism cluster; these patients appear to be endowed with above average intelligence. In a survey of patients in Israel, 37 of 255 patients of working age (not students or retired persons) were engineers and scientists; this is a rate six times higher than that of the general population (Gross 2002). Unlikely as some of these assumptions may appear, they deserve to be examined experimentally.

Eva Jablonka of Tel Aviv University, in a private conversation suggested another possible explanation: that higher intelligence may be the consequence of cultural selective inheritance. For example, families whose children are affected by non-severe genetic diseases, that is diseases unrelated to the hereditary component of intelligence of the children, which limit physical activity, or families who have lost a child because of a severe hereditary disease, would channel their remaining children toward "intellectual" activities. Children in such families will, on average, succeed more than their colleagues under conditions that require intellectual

---

[13] On the internet site, http://www.futurepundit.com/archieves/ cat_brain_evolution.html, in June 25 2005, Randall Parker writes: "Why are Jews so smart? Granted, a lot of Jews want to argue that they are just studious due to their culture. Also, lots of ideologues – particularly on the political Left – stand ready to attack anyone who argues that ethnic and racial groups differ in average intelligence. But the higher average level of Ashkenazi Jewish intelligence is so glaringly obvious that I figure anyone who tries to argue otherwise is either engaged in intellectual con artistry or is ignorant or foolish."

proficiencies, even when their genetic potential is not outsrabding. When "sickly" children are channeled to intellectual activities from childhood, it makes sense that their intelligence quotient would be higher than that of their contemporaries who are involved in other activities and, therefore, they would also produce more progeny than their contemporaries (provided that the gene did not affect their fertility). It is reasonable that such channelling over many generations will lead to the increase of genes for such diseases in the population in correlation to the intelligence ratio of the population.

In summary, it is misleading to treat the subject of "hereditary diseases" or "hereditary intelligence" as if it were pure biological research. The entities we deal with are defined in cultural and political terms. Specific allele frequencies depend not only on their etiologies, but also on factors such as population size and mating patterns, and, of course, the selective value of the alleles and the forces of selection that act on the genetic combinations (heterozygotes, homozygotes, and the interactions between them) under different environmental conditions. Yet following the distribution of single hereditary variables, even those of a well defined hereditary disease, is inappropriate when discussing the history of populations.

As important as the study of the distribution of specific genes is, especially those involved in disease, to understand the structure of populations and the forces acting on them, such a study cannot provide a picture of the history of populations and their origins. The picture is too unstable as the impact of each single variable may drag it in a direction opposite to the previous variable. A multi-gene approach is necessary. Genetic research in recent decades advanced from being a classic bottom-up reductionist science to increasingly becoming a top-down genomic molecular analysis (see, e.g., Falk 2013; Griffiths and Stotz 2013. See also Falk 2014a). With today's sophisticated methods of molecular analyses, culminating in detailed DNA-analyses of variability at the level of single specified nucleotides anywhere along the genome's DNA-nucleotide sequences (genome wide association studies) and of long consistent sequences of nucleotides (haplotypes), a new era has begun. Today, with the sophisticated methods of molecular analyses, that culminated in DNA-nucleotide sequencing (Genome Wide Association Studies) anywhere along the genome, and the analyses of long consistent nucleotide sequences (haplotypes), a new era has begun.

***

The 1940s obviously brought a profound change in the social and cultural perspectives of inhabitants of Palestine. Following World War II and the Holocaust, the demography of the Jewish population in Palestine changed. The Nazi application of racial theory made eugenics taboo in many circles. After Israel's War of Independence, when Palestinian Arabs became a minority within the borders of the State of Israel, the issue of the blood relationship of Palestinian Arabs and Jewish settlers appeared to be marginalized. However, during the 1950s, with the massive immigration of Jews from the Near East and North Africa, the demographic composition of the Jewish population in Israel changed. Interest in the biology of the Jews and especially in the biology of the specific Jewish ethnic groups increased, although many of the questions that were asked were different from the questions that had been asked in the recent past. Not surprisingly, since the beginning of the twenty-first century the "demographic issue" has returned full-fledged (see also Paul 2016).

7.4 The "Demographic Issue"

Dr. Sh. Z. Bychowski, the neurologist working in Warsaw, made it a point of reminding "the Zionists who come to rebuild the life of the people" that

> [i]t is necessary to spread much propaganda in Eretz Israel against the system of "one or two children" that may lead to the annihilation of the race, even for a race that has a prosperous industry, full diplomatic relationships, and great literary institutes. Eugenic literature has devoted many articles to this issue, and specifically to the problem of the degeneration of the French people. It is crucial that also the Zionist literature will acquire knowledge of these issues and will introduce such topics into the sphere of problems that it is concerned with. (Bychowski 1918, p. 299, TRF)

The "Society for the Regulation of Births" was established in Palestine: "Regulation of births means systematic population reproduction; it means the prevention of useless life forms. The regulation of births entails issues like eugenics (the improvement of the race) […]. The reactionary opponents of the regulation claim that the proponents of the regulation of births aim at nothing less than the ruin of the family and the decimation of the population. This is not the case […] it is imperative that children should stop being the products of chance, that their advent should instead be a desired event" (Firsht 1935).

A more biased population-genetics approach to the issue of the specific reproduction of the Jews and that of the various ethnic groups in Palestine was that of Fritz Shimon Bodenheimer (1897–1959), the son of Max Bodenheimer, one of Herzl's close associates. F. S. Bodenheimer grew up in an atmosphere that combined German patriotism with Zionism: As a young man, he proudly served in the *Bonn Husar Battalion* during World War I, but when he encountered anti-Semitism in the army, and especially when he met the Jews of Eastern Europe, he joined the Zionist movement. In his autobiography he emphasized that he explicitly decided to study agricultural entomology in order to further the Zionist activity of settling in Palestine. Bodenheimer was impressed by the figure of the biologist, Ernst Haeckel: "Human wisdom lies in the harmonious union of science and humanism. Modern scientific emphasis on professionalism and technology involves a certain lack of general knowledge, understanding and perspective. A scientist who fails to recognize 'values' is ultimately a mere technician" (Bodenheimer 1959, p. 1). Like Haeckel, Bodenheimer wished to acquire a wide, comprehensive perspective; in a way, he apparently conceived of himself as the modern Zionist-Haeckel. He immigrated to Palestine in 1922 and joined The Hebrew University of Jerusalem in, 1928 as its first zoologist. He is well known primarily through his research in entomology and the dynamics of insect populations. He studied the spread of locust and insect epidemics in Iraq and in Palestine for the British government. Bodenheimer published over 420 articles, books, reports, etc., including those that popularized the sciences, and had a tendency to put forward comprehensive, sometimes sweeping theories.

Already in the 1930s, Bodenheimer preached the importance of intermarriages between the various communities for the biological invigoration of the Jewish people. In his book *The Biological Background of Human Population Theory*, published in 1936, he pointed out that he wrote it in order to make the settlers in Palestine aware of the biological background of humans in general, and of national politics in particular. Bodenheimer agreed that the differences between Jewish ethnic groups were due to the combination of the isolation of the communities, on the one hand, and their dilution with the local population, on the other hand. He conceived of the Yemenites as the "purest Semitic type," whereas all other *eidoth*, "far from being a pure race, are associated merely by a common unique historic fortune" (Bodenheimer 1936, pp. 137–138).

> Geographic isolation of some of the Jewish people in the Diaspora generated in several places a conspicuous distinct type, different from other Jews living elsewhere. That segment of the Jewish race, which we call "Sephardic" – one may assume with near certainty – did not participate in the great mix with Slavic or Tatar blood […]. On the other hand, we do not find among the Ashkenazi the mix with the Negro types, not at such a high rate as found among the Sephardic. The purest Semitic type is conspicuously found mainly among the Yemenites […]. Zionism strives directly, though not with an *a priori* intention, to unify all the different elements, which are prevalent today in the Jewish race, in order to produce one harmonic Jewish type. (Bodenheimer 1936, pp. 138–140. TRF)

Bodenheimer obviously hoped for intermarriages between the – according to him – highly fertile but culturally inferior Sephardim, with the culturally superior, but of low fertility Ashkenazim. Such 'hybrid-vigor' would not only enhance the vitality of the people, but would also provide a solution to the "demographic issue" of the especially high fertility of the Palestinian Arabs:

> Turning to the situation in our settlements in Palestine, the ranks at the lowest cultural level are the most fertile ones among us. Thus, if we disregard immigration, in the future it will be the progeny of these ranks that will increase the Jewish population of Eretz Israel. This is a very positive development, since the settlement in Palestine is threatened by the high rate of natural growth among the Arabs […]. Thus, it is significant for the settlers that among them are members who need little and who are used to living under difficult circumstances, yet are blessed by high child fertility. (Bodenheimer 1936, pp. 141–142. TRF)

Although the Zionist movement by and large ignored the Arab population of Palestine, or tried to minimize the physical and cultural impact of the Arabs, Bodenheimer considered them a demographic threat. Not everyone adopted this approach, above all the physician and anthropologist Elias Auerbach, who already at the beginning of the twentieth century warned against maintaining the attitude, as if the Palestinian Arabs, unlike the Europeans, lacked any worthwhile cultural or national tradition. According to Auerbach, in the future we will undergo a cultural and demographic amalgamation of the country's populations: The simple truth is that the country was never empty, and its character would be shaped by the domi-

7.4 The "Demographic Issue"

nant component in the population. "The Arabic language is superb, and everybody who lives here must learn it" (Auerbach 1911).

Ruppin too was certain of the ancient blood and cultural alliance of Jews and Arabs : This would provide the guarantee against the national envy of local Arabs, as long as the Zionist settlements did not cause the economic and social deprivation of the Arabs. "There is a certain affinity between the Arab and the Jew as there is between the Arabic and the Hebrew languages. It is highly probable that the two peoples would live happily and amicably together even if the Jews were to come in large numbers" (Ruppin 1913, p. 292). Ruppin was primarily worried about the Christian Arabs, who were urban dwellers and owners of great estates. However, the longer he lived in Palestine the more he worried about the unavoidable conflict of the national interests of the Jewish and Arab inhabitants of the country. As Alex Bein, Ruppin's biographer, pointed out: as time passed "the single problem that engages Ruppin beyond anything else is 'the Arab issue' and later in his life he always comes back to it [...]. When Ruppin came in contact with it in his work, his primary response was to avoid injustice and causes of instigation. [...] In the beginning Ruppin strove that the Jews in Palestine would assimilate in the Middle-Eastern Arab world, a world with which they had ethnic and linguistic connections and a common ancient cultural tradition. However, [as he faced] *Memoirs* the complex reality, his longing became more an expression of confusion than a direction leading to a solution" (Bein 1971, p. xviii).

As a matter of fact, Ruppin's attention was diverted from the issue of the essence of the relationship between the various Jewish *eidoth* to that of the Semitic relationship of the Jewish ethnic groups with the Arab inhabitants of Palestine. Life in Palestine convinced Ruppin that some other foundation than that of Zionism as a movement aspiring to give national meaning to the children of the Jewish race must be found. He initiated a desperate search for some link between the loose ends and aspirations relative to non-nationalism – or meta-nationalism – in the form of biological racial cooperation between the peoples of the Near East. He envisioned the bridge that would lead Zionism to the realization of this goal in the communities of Sephardic and Middle Eastern Jews. On April, 12, 1923, he wrote in his diary:

> I think that Zionism is possible only if it provides a completely different scientific foundation. Herzl's proposal was naïve and makes sense only with his complete ignorance about conditions in Palestine. It is our duty to take again our place among the people of the Orient and produce a new cultural community in the Middle East, in cooperation with our racial relatives, the Arabs (and Armenians). I believe that Zionism is now less justified, except for the fact that the Jews belong racially to the peoples of the Middle East. I am now collecting material for a book on Jews that will be based on the issue of the race. I wish to include in it drawings that will present the ancient peoples of the Middle East and the present day population and describe the types that ruled and still rule among the people living in Syria and Asia-minor, I wish to show that these types still persist among the Jews of our time. (Ruppin in Bein 1971, p. 271)

At the end of the day, when he realized the depth of the socio-political problem of the national identity of both the Jews and the Arabs, as one of the founders of "Brith Shalom" (literally: Covenant of Peace; formally: Jewish-Palestinian Peace Alliance, founded in 1925), he hoped for the establishment of a bi-national state in Palestine.

Although many of the Palestinian Arabs were undeniably recent immigrants from neighboring countries or the progeny of those who wandered into the area during recent centuries, Ruppin was not the only one who took notice of Palestinian Arabs and their relation to the ancient Jews. Israel Belkind, one of the founding fathers of the Bilu movement who insisted on "practical Zionism," had reached similar conclusions, namely, the principal of actively immigrating to Palestine and actively laboring there. Belkind settled in Rishon LeZion in 1882, but because of his rebellion against the officers of *The Baron*,[14] he had to leave it and founded Gedera. Eventually he turned all his energies to educational activities. He opened a Hebrew school in Jaffa in 1903 and founded an agricultural school, "Kiriat Sefer," in Shfeia, for the absorption of the children who survived the pogrom in Kishinev. He later became interested in the geography and the ethnography of Palestine, and in 1919 he published his book *Eretz Israel*, in which he described the inhabitants of the country:

> The inhabitants of Palestine are known as Arabs or Syrian-Arabs. It is, however, impossible to view them as Arabs of Arab stock. They really are the progeny of the peoples who were living in this country when it was occupied by the Arabs and who with time adopted the faith of the Arabs, the religion of Islam, and their language. Among them, the Jews were the great majority. [...] For the thousand and three hundred years that passed since the Arabs conquered the country, the inhabitants of the country forgot their nationality and their origins. (Belkind 1919, p. 83. TRF)

In a leaflet that Belkind published in 1928, *On the Arabs in Palestine,* he explained his reasons for those conclusions and presented further evidence: "A common scene in the history is that one people falls upon another and occupies the country." Such occupations may be classified into three kinds: Occupation instigated by "the urge of kings to prevail over peoples and of their craving for foreign power." The second kind is occupiers of other peoples are those who "wish to settle anew in sites not of their own because, for various reasons, they could not stay in their original domiciles. Such an occupation was that of the Land of Canaan by the Israeli people." Although the original purpose of such occupations was to exert pressure and to cause the occupied people to leave their land, history proves that such an aim had never been completely fulfilled. "The result of such an occupation is always that the occupiers and occupied merge to form one people: the one turns into the other without becoming aware of being brought closer." The third type of occupation was different: The Arab soldiers did not leave their country in order to occupy new countries. Their main purpose was to spread their new belief, the belief of Islam. Such occupations may be called "spiritual occupations." The occupied peoples adopted the belief of the conquerors and their language, and also their culture, which further enhanced the imprint of the Arabs on these nations. (Belkind 1928, pp. 7–10. TRF)

Belkind was confident that Palestine was no exception and that what happened in all the other countries in the area would also happen here. "One must therefore ask, whom did the Arabs encounter when they conquered the country from the Byzantines?" He noticed that at the time of the Bar-Kochba revolt (132–136 B.C.E.),

---
[14] Baron Edmund Rothschild was the sponsor of most of the early Jewish settlements at the end of the nineteenth century and the beginning of the twentieth century. He was known as "The Baron".

7.4 The "Demographic Issue"

"hundreds of thousands of people took part in it; thus providing evidence that the Hebrew population in Eretz-Israel in those days was immense." Also, 550 years after the destruction of the Temple, "when Kozro the Second of Persia fought the Romans, there was attached to his army a battalion of twenty five thousand Galilean Jews [...] and when that battalion fought for Jerusalem, a second battalion of twenty thousand Jews attacked the town of Tzor [Tyre]" (Belkind 1928, pp. 12–13).

> These facts prove that we may wrong Truth if we conclude that the Jews completely left their country because of the harassment by the Romans [...]. The country was abandoned by the upper echelons. The intellectuals, the bible students, who preferred religion over land, left the country and wandered eastward [...] while the rich and the merchants turned westward [...]. But the farmers stayed attached to their land. However, after they had been abandoned, without their spiritual leaders, they could not resist first the Christian instigators, and later the Arab conquerors. (Belkind 1928, pp. 14–15. TRF).

Belkind presented further evidence: the names of the settlements that maintained their Hebrew origins in spite of the Roman names that were imposed. "Should we imagine the Arab army commanders browsing the Bible and Talmud in search of the ancient names? No! In reality Jews were the majority of the country's inhabitants" (Belkind 1928, p. 19). Belkind's conclusions were unequivocal: In Palestine we encounter many of our people, who indeed severed common life from the main body of Judaism for fifteen hundred years, yet they are of our flesh. It is on the basis of these facts, Belkind insisted, that we should in the future shape our relations with that people. Obviously, only one attitude can prevail between us, namely that of brethren. Not only as brothers in the political sense, as history destined us to live in a common statehood, but rather as brothers in race, brothers of one nation (Belkind 1928, pp. 23–24).

Furthermore, according to Belkind, the people who returned from the Babylonian exile united with those of the Assyrian exile, so that in the days of the Second Temple the descendents of all the Twelve Tribes united, and "in the days of the Hasmonean Kingdom the segregation of Judea and Israel, that between northern and southern tribes, was annihilated. They approached each other in the lands of their exile, and cooperated in their return from Babylon to Zion, to become one people in Palestine" (Belkind 1928, p. 39). To these, Belkind added the non-Jewish inhabitants of Edom and the Galilee, who had been occupied and converted to Jewry.

It seems that Belkind may have gotten swept away, and such historic or anthropological-cultural claims are not enough to substantiate physical or genetic anthropological claims. Yet, other Israeli archeologists (see, e.g., Broshi 2004) found indications of the relations between the indigenous Arab population in Palestine and the ancient inhabitants of the country.[15] As we shall see, a more critical examination of anthropological and archeological evidence in light of the popula-

---
[15] It must be emphasized that, over the centuries, there was a constant flow of people, mainly from the south who – like the present-day "Out-Of-Africa" migration – were looking for jobs and better living conditions, and thus settled in the land that became Palestine. Also, it is important to note that the recent rapid economic development of the country – compared to neighboring countries – largely due to the Zionist settlers, stimulated intensive immigration from neighboring countries all around.

tion's genetic evidence, and vice versa, is needed: Some direct genetic evidence to support historic claims – the "biology-as-history" approach – is needed just as much as genetic claims – the "history-as-biology" approach – needed historic evidence to substantiate their story.

Yet, in retrospect, whatever the historic evidence and the genetic evidence might indicate, there is some grotesque, tragic irony, less than a century after the disputes of the biological status of the Jews in Europe, that the Zionists who struggled to literally breed a new Jewish race in Eretz-Israel would fall prey to their own efforts to keep separate from the country's Arab citizens. As a mater of fact, they endeavered to kep as advocated by persons like Jabotinsky and Bodenheimer, who endeavored to keep the Jewish people as a distinct race, as well as persons like Auerbach, Belkind, and for a time also Ruppin, who preached about historic-genetic links to the indigenous local indigent Palestinians.

Chapter 8
The Inagathering of Exiles

> […] For Jews were not an accident of race, but simply man's condition carried to its extreme […]. Made homeless in space, they had to expand into new dimensions, as the blind develop hearing and touch. The loss of the spatial dimension transformed this branch of the species as it would have transformed any other nation on earth, […] It turned their vision inwards. It made them cunning and grew them claws to cling on with as they were swept by the wind through countries that were not theirs. […] Reduced to drift-sand, they had to glitter if they wanted to avoid being trodden on. Living in bondage, cringing became second nature to their pride. Their natural selector was the whip: it whipped the life out of the feeble and whipped the spam of ambition into the fit. In all fields of living, to get an equal chance they had to start with a plus. Condemned to live in extremes, they were in every respect like other people, only more so. (Koestler 1946, pp. 355–356).

The decades after the World War II and the establishment of the State of Israel were years of mass immigration. To begin with, the immigrants to Israel were mainly survivors of the Holocaust, whether they came directly from Displaced Persons camps in Europe or from camps in Cyprus, where they were incarcerated by the British authorities for attempting "illegal immigration." Gradually, however, the numbers of these immigrants were reinforced by increasing numbers of immigrants from the Orient – often whole communities – from Iraq and Iran, and soon also from Yemen and North Africa. With the emigration of Arabs from Israel as an outcome of Israel's War of Independence, and with the waves of Jews coming in, Jews became the majority in the young state. This population was even more heterogeneous than before. In the 1950s, only about 30% of the inhabitants were born in Israel, some 20% were born in Poland and about 15% in Romania. Iraqi Jews comprised nearly 15% of the citizens, and North Africans somewhat less. Yemenite Jews comprised about 5% of the inhabitants. There were also some USSR-born and German-born immigrants.

The predicament of individual identity in a foreign country, which obviously concerns the absorption efforts of every immigrant, became extreme and acute (see e.g., Kirsh 2003). Zionist ideology stipulated that the State of Israel would be a melting pot for all Jews. Thus, in addition to the known social and economic issues that immigrants face in all countries, in Israel there was also the fundamental issue of Jewish identity: Who is a Jew? Which ancestral heritage, traditions, religious rituals, languages, or social links identify a Jew? If there was more to this than a halachic issue, it appeared that these were questions about biological links to be answered by scientists, since it had been claimed that the Jews were, in principle, the descendents of the people who inhabited the Land of Israel before their exile (and those who joined them over many generations in the Diaspora), and evidence of common biological ancestry would provide the national ethos with a weighty component. It was, therefore, no wonder that the culture of the Ingathering of the Exiles in the young State of Israel was accompanied by an implicit effort to enlist scientific research to accomplish the task.

The impact of the events in Europe and the rejection of the Nazi vision, on the one hand, and the advances in the study of the biology of populations, on the other hand, were followed by major changes in the interrelationships between the social sciences and the life sciences. The use of the emotionally charged term 'race' acquired new dimensions, not merely in the context of Europeans and Jews, but primarily in that of Black and White relations in America and, significantly, that of the relations between Europeans and Third World Natives. The concept of race, like that of species (and the concept of Jew), reflects the relics of typological notions and of the classification of nature into essential entities as instituted at the end of the eighteenth century in the Linnaean system of classification (see Chap. 3).

The questions "what is a race?" or "who is a Jew?" largely ignore the fact that (within the human species) any marriage circle can be viewed as a potential race. Considering that any female and male of a species may produce progeny with other members of that species,[1] the patterns of marriages or sexual relationships for humans, in general, are resolved not so much by biological determinants, but rather by geographic or socio-cultural affinities and barriers. Today it is common among researchers in the humanities, the social sciences, and even the natural sciences to say that at least as far as humans are concerned, (biological) races do not exist.[2] The biological races that were presumably discovered were in fact the illegitimate product of the classification system imposed on nature. Classification by races is a social construct. As a rule, the use of the term "race" for multiplicity, which is based on the

---

[1] The modern definition of a *species* would be a group of genetically compatible interbreeding individuals of natural populations that is genetically isolated from other such groups: Species have, by definition, a common *gene pool*. In spite of the difficulties in applying this definition in some cases (are pine trees in South America and in the faraway Middle East of the same species? Are the mountain goat and the domesticated goat (that do produce fertile progeny when hybridized by men), two different species? Including all human beings in one species, *Homo sapiens*, is not a problem, although not more than a century ago there were researchers who considered some African tribes and the Europeans to be different species. See, e.g., Gould (1981). See also Wade (2014).

[2] This does not exclude the possibility that social relationships may eventually end up in biologically meaningful segregation. Already many years ago (see e.g., Montagu 1974. One chapter in his book is entitled, "Are 'the Jews' a 'race'?") authors doubted the very term Jew as a race. See Barkan (1992).

typological mind-set, has been pushed aside in the scientific parlance and replaced by "population" in terms of statistics (see Chap. 3).

Although genetics flourished in the first forty years of the twentieth century, it was only towards the 1940s that population genetics was established, and human genetics became a science only in the 1950s. Starting in mid-century, variables of human population genetics increasingly replaced physical anthropological characteristics in discussions of the dynamics of human populations and the variability of human populations as the product of their evolution.[3]

But the introduction of genotypic markers to describe the evolutionary historic relations between "populations" did not eliminate attempts to correlate differences in character-patterns or intellectual capabilities of ethnic groups with properties such as skin color or facial patterns. Even though it became obvious to scientists that there was no inbuilt correlation between genetic factors involved in behavioral and mental traits and hereditary factors that affect the pigmentation of the iris of the eye, hair curling, or skin color, the segregating classification of people according to ethnic group or birthplace continued. Primarily on the basis of external phenotypes and then on the geographic origins of a persons' immediate relatives, it was claimed that the environments that selected for, say, dark skin color also promoted certain patterns of behavior or intellectual capacities (or put differently, did not promote properties valued by white Europeans). However, the more genetics became molecular, the more it became clear that the expression of personal properties depends on so many variables that they may only rarely be reduced to specific hereditary variables of individuals, and to speak of general properties of populations would be absolutely meaningless.

8.1 Medical Anthropology and Population Genetics

The differences in the frequencies of hereditary diseases in Israel among persons coming from different communities were conspicuous and called for examination. The physician and medical administrator Chaim Sheba (1908–1971) pioneered the study of this subject already at the end of the 1930s by exploring the differences in

---

[3] In recent years, repeated attempts have been made to overcome the reluctance conditioned by socio-cultural attitudes about using the term "race" for humans. The criteria should not be those of identity of characteristics, but may rather be those of historical – even prehistoric – partnerships: Races are ancestor-related breeding populations, whose members share common genomic patterns. As far as modern humans are concerned, distinct breeding populations are not so much an issue of geography as of socio-economic relationships. See, e.g., Part Two of Koenig et al. (2008); and Sesardic (2005). However, special attention must be given to Gannett's (2004) reaction to Andreasen's (2004) reification of the race concept and her criticism of the adherence to dichotomies between science and society, facts and values, nature and culture, and the biological and social. Gannett (2013, p., 138) quotes Griffiths' words, "Kinds are the realist interpretation of Goodman's [*Facts, Fictions and Forcast*] 'projectible properties'. They represent correlations between properties, which our background theories suggest, can be relied upon to hold up in unobserved instances."

the distribution of diseases among people of different ethnic origins. Thus he proposed introducing "medical anthropology" as a branch of medicine that would follow the origins and relations between people according to the distribution of diseases among individuals of known hereditary backgrounds. Sheba acknowledged the nominal traditional story of the history of the Jewish people as related in the Jewish texts. He turned his attention to the peoples of the Biblical Eretz-Israel, whom he called *Homo israelensis,* and attempted to follow their subsequent fate in the Diaspora. Although the Bible served as his primary source for elucidating the chain of descendents of the Jewish nation, he suggested interpreting biblical stories and other descriptions of Jewish mythology in the light of modern science, especially human biology and medicine. This way he claimed, for example, to have discovered lost Jewish tribes, such as the inhabitants of Sardinia, based on the high frequency of the factor for thalassemia found among them, as well as in some Jewish communities.

In a lecture at a conference on "Epidemiology of Cancer and other Chronic Diseases among Migrants to Israel," Sheba reported that when his Zionist activity brought him to Palestine in 1932, his attention turned to a certain disease that erupted in the springtime among non-European Jews, mainly of Iraqi and Kurdistani origins. In the spring of each year, the erythrocytes of these patients break up (Sheba 1971). The disease, called *favism* because its eruption was linked to the bloom of *Vicia faba,* the broad bean, was also called "Baghdad Spring Disease"– for the period when the air in Baghdad is saturated with the broad bean pollen. Indeed, a factor in the broad bean was found to cause hemolysis in persons deficient in a certain enzyme. Sheba took on the task of finding out the origins of the immigrants who suffered from the disease: "I mention all this because when we first encountered the high frequency of glucose-6-phosphate dehydrogenase (G6PD) deficiency in Babylonian (Iraqi) Jews, [we found that] an identical situation applied in Italy. All the families in which favism occurred on the Italian peninsula could be traced to Italians of Sardinian, Western Sicilian or Maltese origin" (Sheba 1971, p. 1336).

For many years Sheba also followed the epidemiology of other diseases mainly restricted to one ethnic group or another of Middle Eastern origins. He interpreted Jewish history on the basis of the distribution of the diseases:

> If we try and put this data together, I would propose the following historical account. The Libyan Jews and the Armenians, both having the highest gene frequency for FMF, must have left the common Semitic gene pool in and around the 8th century B.C., most likely from the Kingdom of Israel. [...] The Yemenites were refugees from Judea to Eilat and down to Yemen before the major expulsion of about 26,000 families to Babylon in 586 B.C. [...] The Iranian isolate with one DJS case for every 1,200, as against 1:40,000 in the Iraqis, can only be explained by the fact that the deportation of King Joachin (of the Tribe of Judah) in 597 B.C., with his entourage including Kish, [...] must have taken with them all there was of this mutation into the Iranian exile. [...] (Sheba 1971, pp. 1336–1338)[4]

To Sheba, the distribution of these (relatively) common diseases in Jewish communities and neighboring peoples, reflected the original distribution of these

---

[4] FMF refers to Familial Mediterranean Fever disease; DJS refers to Dubin-Jones Syndrome.

diseases and served as indicators of the allotment of biblical Jewish populations and of their mating customs. From his perspective, the common presence of these markers also among the Gentiles of these countries indicated inter-ethnic mating.

G6PD deficiency is inherited as a "sex linked" trait, i.e., as a property whose gene is located on the X-chromosome. Females carry two X chromosomes, one from each parent, and transmit one of the pair to each of their offspring. Males carry only one X-chromosome, inherited from their mother, and a Y-chromosome inherited from their father. Thus for a female to be G6PD deficient, she must have received an X-chromosome carrying the deficiency allele of the gene from both her father and her mother. A son may become G6PD deficient when his mother carried an X-chromosome with the deficiency-allele (even if she herself was not affected – being a heterozygote – she carried one ordinary and one affected allelic version of the gene), but not when his father was affected (a son inherits a Y-chromosome, the "partner" of the maternal X-chromosome, from his father). Accordingly:

> Why do the European (Ashkenazi) Jews lack or practically lack G6PD deficiency, DJS, phenylketonuria, and the thalassemias? For G6PD deficiency which is an X-linked mutation, [...] if you read "The Jewish Wars" of Josephus Flavius, you will learn that the Roman exile into Italy consisted of males only who were sold as slaves or thrown to the lions in the arena. [...] Thus, under the conditions of slavery the Semite males had to marry, and obviously convert, the Japhethite females. Therefore the core of these Jews who moved into Europe in the paths of the various invaders of Rome from north of the Alps did not, in most cases, have wives of Semitic origin; and thus these women did not carry the mutant gene for G6PD deficiency to their male offspring at all. (Sheba 1971, p. 1338)

In other words, the origins of the Ashkenazi ethnic group are Israelite slaves expelled to Rome. Those selected to become slaves (and who survived) were obviously the healthy and strong individuals, free of an anemia that may erupt in contact with broad beans, that is, not carrying the "defective allele" of the gene on their single X-chromosome. These slaves eventually produced children with non-Jewish women whose X-chromosomes were obviously free of the defect-inducing allele. Although this reconstruction of the origin of the Ashkenazim from converted non-Jewish mothers is contrary to the current Jewish Halacha, at least it alleviates the current Ashkenazi of the G6PD deficiency.

> If we now look at the 4,000 years of migrations of the Jews, we can say that they shared the genes, which were theirs in the Persian Gulf, with the Phoenicians, whose traces we can find in the Sardinian lowland inhabitants, in Western Sicily and maybe in Malta and in Minorca. [...] and the Philistines who were also Greek, became "Mediterranean" only by the influx of mutations prevalent among the Semitic Greek islanders. (Sheba 1971, p. 1339)

Contrary to the Hebrew slaves who were not affected by the factor for the disease, the community of exiles in Babylon transmitted the factor to the Phoenicians through intermarriage, and these later founded the populations of the western Mediterranean islands. The Greek islanders, however, got the gene through the Philistines. Sheba overlooked any possibility that other factors might have brought about the distribution of the alleles of the gene. The improbability of two rare mutational events seemed to him to be overwhelming. Thus, he summarized his simplistic biologist's interpretation of the Jewish history of migration:

> The variety of mutations known to exist among Jews perplexed us, in view of our uniformity in language and religion, until we began to look at the genetic disorders in terms of our history. [...] Whoever wishes to dig deeper into the genetic, and even environmentally influenced "*Homo israelensis*" of today, has no alternative but to spend part of his time in reading the Old Testament and then a wealth of information accumulated in Aramaic and Latin as well as in the middle ages in the glorious period of Spanish Jewry. (Sheba 1971, p. 1340)

Sheba's associate in these studies, Arieh Szeinberg, also dealt with the high frequency of G6PD deficiency in Jewish and non-Jewish Mediterranean populations. Szeinberg noted that up to 60% of the Kurdistani carried the deficiency allele, which was at variance with the low frequency observed among Ashkenazi. But he was aware of the possibility that another factor of population dynamics, other than the proposed long-term historical relationships of these populations, may have been a determinant in the "defective" allele's widespread distribution:

> In the early studies of this trait, it seemed to us that it might constitute a good tracer for the ancient Hebrew gene pool, on the hypothesis that this gene was very frequent in this ancient population, and that the present intercommunity differences have been established through a dilution effect. [...] However, with progressive accumulation of laboratory data on the prevalence of G6PD deficiency in the populations of the Middle and Near East, this view became untenable since the G6PD deficiency of the Mediterranean type is found in relatively high frequencies in most populations of this area. [...] Most investigators agree that this distribution, and also microscale differences between close communities (e.g., in various localities of Sardinia), may be explained by a positive selective value of G6PD deficiency in the protection against malaria. (Szeinberg 1973, pp. 1173–1174)

In other words, in some environments, like that of the Mediterranean shores, the deficiency of the allele for the synthesis of G6PD may be somewhat advantageous so that rare independent mutations could have been selected for in these communities. Once DNA-sequence analyses were introduced, such issues were easily resolved.[5]

Sheba's "epidemiological historic" argument is based on the Mendelian understanding that a genetic change – a mutation of one allele to another allele (in our case, an allele that allows healthy life, mutated to an allele that entails a propensity to disease) – is a rare, random event. When such a mutation does occur in a small relatively isolated population, it spreads by mating between members of the community and is absent in neighboring populations. Consequently, while in one population several affected individuals may appear in almost every generation, in neighboring populations usually no affected individual appears. Thus, the appearance of an affected individual in a neighboring, or even in a distant population, would suggest some covert common history, such as an occasional mating between members of the communities, rather than another rare mutation.

However, one must keep in mind that a disease at the phenotypic level might be due to different causes at the genotypic level, like the involvement of another gene in the metabolic or developmental pathway leading to the given disease; or to mutations

---

[5] Note that the probability of repeated events of rare mutations is not zero. Mutations in a gene may occur repeatedly, and the probability of observing such repeated occurrences increases the larger the population and the longer the time span observed.

in different nucleotides of the same gene that were not discerned at the clinical or the laboratory level as being distinct, keeping in mind that relevant genes consist of a sequence of multiple nucleotides.[6] Furthermore, circumstances of natural selection may account for the spread of "similar" (at one or another phenotypic level) mutations in a population due to some small advantage to the rather numerous heterozygous individuals, thus balancing the very deleterious effect to the few homozygous individuals. Also, drift of a defective allele with a closely linked gene on the chromosome that is positively selected, may lead to the presence of an unwanted allele in a population, disproportionately to its own selective value; or even a random event, such as the founding of a new population by a small number of individuals (say, boat survivors on an isolated island) who happened to carry (disproportionally) the relevant allele in one or several of the founders of that popular ("founder effect"). In other words, the distribution of a "hereditary disease" may reflect the special conditions of the populations involved as much as it reflects its common genetic history.[7]

J. B. S. Haldane (1892–1964) speculated as early as the 1930s that malaria was a principal factor in shaping the history of humanity. Homozygotes for an allele of the major gene for haemoglbin molecule (S_2S_2) suffer from severe sickle cell anemia. Notwithstanding, in malaria infested areas, such as Central Africa, heterozygotes for the hemoglobin mutant (S_1S_2) were found to be more resistant to malaria than the non-sickle cell individuals (S_1S_1). Thus, population-wide equilibrium is eventually reached between the (severe negative) selection against the relatively few homozygotes (S_2S_2) who suffer severe sickle cell anemia and the (slight positive) selective advantage of the abundant heterozygotes (S_1S_2) who enjoy some resistance to the burden of malaria (Haldane 1949; but see also Lederberg 1999). The presence

---

[6] DNA molecules are helices constructed of double strands of anti-polar sequences of nucleotides. Each nucleotide is composed of a desoxyribose and phosphate moiety, connected to two adjacent similar moieties of the strand, and to one of four purine- or pyrimidine-residues. The strands are held together by specific weak hydrogen bonds between the purine residues and the pyrimidine residues of the counter-polar strands: adenine (A) with thymine (T); guanine (G) with cytosine (C).

A sequence of nucleotides codes for a sequence of amino-acids (a peptide) that is synthesized on the ribosomes (via an intermediary, the messenger-RNA copy of the respective nucleotide sequence of one of the complimentary DNA strands: G for C, C for G, T for A and A for U. U is the RNA-analog of DNA's T): each specific triplets of nucleotides codes for the twenty amino-acids that comprise all proteins. The sequence of nucleotides (actually that of the copied messenger-RNA sequence) that codes for a peptide may be conceived as an approximation of the concept of the gene as a structural entity (see Falk 2010). Mutation in any of the nucleotides that code a given peptide may result in a similar effect at the phenotypic (e.g., clinical) level. For more details see any textbook of molecular genetics.

[7] There are many factors that act on natural populations in opposing directions, thus one may view a natural population as an entity that is striving toward equilibrium. A basic situation would be that of a large population with random mating, no selection, and no migration. As early as 1908, George Hardy and Wilhelm Weinberg independently showed that if the proportions of the pair of alleles of a gene, A and a, are any value p and q respectively ($p + q = 1.0$), then within one generation the (large, random mating, non-migrating) population will reach an equilibrium of AA, Aa and aa at p^2, $2pq$ and q^2, respectively. Thus, the *Hardy-Weinberg equilibrium* provides an anchor for analyzing the impact of forces working in any population in terms of its deviation from the equilibrium (see, e.g., Li 1955).

of the malaria-spreading mosquitoes on the one hand, and the relative immunity to malaria of heterozygotes with the sickle cell allele, on the other hand, rather than inter-human relationships were crucial (in this case) in drawing the map of the history of human settlements.

In the 1950s it turned out that the sickle cell disease was caused by the replacement of a single amino-acid in the β-hemoglobin molecular moiety of the gene S_2.[8] Such a balance of homozygotes, who die of the handicap, and of heterozygotes, who develop only minor pathological symptoms and survive having been infected by malaria-carrying parasites in some regions of Africa, resulted in up to 40% of the population being heterozygotes, despite the price that one in four progeny of the mating of two heterozygotes may die of sickle cell anemia.

The mutant allele is maintained in the population at an equilibrium level so that the fitness advantage of the many population's heterozygotes balances the fitness loss of the few homozygote individuals, independent of the history of the mutation's origin.[9] Although this has long been the most detailed analysis of a case of *balanced polymorphism* of a gene, accumulating evidence indicates that quite a number of genetic diseases are maintained in populations by the equilibrium between their small advantageous effects on many and their severe deleterious effects on few others, even when the precise physiological and molecular details are not always known. In such cases, it is to be expected that a hereditary disease common at the clinical level may be due to different, rare, independent mutations at the molecular level of the same gene in each locality. Malaria was a very common killer in the Mediterranean basin, in Greece, Turkey, Cyprus, Sicily, Sardinia, as well as in Kurdistan and Iraq (and in the Far East, especially in Indo-China). In all these countries, "genetic diseases" involved in the breakdown of erythrocytes (anemia) are common. Thalassemia – the ancient Greek term for "near-the-sea" – is one of these; G6PD deficiency is another. There are good reasons to claim that all these have been maintained in the local populations, despite their deleterious effect on some individuals, because of the slight selective advantage that they conferred on many individuals of the populations of the region.

The further biochemical and later molecular analyses advanced, the more it became clear that *different* mutations are responsible for the *same* disease in different populations. This shows that the distribution of diseases may be an issue of Darwinian population dynamics rather than that of historical phylogenetic relationships. The people of Sardinia are not the descendants of exiled Jewish slaves, but

---

[8] This was the first "molecular disease" described. It turned out that the sickle cell defect was due to the replacement of a single amino-acid, glutamine, by valine at position 6 of the peptide, in the β-hemoglobin moiety. This caused the formation of the (defective) hemoglobin-S instead of hemoglobin-A. Eventually it turned out that, at the DNA level, this was a mutation that changed a specific adenine into thymine (turning the coding-triplet GAG into GTG).

[9] Whereas in Africa the frequency of the sickle cell allele is maintained by the balance between its deleterious effects and its beneficial effects, in America, where the need for protection against the impact of malaria became negligible, the equilibrium in African-Americans had been disturbed and, over the years, selection has already reduced the frequency of the allele for the defective hemoglobin.

rather victims of malaria stricken areas, like the people of the eastern Mediterranean. Long before detailed DNA sequencing confirmed this pattern, support for the differential selection hypothesis was provided by Sardinians, who lived in small and closed communities that rarely intermarried. While thalassemia is frequent in the low-land swampy regions along the beaches, it is hardly known in the villages in the hills only a few kilometers away.

8.2 Common Relatives *versus* Common Genes

Attempts to establish traditional claims relating to Jews' typical facial patterns and character outlines on scientific-biological foundations were repeatedly made in the eighteenth century. If indeed there are typical inbuilt Jewish characteristics, it was claimed that one may detect them independently of the cultural or religious contingencies of the populations. For example, Carl Heinrich Stratz, a Dutch philo-Semitic researcher studied the Jewish facial features in various exotic peoples and tribes in an "ethnographic-anthropologic study" entitled "Who are the Jews?" (Stratz 1903). He presented pictures of Japanese individuals, members of tribes in Papua in the Pacific, and individuals in the jungles of Brazil with "Jewish facial features." Thus he related a tale of an isolated village in central Brazil, whose inhabitants' physiognomy is so Jewish that travelers nicknamed one tribesman "Itzik," a common Jewish name in the *shtetel* (Fig. 8.1).

When Wilhelm Johannsen called for discriminating between phenotypes and genotypes (see Chap. 3) in 1909, it appeared that the foundation was laid for a reductionist definition of trustworthy biological parameters that discern between *nature* and *nurture* with reference to specific characteristics in individuals, and more so, with regard to characteristics in populations. But our prejudices went on to haunt us. The reductionist dream of distinguishing between nature and nurture

Fig. 8.1 "Jewish facial form" in a Bakaïri – Papuan native (Stratz 1903)

repeatedly turned out to be illusory, and evidence showed that it was only an epistemic device to operatively isolate the impact of single genetic factors on the basis of phenotype. As emphasized repeatedly, we must conceive of life as a complex interactive system (see Falk 2009, 2011; Keller, 2010. Also see Mitchell 2009). Gradients in the frequencies of alternative alleles of one or two genes, presumably neutral, that is free of environmental input, turned out to be too small to decide on problems of common biological origins: Social and cultural factors affect the mating patterns of populations as much as do, for example, topography, disease, even climate. The most detailed genetic study cannot draw maps of the origin of populations and determine their historic relatedness. At most, it may uphold such claims, or sometimes refute them – if we make certain helpful assumptions.

The extent to which politicians and social reformers exploited biological arguments in general, and eugenic arguments more specifically, to advance their purposes (see ample examples in Bashford and Levine 2010), soon led scientists to search for ways to present populations and their interrelations in terms of abstract variables. But it was primarily the Synthetic Theory of population genetics, established in the 1940s, that contributed to the modern understanding of the forces acting in evolution. However, fortunately, from the moment this theory of population dynamics was established, it provided the basis for modifications (see, e.g., Crow and Kimura 1970 *versus* Jablonka and Lamb 1995. See also Laland et al. 2015).

Karl Landsteiner discovered the ABO blood group system in 1900, and its value for anthropological and eventually genetic-population classification was recognized in 1919 when Ludwik and Hanna Hirszfeld, two Polish Jewish physicians, published data collected on the battle field during World War I in Macedonia on the "Serological Differences between the Blood of Different Races" in an attempt to preserve a pure race concept. We do not know of any natural or artificial factors that may change a person's blood type, which makes these phenotypes ideal indicators of *genotypic* variability. The inheritance of ABO blood types is principally explained by three alleles of one gene. We may simplify matters and call the alleles *A*, *B*, and *O*. A person inherits one allele of each gene from each parent. If an *A* allele is inherited from one parent and a *B* allele from the other parent, the person will be of genotype as well as the phenotype AB. If both parents transmit the *O* allele, then the offspring's blood type (and genotype) will be O. However, if one parent contributed an *O* allele and the other an *A* allele (or a *B* allele) the phenotype of the progeny will be of blood type A (or B, respectively).[10] Blood types thus often enable deduction from the phenotype to the genotype. Early on, when each of the three alleles was ascribed to a pure race, the frequencies of the blood types of a population were assumed to reflect the racial mix of the population in question (Bernstein 1925).

In the beginning, only the A, B, AB, and O blood types were recognized. Soon, however, more blood types were discovered. In the 1940s, Rh– and Rh + were added. It turned out that these were much more varied, and today Rh blood-groups

---

[10] Allele *O* is recessive with respect to alleles *A* and *B*, both of which are dominant over *O*. Similar principles hold in the inheritance of other blood groups though the multiplicity of alleles and recessivity-dominance relationships between them may complicate the picture.

8.2 Common Relatives *versus* Common Genes

are divided into numerous sub-types (C, D, E, etc.), which allow (together with up to a dozen other blood types) a much finer classification of humans into detailed genotypes. However, although differential diagnostic typing known today is quite extensive, their possible combinations do not allow a unique racial definition of a person. The most we may assert is that a given person's combination is rare (or frequent) in a given *population* that had been defined by some other criteria, or rather that a person with a given combination could only rarely be the progeny of a person of another population.[11] Over the years, many more variables that reflect the genetic polymorphism of populations were added. Of special importance were the allelic polymorphisms of many enzymes. Small differences in many enzymes (usually so small that for all practical purposes they do not significantly affect the enzyme's activity) may be discovered by running the enzyme extract in an electrophoretic field (Harris 1966; Hubby and Lewontin 1966).[12] With time, more kinds of markers were observed, such as those of the variability of fragments of DNA cut by specific enzymes at certain sites, RFLP (restriction-fragment length polymorphisms). Nowadays, since the full DNA base-sequences of many genomes are known, polymorphism can be studied at the level of single DNA base-pairs at specific sites, known as SNPs (single nucleotide polymorphisms).

Thus, although gene frequency studies increasingly replaced the old anthropological studies on skull-shape, hair curling, eye color, and other morphological characteristics starting in the 1950s, the classification of persons by one or a small number of trustworthy indicators of genetic variables became very problematic for the positioning of populations in relation to each other, especially when there was no notion as to the (biologically relevant) criteria that specify the population and the forces that affect the distribution of its variables. Yet, the more variables one examines to describe a population, the more confidence one gains regarding the steadiness of the parameters of that population. Furthermore, following numerous variables may provide indicators to the factors that affect the variation of these parameters: If there is a correlation between the variability of markers of different functions in a population, it is reasonable that their distribution is independent of their function, and may be due essentially to the common histories of the members of the population. If, however, the distribution of some marker differs significantly from the (correlated) variation of numerous other markers, this may indicate that its distribution has been affected essentially by its specific function. If there is no correlation at all between different variables, this may indicate that the population went

---

[11] These statements proved to have enormous practical value because they may allow for unequivocally *excluding* claims of identity. Hence the role that blood typing plays in forensic paternity cases. Furthermore, even when it is not possible to exclude paternal relationships, it is often possible to express in statistical terms the *probability* of a person carrying the given blood type.

[12] Electrophoresis – a laboratory method that measures the movement of a material like a drop of an enzyme in a semi-solid (gel-like) horizontal plate by running an electric current along the plate. Each component of the tested matter will move along the field as a function of its electric load and its attachment to the material of the plate. After some time, the current is broken and the distance each of the components moved is revealed by specific chemical staining.

through a process such as a bottleneck of very few individuals, and random drift determined the present frequency of variables.

The human geneticist Arno Motulsky summarized the proceedings of the First Symposium on Genetic Diseases among Ashkenazi Jews in December 1979, noting that "the newer genetic markers studied in Ashkenazi populations clearly suggest a Mediterranean origin of current Jewish populations." Data collected in the past on blood groups, which seemed to indicate a large proportion of admixture of Jews in the surrounding populations where they lived, "may have been [due to] exceptional characters that underwent rapid changes in gene frequencies because of selective advantages in the European environment (convergent selection)" (Motulsky in Goodman and Motulsky 1979, p. 425). Motulsky preferred the newer results that seemed to correspond better to his notion that the Jews had preserved a considerable degree of cohesiveness and genetic isolation.

An additional cause for such non-random distribution of single genes between populations might be the "bottleneck" effect. I have already discussed the special case of the founder effect, where a new community is established by a limited group of members of the original population who happen not to reflect a "representative" sample of the original population. The new community may exhibit extremely different, specific gene frequencies when compared to those of the mother population. Considering the history of wandering of Jewish individuals and communities, and the splinters of persecuted communities, such bottleneck effects of minute non-random samples of populations are expected to be the rule rather than the exception. Indeed, Motulsky and his colleagues tested the plausibility of such effects on the distribution of diseases in Jewish communities (e.g., Motulsky 1980, 1995). Although modern analyses at the level of DNA sequencing greatly increased the level of resolution, it must be kept in mind that all these procedures are explanations of how the current frequencies of diseases in different communities *may* have been established, and most of these explanations are based on the assumption of one common ancestral population.

Many Jewish communities were culturally isolated from the Gentile periphery and there was a high rate of inbreeding of intra-community, even intra-family mating. Consanguineous mating was especially frequent among Iraqi and Persian Jews, reaching up to 30% of marriages among Kurdistan Jews (see Goldschmidt, Ronen and Ronen [1960], Ronen, Ronen and Goldschmidt, "Marriage Systems", and "The Population of the Kurdish Jews" [Cohen, et al. 1963] in Goldschmidt 1963, pp. 340–349. See also Tsafrir and Halbrecht 1972; Cohen et al. 2004). This would theoretically lead to an increased frequency of persons affected by genetic diseases as a result of both parents carrying the relevant gene, even if the frequency of the relevant alleles was not higher than elsewhere.[13] As for alleles of genes involved in providing increased resistance to malaria, it was no wonder that in these Kurdistan and neighboring communities frequencies were high. However, to the surprise of

---

[13] Each population carries at least some mutant alleles from the past. Given the frequency of an allele in a population, the frequency of homozygotes for this allele is linearly related to the rate of inbreeding. Put differently, in two populations with the same gene frequencies, the one with a higher inbreeding coefficient will have more individuals homozygous for the alleles of that gene.

Goldschmidt and her associates, there were not the expected high rates of persons affected by other genetic diseases (not related to malaria infection) in spite of the communities' excessive inbreeding coefficients, contrary to the experience in Western communities. Also, other phenomena, such as reduced fertility, which would be expected among members of communities with high inbreeding coefficients, were not observed in these Eastern communities (*vide supra*). Goldschmidt suggests that this unexpected low frequency of affected persons was the consequence of the many generations of high inbreeding rates in these closed, relatively small communities. The frequent appearance of affected homozygotes over many generations in the past effectively "cleaned" the deleterious allele, so that today these communities are (almost) free of these alleles (an unconscious eugenic act?).[14]

On the other hand, effects like the "founder effect" that are more conspicuous in small isolated populations resulted in the presence of some diseases. This is suspected to be the case for the relatively high frequency of phenylketonuria (PKU) among the Yemenite Jewish community (Adam 1973). Tracking the family relations in 25 families with affected members who were spread over Yemen revealed that all were related to the same founder person in Sana in the eighteenth century. It is, however, not clear that the mutation to a PKU-deficient allele occurred in this Jewish founder person, or whether this person was a non-Jew who was the founder only in the sense that she/he was the one who transferred the gene by intermarriage into the Jewish community. According to Paul and Brosco (2013), more than five hundred mutations have been identified at the gene locus, and this genetic heterogeneity is associated with considerable clinical heterogeneity. Most people diagnosed with PKU are "compound heterozygotes." The frequency of PKU in this community is among the highest known and molecular tests eventually carried out among these persons showed that all carry the same mutation, which is unique to them (Avigad et al. 1990; Shiloh et al. 1992). Such data support the historians' claim of the isolation of this community from other Jewish communities, in spite of the lack of data on the non-Jewish communities for comparison.

Familial Mediterranean Fever (FMF) disease, prevalent among Jews of Libyan descent (Pras et al. 1992), has been well known also among Armenians, Turks, Syrians, and Egyptians, and in other ethnic groups in the Near East. It is rare among Ashkenazi Jews and even among Gentile Italians and Greeks (McKusick 1979). Such a distribution may suggest a common origin of these people, but clinical examination reveals that among Armenians and Ashkenazi Jews, the disease varies from that of the other groups. It appears that for some reason the clinical pattern of the disease is common along the Mediterranean shores, irrespective of the inhabitants' origins. Are there some conditions in the Mediterranean regions that promote the frequency of the FMF alleles that result in the clinical disease? Is the frequency of the alleles in these populations due to the equilibrium between the small advantage of many and the big disadvantage of a few?

All these issues acquired new dimensions with the introduction of DNA-sequence analyses. Are there at least two independent FMF mutations at the DNA-sequence

---

[14] Bonné-Tamir suggested a similar idea in 2010.

analysis level? Although researchers acknowledged "a certain role of selection," some have returned to Sheba's medical anthropological notion and have directed special attention to the "Apparent ancestral ties between Sicilians and Arabs; Jewish and Armenian FMF carriers are especially noteworthy and probably reflect Sicily's historic status as a crossroads of Mediterranean commerce" (Aksentijevich et al. 1999, p. 961). Others even bring maps suggesting the likely distribution of the two main mutations from Israel along the Mediterranean and further on to the Chuetas of Mallorca, who are believed to be descendants of converted Jews (Ben-Chetrit and Levy 1998, p. 660).

It is, of course, impossible to exclude the speculation that the migration of the mutation was due, for example, to the mating of a Jewish bride from Israel to a Sardinian inhabitant or to a lonely sailor whose ship was wrecked on the coast of Mallorca, and that their progeny settled there. Notwithstanding, the implied assumption of some is that such maps reflect the trail of the wandering of the children of Israel.

It is, therefore, important to repeat that the distribution of "hereditary diseases" is the less trustworthy indicator for the historical migrations of populations. True, hereditary diseases may be conceived as representatives of classical Mendelian characteristics in humans, but the very fact that diseases are markers that have selective value in the relevant populations suggests that the distribution of these genes is also related to processes involving the specific living circumstances that prevailed in the respective areas and, thus, provide a biased picture of the origins of the populations in historic terms.[15]

Although it appears that local Ashkenazi communities were by and large less isolated from each other than the Sephardic and Middle Eastern communities, there are indications that they too were often small, isolated breeding entities for many generations. Under extreme conditions of riots and plagues, when many communities shrink, high rates of consanguinity would result in increased frequencies of homozygotes for the "hereditary diseases" specific for each isolated community, which would further distinguish the Jewish community from its neighboring Gentiles.

We have already mentioned Tay-Sachs disease, which causes a newborn who has inherited the allele of the defective enzyme hexosaminidase-A (Hex-A) from both parents to be physically and mentally impaired and to die early. The rate of newborns with this disease is especially significant among the descendents of Ashkenazi parents originating in a specific region of Poland and Lithuania. Because of the prominence of these Jewish communities, Tay-Sachs has been considered the prototype of a Jewish disease (see, e.g., Bychowski 1918, p. 299). A mutation in the gene probably occurred in a Jew in one district of Poland and considering the demography of the community, it turned out to be essentially a community indicator scourge. In such a small, closed community, there might be considerable fluctuations in the frequency of the allele for the disease, and random drift might result in a quite high frequency of the disease (irrespective of its selective fitness) in some

---

[15] Notice that Mendel used his pea characteristics as good markers for elucidating the rules of inheritance of individuals in the family, and not of their distribution and inheritance in populations.

8.2 Common Relatives *versus* Common Genes

communities. Motulsky suggested that the distribution of the disease reflected the short-term history of the Ashkenazi Jews in the last 300 years. The distribution of the disease corresponds to the story of the origins of present day Ashkenazi from a very small assemblage of refugees of pogroms that annihilated the Jews a couple of centuries ago (Motulsky 1995). Natural selection may, with time, reduce the frequency of the Tay-Sachs allele in these populations that happened to produce many carriers by random sampling events, but this takes many generations, unless positive eugenic means (artificial selection) are carried out to negate these circumstances (see further on and Chap. 9).

Other diseases, such as essential pentosuria, Gaucher's disease, and familial disautonomy, can also be identified rather faithfully in Ashkenazi Jewish communities in specific areas. Bloom's disease that develops in homozygotes for alleles of a gene involved in DNA-damage repair, turned out to be limited to Jews from specific regions in Poland, around Lwow and Cracow (German 1969). The mutation found in Jews is unique and distinguishable from, say, that of a similar disease in Japan. It is a reasonable assumption that the high frequency of these diseases among Ashkenazi Jews is the consequence of the 'founder principle,' namely, unique mutational events that occurred in recent centuries somewhere in Europe and has been maintained by drift in the small breeding populations. Still, additional socio-cultural factors may have contributed to the public consciousness of ethnic diseases. Tay-Sachs's status as a 'Jewish disease' was established by its relative prominence in middle class Jewish circles that had access to advanced health services, like those of the East Coast of the United States (New York, Baltimore, etc.). The misfortune of several children affected by Tay-Sachs in the family of Rabbi Joseph Eckstein motivated him to establish *Dor Yesharim* (upright generation, Psalms, 112:2),[16] an organization that performs premarital tests for Tay-Sachs, and by now some dozen or more other hereditary diseases, primarily in the Ultra-Orthodox Ashkenazi community, in which marriages are arranged by matchmakers (the discrete test results are known only to the matchmaker, who regulates marriages accordingly) (See Sagi 1998, and further discussion in Chap. 9). But Tay-Sachs as well as the other diseases was also found in Sephardic communities and, of course, in non-Jewish communities even though, in most cases, these are affected by other alleles of the relevant genes (Neufeld 1992; Frisch et al. 2004. See also the distribution of another "Jewish" recessive disease, cystic fibrosis, within and between ethnic groups in Israel, in Kerem et al. 1995). Still, there are several researchers who insist that under the specific living conditions of the Jews in Poland, there was some selective advantage to the heterozygotes, although no one knows what this was (see Motulsky 1995). Support for such claims may be found in the fact that two other hereditary diseases

---

[16] *Dor Yeshorim: Committee for Prevention of Jewish Genetic Diseases*, established in Brooklyn, NY in, 1968. *Yeshorim* is the Yiddish pronunciation of the Hebrew *Yesharim*.

In 2013, the parents of Eden Gold, who suffered from mucolipidosis type 4 (ML4), established *JScreen*, a national public health initiative that offers a screening panel of 40 diseases common in the Jewish population (including the roughly 20 diseases that Ashkenazi Jews are at increased risk for) to Jews and also the general population.

of similar metabolic defects (sphingolipoid storage diseases), Niemann-Pick disease, and Gaucher's disease are also relatively frequent among Ashkenazi Jews. Such a correlation between the three diseases of the same metabolic pathway may indicate that a factor, such as nutritional habits, could have contributed to some reproductive advantage of the heterozygotes for these alleles that compensated for the failure of the homozygotes.[17]

Since tuberculosis was for a long time a lethal disease in Europe, it has been speculated that the hereditary pattern of the lipid storage disease somehow imparts to its carriers a relative resistance to tuberculosis. According to this logic, just as malaria was responsible for the widespread distribution of thalassemia and G6PD deficiency in the East, so was tuberculosis, under the East European ghetto conditions, responsible for the widespread distribution of the lipid storage diseases among Ashkenazi Jews: "Because tuberculosis was probably a major killer in the densely populated urban ghetto, it has been suggested that resistance to this infection might have been the selective factor that made these lipid-storage diseases common in Jews" (Beutler 1993, p. 5386).

Doron Behar and colleagues, who are of the era of DNA analyses, on the other hand, wondered whether the basis of more than 25 genetic diseases present in Ashkenazi Jewish populations may be due to "accentuated genetic drift resulting from a series of dispersals to and within Europe, endogamy, and/or recent rapid population growth" (Behar et al. 2004a). They followed a series of molecular markers not related to diseases of single nucleotide polymorphisms (SNPs) and microsatellites (multiallelic short tandem repeats – STRs) in the effectively haploid non-recombining portion of the Y-chromosome (NRY, see Chap. 9) within Ashkenazi Jewish communities and their non-Jewish host populations in Europe.[18] Ashkenazi populations were found to be genetically more diverse at both the SNP and STR level than their European non-Jewish counterparts, but they are greatly reduced within haplotype STR variability (especially in the presumably more ancient hapolotypes). This pattern in the Ashkenazi populations, they claim, indicates a reduction in male effective population size.[19] This could result from a series of founder events, or because a small number of individuals have survived in Europe in a

---

[17] An interesting argument was raised in 1962 by Louis B. Brinn, a Jewish physician in New York, in the medical journal *Harofe Haivri*. According to him, with respect to these diseases in the Jewish communities, it was not necessary for the advantage of the heterozygotes that the frequency be higher among Jews than among non-Jews. It would suffice that the heterozygotes' *dis*advantage among Jews would be less than that among non-Jews in order for the disease frequency among Jew to be higher than among non-Jews. The problem with such explanations is that, until today, we have no indication as to the reproductive advantage/disadvantage of heterozygotes for these phenomena (see Post (1973).

[18] Haploid segments are more bound to change than diploid segments. SNPs are due to rare mutations, STR polymorphisms are due to rather frequent intra-chromosomal recombination events. See also Chapter 9.

[19] Effective population size is the mathematically adjusted number of individuals in a population that are actually involved in reproduction in a given generation (so that populations with varying parameters, such as sex-ratio, inbreeding, etc., may be compared).

8.2 Common Relatives *versus* Common Genes

particular generation – a bottleneck – and high rates of endogamy (during bottlenecks, STR haplogroups are expected to be lost at a much higher rate than SNP haplogroups. This may serve as indicator for a differential diagnosis). This reduced effective population size may explain the high incidence of founder disease mutations, rather than variations in selective values of the mutated alleles in heterozygotes and homozygotes (Behar et al. 2004a, b).

Others attempted to deny altogether the "founder effect" as an explanation for the high frequencies of disease-related recessive alleles of genes among Jews. Harpending and coauthors (*vide infra*) claim that it was not bottleneck effects that brought about the high frequency of the hereditary diseases, but rather a process of selection in the communities of a hereditary component of intelligence (as measured by IQ tests). According to them, the increase in the genes related to these diseases was the by-product of such selection (Cochran et al. 2006. See also Wade 2014, pp. 202ff. for an extensive discussion of this work).

***

There have always been two fundamental assumptions of physical anthropology regarding the roots of Jews:

a: A Jewish population of distinct national and socio-religious uniqueness existed in Palestine or Eretz-Israel for centuries B.C.E. In genetic terms, this meant that there was a population with a common discernible gene pool, although it was probably heterogeneous *per se* ("Twelve Tribes"!), and also that it was the result of partial amalgamations with several neighboring gene pools.

b: The Jews of today are the descendants of that gene pool, although the common pool was split into several partly distinct sub-populations or ethnic groups. Furthermore, there was a certain degree of dilution with pools of the populations among whom the Jews lived in the Diaspora (see Chaps. 5 and 6). Such processes occurred by choice or by force: through intermarriage or proselytism of individuals or of whole communities; or by rape or by the alleged "lawful" means, such as the *droit du seigneur* – the right of the lord of a medieval estate to take the virginity of his serfs' maiden daughters.

Search for common relatives has always been also a search for status in human history, or as expressed by the geneticist Steven Jones in a BBC program, "nowadays a search for genes became a search for nobility." In light of assumptions a and b, religious groups that tried to emphasize their common origins with the ancient Jews, the inhabitants of biblical Palestine, or at least of the Aryan roots of early Christianity, would appear again and again over time. Even today, there is no lack of people who claim such nobility.[20]

In Zimbabwe, in southern Africa, resides the tribe of the Lemba, many of whose members are said to differ in their physical appearance from members of neighboring tribes. Their culture and language contain dialect elements that also set them apart from neighboring tribes.

---

[20] See, for example Mourant's confession (quoted further on in this chapter) regarding the sources of his interest in the blood groups of Jews. See also Goldstein (2008, pp. 44–45).

According to the (unwritten) tradition of the Lemba, they stem from the tribe of Judah. They claim that their customs and traditions testify to their links to Zion, and to their being the descendents of the Lost Ten Tribes and of the Jews of Yemen. According to their tradition, their forefathers in Arabia went on to Africa, where they founded a city named Sena (Sanaa?). Indeed, the frequencies of the alleles of many genes among the Lemba reveal that they appear to be more similar to the people of Yemen than to those of the Lembas' neighboring Bantu tribes (Jones 1997).

Eventually the Lemba were included in the "genetic archeology" tests of Jewish communities. Examined for the "Jewish Y-chromosome," they were found to carry remarkably high frequencies of CMH sequences, which are typical of Jews, together with sequences typical of the neighboring Bantu tribes. Most researchers agree that the Lemba do represent a population whose nucleus is of "foreign blood" that was introduced into Africa by outsiders.

This does not mean that the Lemba descended from the migration of a tribe, Jews or others, with specific national, cultural, or biological identity. It is feasible, for example, that some (Jewish?) slave traders from the Arabian Peninsula begot progeny with local women (these merchants may also have left behind some customs and linguistic expressions). The evidence available does not substantiate these theories, and historical evidence would be needed to support the theory raised by the genetic data, no less than biological data are needed to support the historic version.[21] Whether the Lemba represent a trunk of the tree of the ancient Jews that was diluted by intermarriage, or simply by occasional twigs of Jews who intermarried with locals, is a question that geneticists cannot answer.

It must be emphasized again that genetic investigations cannot reconstruct the history of an ethnic group, though differences in the frequencies of the alleles of genes in populations may refute or substantially support historic narratives. Comparing the frequencies of the alleles of genes in a Jewish population (assuming there exists a reliable instrumental definition of "Jewish") with alleles of these same genes in the population among whom the Jews live, or for that matter comparing the frequencies in one Jewish community with those in another Jewish community elsewhere, *may indicate* a possible past blood relationship between the populations, but this cannot be considered evidence for *what kind* of relationship it is.

Convergence to common ancestors might be "vertical," like branches of a tree (Fig. 8.2) – or it might be the result of "horizontal" intermarriages based on geographical and cultural relationships. When a rare phenomenon is detected, such as a hereditary disease due to a specific mutation (at a specific DNA nucleotide), since it is quite improbable for that event to occur twice, its presence in more than one com-

---

[21] For a discussion of the plight of the Lemba, see Goldstein (2008, pp. 40–60). The *Eugenical News*, 15 (1930, pp. 142–143), reviews the book by the Jesuit priest, J. J. Williams, *Hebrewishness of West Africa*, which claims Jewish infiltration into Central Africa, and presents claims of evidence of Jewish traits and beliefs among the Masaii, Ashanti, and Ivory Coast Africans: "The collection of a mass of data 'cumulatively' supporting the theory of a trek of Jews from Jerusalem to the Niger."

Fig. 8.2 Haeckel's tree life in *The Evolution of Man* (1879)

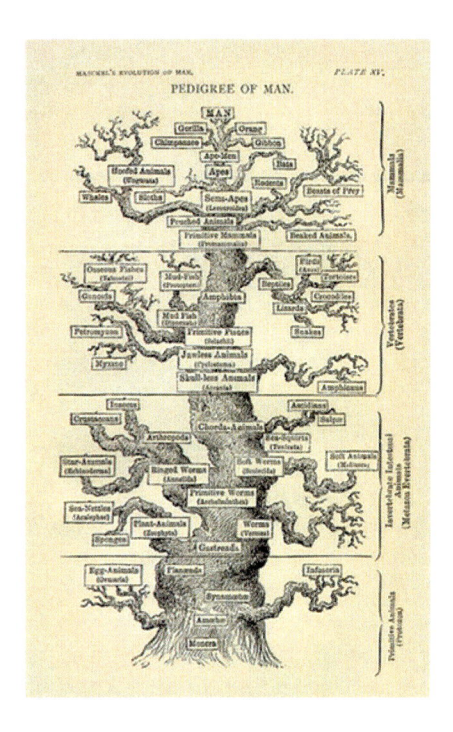

munity may indeed indicate some biological relations between the communities,[22] but only the distribution pattern of many genes (or molecular markers) may indicate population-wide processes, rather than occasional contacts. Independent evidence is required to determine whether the connection between communities is due to, say, a split of an original population into two, or due to separate contacts by rare intermarriages.

A decision of whether members of two communities are Jews or whether they merely carry biological markers common to Jews (however defined), cannot be based on biological data, or determined by anti-Semites nor by Jewish Halachic authorities. The question of whether or not members of a community are Jewish is raised only with respect to communities that deviate in their life style and cultural habits from those who consider themselves (or are considered by others) to represent the so-called accepted Jewish consensus. The decision with respect to the Lemba has been negative (so far). With respect to other communities, such as the Bnei-Israel in India in the 1960s and the Ethiopians in the 1990s, the decisions, which were based on political-religious considerations of the Israeli establishment, were positive. With respect to most other communities, the issue has never come up.

---

[22] See, for example, "The Lethal Gene that Emerged in Ancient Palestine and Spread Around the Globe," in the December 2011 print of *Discover Magazine*, regarding the *BRCA1*.185delAG mutation "that causes breast cancer."

Researchers, as a rule, adopt positions based on non-biological grounds and then endeavor to establish biological grounds as much as possible.[23]

Is there a way to determine the relative impact of each of the factors on the gene pool of a community? Is the history of present day Jewish *eidoth*, or communities, sufficiently circumscribed to allow assigning to them a pattern of diverting branches of a tree, as assumed by applying the *Structure* computer program for building the best phylogenetic branching tree (see, e.g., Bolnick 2008, pp. 74–80)? Or might it be probable, or at least not impossible, that the genetic pattern is more like that of a twisted lattice with various levels of complexity and interaction, including a level of multiple roots, rather than the traditional model that converges on a single root?

8.3 The Genetics of the Israeli Melting Pot

The enormous variability encountered in all aspects of life with the mass immigration to Israel in the 1950s was especially pronounced for those communities and ethnic groups whose culture and habits deviated from the main body of those who defined themselves as the normative Jews. There was demand and hope that it would be possible to identify a typical Jewish biological component, if not in each individual, then at least in every community. Frustration among experts in the disciplines of society and politics concerning the question of who is a Jew directed the task of answering this kind of question to the life scientists. Many willingly accepted the challenge, not only as a contribution to the national effort, but also as a unique opportunity to study fundamental issues concerning structure and dynamics of human populations.

Up until the 1970s and to some extent also later, research of the genetic structure of populations and their dynamics centered on the study of one or two variables at a time: a gene related to one or another blood group or one or another genetic disease, or the presence of one form of enzyme or another. Since different forces may act on the distribution of different genes, it is not surprising that different researchers arrived at different conclusions regarding the genetic relationships between the various communities. Ironically, in retrospect, one may even argue that the old-style physical-anthropological characterizations had a certain advantage over modern

---

[23] Anthropological studies, relying on cultural and physical data for determining Jewishness (or the existence of the "Ten Lost Tribes") have not disappeared (recall footnote 11 in Chap. 1). Rabbi Eliyahu Avichail followed the footsteps of tribes that he claimed were isolated communities who had lost their link to the Jewish world, contrary to other "Gentiles who attempt to be absorbed by Judaism." He identified communities in South East Asia, Central Asia, and South America on the basis of their cultural traditions as well as on "a biological argument" [personal conversation]. He has even brought some communities to Israel. To the best of my knowledge, no genetic tests were carried out among them. Avichail is not the only one involved in identifying "lost Jewish tribes." See, e.g., a review by Phillips (2002) of Hillel Halkin's (2002) *Across the Sabbath River: In Search of a Lost Tribe of Israel*. Nurit Kirsh (2003, p. 651) has called attention to the mission-oriented use of rhetorical expressions, such as "tribe," in these people's exhortations.

characterizations of population dynamics, which followed a single gene distribution. The former – inadvertently – followed variables that were determined by many genes (and environmental factors).

Almost from the beginning, the study of the genetics of Jewish communities was based on *two* concepts: one led by Elisabeth Goldschmidt, who seized the unique opportunity to understand fundamental issues concerning human populations, or as she formulated it, to study *the genetics of migrant and isolate populations*, the title she gave to the conference she led in the early 1960s with Chaim Sheba, whose name is associated with the second concept that is based on the belief in the power of genetics to resolve the *archeology of the Jewish people* and portray the *Urjude* (see note 3 of Chap. 1).

In the 1940s, Joseph Gurevich of the "Hadassah" Hospital in Jerusalem had already pioneered the systematic collection of data on the distribution of blood types among members of various communities of Jews in Palestine (Gurevitch et al. 1951, 1953, 1954, 1955a, b, 1956; Gurevitch and Margolis 1955). Indeed, over the years, a considerable dossier of data was collected on the frequencies of various blood groups of members of the Jewish communities and among local non-Jewish communities. Arthur Mourant and his colleagues reviewed and published all these data in 1978 in their book, *The Genetics of the Jews* (Mourant et al. 1978). Mourant had been involved for many years in the study of the distribution of blood groups. In the introduction to his book, he related that his interest in the genetics of the Jews originated in his early childhood; his first school teacher was a British Israelite, who believed that all the British people were descended from the "lost tribes of Israel." Consequently he too, and the people around him believed for many years that they were lost Jews. Thus, Mourant noted that "though I can now see no evidence that I have any Jewish ancestors, I have maintained a deep interest in the Jewish peoples" (Mourant et al. 1978, p. v).

Mourant and his associates opted for the simplest two-dimensional graphs to present population frequencies of ABO blood types (and other blood types). On one axis, the frequency of the allele for type A is presented; on the other axis, that of type B (that of type O is the complementary frequency to 1.0) (Fig. 8.3, and Mourant et al. 1978). Connecting the Jewish values (black circles) and the corresponding non-Jewish data (white circles) with full lines and connecting the different Jewish data with dashed lines, allows relating to the length of lines as a measure of the proximity of the Jewish communities to each other, as well as of the corresponding Jewish–Gentile relatedness. Obviously, the more the Jewish communities resemble each other, the shorter the dashed-lines, and the more the Jews resemble the local non-Jewish population, the shorter are the full lines. Upon examining Mourant's graphs, it is difficult to claim that the Jewish communities are very coherent or very much related to their corresponding non-Jewish communities. Yet Mourant et al. summarize the total picture presented in his graphs:

> Looking at the complete blood-group picture of either the Ashkenazim or the Sephardim separately, one may observe that neither of these populations resembles closely the peoples among whom they now or recently have lived, and the range of variation between separate samples of the Ashkenazim compared with one another or the Sephardim compared with

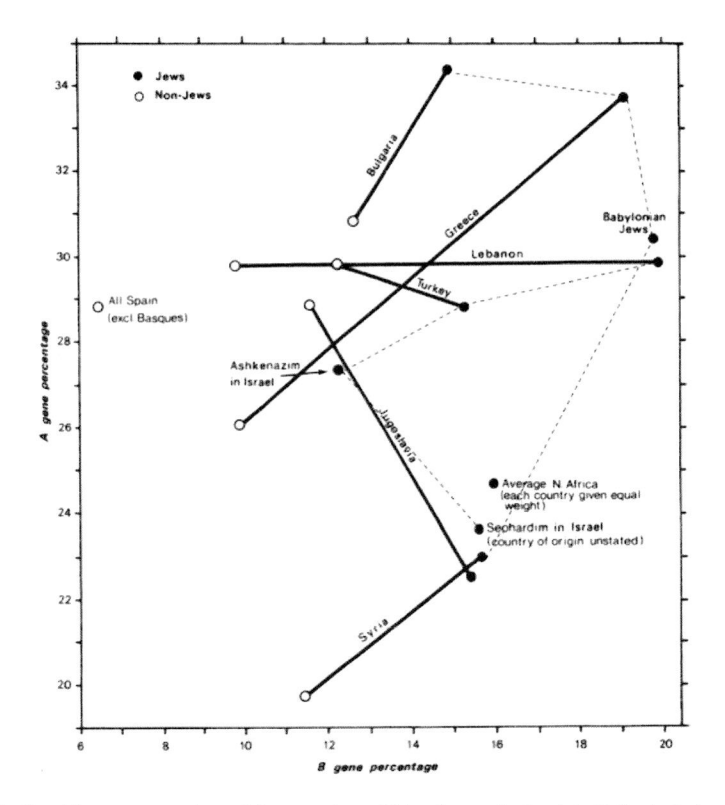

Fig. 8.3 Graphic representation of frequencies of blood type in Jewish (full circles) and non-Jewish (open circles) communities. Full lines connect Jews and non-Jews of similar sites, broken lines connect Jews of different background (modified from Fig. 4 of Mourant et al. 1978)

> one another, is so small that we can be sure that each is essentially a single population group.

> When, however, we compare Ashkenazim with Sephardim we find that there are indeed systematic differences between them. But these are so small that we can hardly avoid the conclusion that the two populations have a common origin, and a common original blood-group picture, only slightly modified in one direction or another by their different histories since separation. (Mourant et al. 1978, p. 51).

On the basis of his data Mourant also rejected the Khazar origins of the Jews, although he did not exclude the possibility of Khazar contributions to present day Jews. According to him the data did support Palestinian origins for both Sephardic and Ashkenazi Jews, in accord with the biblical legend. Even when Mourant and his associates observed deviations from this image of common origins, they tried to connect them to the biblical story. Thus, "The Jews of Kurdistan, Persia, and the

8.3 The Genetics of the Israeli Melting Pot

lands to the north and east of these, although they show a considerable scatter of gene frequencies," do fit fairly well into the regional non-Jewish picture. "Their traditions suggest that they are descended, at least in part, from the 'lost tribes' [...]. Their rather low frequency of African marker genes suggests that they have long remained genetically separate from the more southerly Jews of the Babylonian captivity and of the Dispersion into north Africa and Europe" (Mourant et al. 1978, p. 57). With respect to other ethnic groups, the Jews of Yemen significantly deviate from all other Jews of the Orient and are very similar to the Arabs of the Arabian Peninsula: There can be little doubt that substantial numbers of Jews migrated to the south at about the time of the destruction of the second Temple, but they seem to have become merged genetically into the more numerous Arab community. Nevertheless, for a time those who had remained Jews by religion prevailed culturally and set up a Jewish Himyarite kingdom in southern Arabia (Mourant et al. 1978). We will encounter again this kind of approach, which first throws the darts and then draws the target around them, a practice that is not limited to politicians.

The Israeli demographer, Helmut Muhsam, further elaborated on these attempts at a statistical and presumably unbiased method for guessing about the gene pool of the original Jews. Muhsam compared the data of the frequencies of genes in various Jewish ethnic groups – *eidoth* – with those of the frequencies in their "genetic environment," namely in the Gentile populations among whom the Jews of each *eidah* lived in the Diaspora in recent centuries (Muhsam 1964). His assumption was that all Jews were of one single, original population that upon exile became fragmented, with each splinter being diluted by the respective local population, such that the gene frequencies in the various *eidoth* would be an intermediate value between that of the original Jewish population and that of the local non-Jewish population among whom they lived. Graphically, the lines drawn from the given frequencies of the "genetic environment" population towards the corresponding Jewish populations would tend to converge, at least roughly, on the frequency of the original pre-exile Jewish population. Mourant's data of the distribution of the ABO blood-groups in 36 Jewish *eidoth* and their corresponding "genetic environment" were the most comprehensive data that Muhsam could use. Unfortunately, the vectors showed anything but convergence (Fig. 8.4). Muhsam admits that the general picture "can be considered to be largely the opposite of the starlike structure, which would be expected if all *eidoth* stemmed in fact from a common origin: the rays of our star seem rather to diverge into all directions than to point to a single point or, at least a limited area" (Muhsam 1964, pp. 15–16). Muhsam, however, was not prepared to agree with a conclusion that "makes it very unlikely that all Jewish *eidoth* included in the study stem from a common origin." So he offered "at least three different explanations of this apparent irregularity [...], each of which alone is sufficient to account for the deviation of the empirical data from the expected starlike structure, without contradicting the basic assumption of a common genetic origin of all Jewish *eidoth*" (Muhsam 1964, p. 17). Muhsam wrote a five-page treatise presenting possible factors that may have unfavorably affected the results, over which he had no control, so that the "real" links were not detected. He concluded:

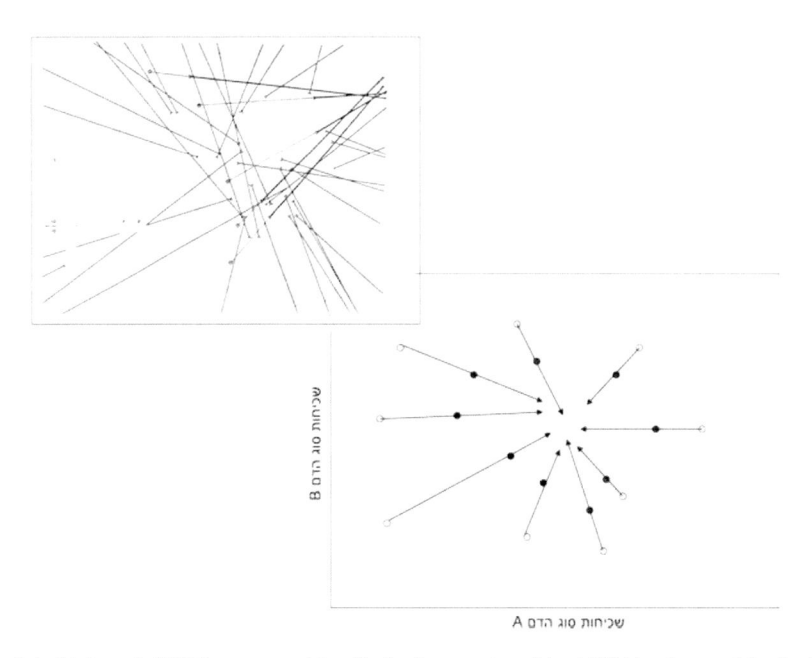

Fig. 8.4 Muhsam's (1964) attempt to identify the frequencies of the ABO blood type of the Jewish forefathers: vectors from "Gentile environments" to the corresponding "genuine" Jewish *eidoth*. *Lower right*: expected; *upper left*: observed

> [I]t would not require us, in view of the observed data, to abandon the assumption that each *eidah* is a mixture of a 'genuine' Jewish group and its environment, where the Jewish group involved is called 'genuine' because it stems directly from the original pool. [...].
>
> [...] in view of various historical facts such as wide-range migrations and the formation of isolates for many generations, the simple model may not be able to explain the relationship between each *eidah* and a hypothetical common genetic origin. [...] It is hoped that the analysis of additional traits, taking full advantage of the possibility to extend our model into a multidimensional attribute space will throw further light on the problem. (Muhsam 1964, pp. 21–22)

Put otherwise, if the conclusions of the biological research agree with the tradition (or the preliminary assumptions) – well enough. If not, the data are faulty, not the assumption.

Leo Sachs and Mariassa Bat-Miriam's study of the finger-print patterns of Jews and non-Jews of different origins was one of the first efforts to switch from anthropological to genetic variables to establish conclusive evidence for the common biological foundation of Israelis, in addition to that of blood type. This paper was intended to be the first in a series on "Genetics of Jewish Populations," but there was never a sequel to this single paper. The genetic basis of the finger-print pattern is quite complex. Yet according to Sachs and Bat-Miriam they succeeded in detecting in different communities an element common to Jews, wherever they come from,

8.3 The Genetics of the Israeli Melting Pot

which was distinct from that of Europeans, but similar to that of the Mediterranean people (Sachs and Bat-Miriam 1957).[24]

Goldschmidt, on the other hand, exploited the unique opportunity to study in quasi-laboratory conditions the processes in human populations that affect specific gene-related diseases in the populations that had lived for many generations under specific conditions and who were abruptly moved to completely different environmental conditions. Furthermore, many of these communities were small isolated populations with strong inbreeding habits that suddenly were exposed to intermingling populations. Thus, Israel offered geneticists unique opportunities to study juxtaposed migrant and non-migrant populations, some of whom were trying to maintain their genetic isolation, others who were not, and some who were establishing different isolates, all under conditions of convenient demographic and medical care:

> The study of isolate and migrant populations is valuable to the geneticist because such groups offer a "laboratory" in which certain variables can be observed to operate largely without contamination. [...] For this reason, the population of Israel offers the geneticist a unique opportunity. There, within the present population, one finds native-born Jews, both isolate and assimilated, urban and rural; immigrant isolate groups from various parts of the world that maintain their isolate characteristics; migrant groups that became isolates after immigration; and isolates that assimilated upon entering Israel. And for all these groups accurate demographic records are available. [...] Application of the tools of modern genetics to populations of ancient origin may be as valuable for the questions they raise as for the answers they provide, but they contribute significantly to our understanding of many genetic traits in a variety of ethnic groups. (From the dust-cover of Goldschmidt 1963).

In the exhibit that accompanied the conference, Goldschmidt presented the results of research that was based on experiments that had been carried out in Israel, mostly by clinicians. This was an early attempt to call attention to the genetic aspects of medical research in the country and to direct the attention of other researchers to the unique opportunities for analyzing specific genetic diseases in humans.

Indeed, during the following years, human genetic research developed dramatically. Several other international conferences dealing with the genetics of Jewish ethnic groups were convened in Israel. Much of the developments in this field of research depended on the significant developments in bio-medical and genetic research worldwide, but human genetics research in Israel led the way. Eventually the Sixth International Congress of Human Genetics in 1981 convened in Israel, and as a token of appreciation for the contribution of Israeli scientists to human population genetics, Bat-Sheva Bonné-Tamir of Tel Aviv University was elected its Secretary.

---

[24] In October 1921 Arthur Ruppin of Jerusalem turned to the criminal identification services of the police in Berlin with a request to examine and classify some 10.000 identification-sheets of the service that had been accumulating over the years, and on the basis of this data, to determine where the Jews versus non-Jews belong. The project was never carried out because of the lack of manpower (Doron 1980, p. 416).

Even though the declared aim of this research was to make shrewd use of the specific opportunities that Israeli demography offered to promote general interest, there was always consciousness of the prospect of establishing national unity; our blood ties never escaped the researchers' attention. Nurit Kirsh made an explicit attempt to examine how mobilized Israeli researchers were to the national tasks of the State of Israel during its first twenty years. She tried to examine quantitatively the extent that Israeli geneticists used their findings in the 1950s and 1960s as a basis for their national identity and their common historic roots (Kirsh 2003, p. 632). Although her samples are rather small, she demonstrated that the Israeli researchers of human genetics and Israeli physicians (and to a lesser extent, the non-Israeli Jews) emphasized the sociological-historical aspects of their studies rather than the genetic ones. Furthermore, as a rule, Israeli researchers tried to avoid formulating conclusions that conflicted with the Zionist tale and endeavored to reach conclusions that would support it (Kirsh 2003, p. 646). This tendency to examine and to confirm the historic formal narrative of the Jews in the Diaspora as being the direct progeny of the Biblical Children of Israel was further grounded and received wider attention with the establishment of "genetic archeology," the term suggested by Sheba.

The summary of a paper by Bonné-Tamir, published in the Hebrew science journal *Mada*, entitled "A New Perspective on the Genetics of the Jews" was typical of this tendency (Bonné-Tamir 1980). According to the author, "studies carried out before the 1970s emphasized the genetic differences between Jewish *eidoth*; on the other hand, later studies stress the similarity and the paucity of the contribution of 'strangers' to the gene pool typical to the Jews." Bonné-Tamir emphasized repeatedly that in spite of methodological and theoretical difficulties, "one of the conspicuous findings is the genetic relationship between Jews of North Africa, Iraq, and Ashkenazim. In most comparisons, they form one entity, whereas the non-Jews (Arabs, Armenians, Samaritans, and Europeans) are significantly different." She pointed out that the Yemenite and Cochin Jews were more distant, but this too could be explained as due to the geographical distance between them and the long historic isolation from each other. The author appropriately pointed out, however, that "a small genetic distance" does not necessarily indicate "biological closeness," or common genetic origins (Bonné-Tamir 1980, p. 185 f.).

> Our purpose in studying the differences and similarities between various Jewish populations was not to determine whether a Jewish race exists, nor was it to discover the original genes of "ancient Hebrews," or to retrieve genetic characteristics in the historical development of the Jews. Rather, it was to evaluate the extent of "heterogeneity" in the separate populations, to construct a profile of each population as shaped by the genetic data, and to draw inferences about the possible influences of dispersion, migration, and admixture processes on the genetic composition of these populations. (Bonné-Tamir et al. 1979b, p. 325).

Other scientists too expressed their opinions. David Goldstein introduced his book, *Jacob's Legacy*, by noting that "had I no Jewish heritage, this work would likely have never led me into Jewish genetic history. Events, however, primed me to look for ways to translate my professional activity into some kind of connection to

8.3 The Genetics of the Israeli Melting Pot

my own ethnic background" (Goldstein 2008, p. xii). Nonetheless, even in retrospect, after fifty years of research, Bonné-Tamir denied that in the process of her work any ideological notion directed her or her colleagues, and insists that they did not try to determine whether a Jewish race exists, or to discover the original Jews. Yet, Bonné-Tamir and her colleagues were "rather surprised" to discover the extent to which the populations of Ashkenazi, Sephardic, and Iraqi Jews show indications of relationships in spite of the known morphological and anthropological differences between them. Bonné-Tamir and Avinoam Adam confess that "[w]ith very few exceptions, it is practically impossible to trace, or quantify separately, such intracommunity diverse genetic components, which make up most of the present-day-groups" (Adam and Bonné-Tamir 1997, p. 433. See also Bonné-Tamir 2010). Still, these authors persisted in attempts to establish the biological foundations of the Israeli melting pot: "We are all Jews by descent, related to the ancient dwellers in this land; there were circumstances and environments that brought about our present genetic diversity. It appears that the Israeli melting pot is about to bring us to live under conditions where the biological differences that had evolved during our years of exile will seem a hump on the Jewish genetic pool and will disappear."

Some 11.5 million of the 14 million Jews in the 1970s were identified as 'Ashkenazi,' 1.5 million as 'Sephardic,' and another million as 'Middle-Easterners.' Still, it was accepted that contrary to the relative genetic homogeneity among the "Ashkenazim" wherever they lived, there was great variability among the communities of "Sephardim" and "Middle-Easterners."[25] Data supported a rather intensive rate of mixing and wandering of communities among the Ashkenazim, whereas most Jews of the East, North Africa, and the Mediterranean were divided, at least in recent centuries, into small or even minute unities effectively isolated from each other. This became most conspicuously apparent with reference to genetic diseases (see Chap. 6). Genetic studies that were carried out especially in communities that deviate from the so-called 'Israeli standard,' may be taken as another attempt to examine how far the cohesion of Israeli communities extends at the genetic level. This was the case with the Yemenite Jews, and with those from Cochin, and later with the immigration of the Ethiopian Jews. Even though the researchers understood that it was not the biological data that should determine the immigrants' identity, it seems that they still hoped that their research would explicitly or implicitly support the existence of a firm biological foundation for admitting these communities into the Israeli melting pot.

Evidence of the complexity of the subject may be derived from the treatment of the veteran communities of the Yemenites. They are lumped together, although they actually comprise demographically three relatively isolated populations: that of

---

[25] Not all agreed. See, e.g. Montagu 1974, p. 325, who claimed that the Sephardim of today comprise physically a much more homogeneous group than the Ashkenazim, and that the Sephardim "preserve rather credibly the racial pattern of their Palestinian ancestors." According to Montagu, it makes sense that the Sephardim are also less intermingled than the Ashkenazim; however, he doubted the extent to which they preserved the "racial" character of their Palestinian ancestors.

northern Yemen, that of southern Yemen (Aden), and the Habbanites from Hadarmaut. Yemenite Jews have often revealed a truly warm devotion to Judaism in spite of sustained centuries of persecution, and there is historical evidence that there were times of intensive proselytizing to Judaism of local tribes. In the third century C.E. there was even a royal family among the proselytes. Jewish kings ruled Yemen and fought the Ethiopian intruders until their final defeat in the sixth century.

Yemenite Jews became a persecuted minority among the Muslim majority, and they were prominent among the immigrants to the Holy Land during every period, including during the great immigration in the, 1880s. Large groups of Yemenite Jews immigrated in the first decades of the twentieth century (see Chap. 6), and there was a significantly large final immigration of Yemenite Jews in the first years of the State of Israel.

Although they maintained a quite effective isolation, as evidenced by certain hereditary diseases that are almost completely restricted to them, there is evidence that already in the Middle Ages there were contacts and intermarriages between Yemenite Jews and Jews from communities in Mesopotamia, Egypt, and Morocco. Nevertheless, the issue of the Yemenites' biological relation to other Jews was repeatedly raised by those who considered themselves to be the genuine Jews. Can the Yemenites be accepted as their blood-relatives? Different investigators reached diametrically opposite conclusions: Salaman concluded that they were not Jewish, whereas Bodenheimer considered them to be the most genuine representatives of the ancient Jews. Finally, in a 1999 article in the *Yemen Times*, two historians, Mohamed El-Kudai and Mohamed Ben-Salem, claimed that "the historic and social reality confirms that the Yemenite Jews are an integral part of the Yemenite nation. These people proselytized and adopted Judaism in their homeland, which at the time enjoyed religious tolerance."[26]

Although the Jewish status of communities from all parts of Yemen is not contested in Israel anymore, another community, that of the Ethiopians, has attracted a great deal of attention in recent years. According to their tradition, the people of Beta-Israel, the name they call themselves, arrived in Ethiopia with Menelik the First, the son of King Solomon and the Queen of Sheba. Some investigators see them as the descendents of the Jewish community in Upper Egypt, others, as the progeny of an African tribe who converted to Judaism when Yemenites ruled Ethiopia during the early Christian era. The religious authorities seem to be split in their ruling of their Jewishness.

One of the most efficient genetic markers used to be that of the highly polymorphic pattern of Human Leukocyte Antigens (HLA). Comparing the distribution of HLA markers in Jews of Ethiopian origin and that of the non-Jewish Ethiopians, Blacks from South Africa, Yemenite Jews, Cochin Jews, Ashkenazim and non-Ash-

---

[26] *Zionism in Yemenite Eyes* (in Hebrew), *Haaretz*, October, 15, 1999, p. 8B. For a recent discussion of the reactions of the official Israeli and American Jewish authorities to the Yemenite Jews and their longing to immigrate to Israel, see Meir-Glitzenstein (2012).

kenazim, researchers noted that every population had its own typical antibody distribution. The pattern of the Ethiopian Jews resembled considerably (but was not identical with) that of the non-Jewish Ethiopians and both varied from that of South African Blacks, as well as from that of the Ashkenazi pool. Thus, the authors concluded that the gene pool of the Ethiopian Jews indicated a mixture of Mediterranean and African sources (Brautbar et al. 1992). Such a statement reflects the predicament of the authors attempting to maintain an unbiased position, in spite of dealing with a contested socio-political issue: Are these people Jews who absorbed "Ethiopian blood," or the other way round, namely, Ethiopians with some Arabian or Jewish input?

Even though the Israeli experiment of creating a melting pot to form a single cultural and social entity apparently did not succeed and has been replaced in recent decades by the notion of a multicultural Israeli society, it seems that the geneticists who primarily followed the processes within communities can report today a considerable level of success in the processes of biological amalgamation. Tirza Cohen, one of the pioneers of research of heredity in Jewish communities, and her colleagues report that towards the beginning of the twenty-first century, there was a significant decline in consanguine mating, which was extremely high in the 1950s, especially among Jews from Middle Eastern countries, to the extent that nowadays inter-community mating comprises about a third of all mating in the population (Cohen et al. 2004).

8.4 From Single-Genes to Systems Polymorphisms

Towards the end of the twentieth century, even before the accomplishments of DNA-sequence-mapping studies, interest was increasingly turned to following genetic variability in genes other than blood groups. In 1966, Lewontin called attention to molecular polymorphism in enzymes and its applicability to the study of genetic heterozygosity in natural populations (Hubby and Lewontin 1966. See footnote 12). Thus, gel-electrophoresis detected in the extract of the enzyme adenosine deaminase (ADA) two forms that migrate at different rates in the electrophoretic field, ADA^1 and ADA^2, which represent two alleles of the same gene. As it turned out, ADA^2 is very rare in Northern and Central Europe, while very common in Mediterranean nations. Both Ashkenazi and non-Ashkenazi Jews have a high frequency of the Mediterranean allele ADA^2. It makes sense to speculate that this points to the ancient gene pool of the Jews that has not been completely diluted by local alleles or selected by local conditions (note, however, that there are no data on East European non-Jewish ADA-allele frequencies, which is where most Ashkenazi Jews come from) (Szeinberg 1973). However, it is also plausible that members of Jewish communities with a similar religious-cultural background maintained some level of intermarriage, irrespective of whether they really had common forefathers or not. On the other hand, once we study the polymorphism of another gene, PGM_1

for the enzyme phosphoglucomutase, we find a low frequency of the PGM_1^2 allele among Ashkenazim, like that among non-Jewish Europeans, and a relatively high frequency among Middle Eastern Jews and non-Jews alike. Does this indicate different patterns of mixed Jewish and non-Jewish blood in different sites, or does it indicate local differences in natural selection of alleles? Arieh Szeinberg, who studied the distribution pattern of numerous enzyme-polymorphisms, concluded that the results were "compatible with the belief that all the main Jewish communities stem from a common gene pool, and that the influx of foreign genes was not significant. All the differences found among these communities might be explained by the existence of local selective forces which left an imprint on the genetic make-up of each group" (Szeinberg 1973, p. 1176).

If one accepts the premise of the common origin of Ashkenazi and non-Ashkenazi Jews, and the premise that the ancient gene pool has not been diluted to a significant extent among Ashkenazi Jews, as suggested by ADA polymorphism, then the similarity with most European populations in the distribution of PGM genes strongly suggests that this polymorphism is markedly influenced by selective forces. (Szeinberg 1973, p. 1174).

In other words, the dissonance between the data of the distribution of the alleles of different genes and the fundamental assumption of a common denominator for all Jewish communities may be settled if we add an additional assumption, such as a reference to the selective value of the different enzymes. Szeinberg was aware of the problematics of his argumentation and mentioned that many assumptions still lack factual basis. He went even further and noted that contrary to the conclusion that "most genetically informative properties support the Mediterranean origin of the Ashkenazim," it is precisely the distribution of ABO blood groups that demonstrates a similarity between Ashkenazi and non-Jews in Europe. Put differently, the ABO blood groups indicate links to the people among whom the Jews lived. Remember that researchers perceived blood type polymorphisms to be reliable markers for following the histories of populations and of their admixture, because they were presumed to be neutral with respect to natural selection (see also Chap. 7).[27]

As it turned out, polymorphisms of blood types may also have differential selective values, and in order to reconcile the findings with his assumptions, Szeinberg suggested that indeed, natural selection played a part in shaping the frequencies of the blood groups (Szeinberg 1979). The studies of Luigi Cavalli-Sforza and Dorit Carmelli (1979) were more equivocal with respect to the role of natural selection in shaping the gene frequencies of the Jews. They suggested that the differences between the Ashkenazi Jewish ethnic groups, on the one hand, and the similarities to these populations among whom they lived, on the other hand, stem from intermarriages. They estimated that these were responsible for dilution by up to 40% of non-Jewish genes into the Jewish gene pool. Obviously, the disparate conclusions of

---

[27] To be precise, it was the ease of collecting data that made blood types suitable for the study of the dynamics of population gene pools, as well as the high reliability of the field-collected data.

8.4 From Single-Genes to Systems Polymorphisms

the different groups of investigators stem from their assumptions about the factors that they believed had been active in specific communities over generations.[28]

As Mourant stressed, it is difficult to avoid mistakes in determining the uniqueness of the Jews among the non-Jews when the distribution of the blood types followed is small. Indeed, Bonné-Tamir and her colleagues systematically introduced multiple markers to examine the assumption of the existence of a Jewish gene pool that crosses ethnic groups and eras, and to study the variables in a large number of blood types, enzymes, and antibodies (Bonné-Tamir et al. 1979a). They claim to have found in Israel closer relations between Jews of different origins than between these Jews and non Jews, and more specifically, "Ashkenazi Jews are essentially one uniform and homogeneous group [...]. When compared to European populations among whom they have lived before migration to Israel, different degrees of closeness in frequencies are demonstrated at different loci; however, the overall picture based on all markers is one of distance of the Jews from the European populations" (Bonné-Tamir et al. 1979a, p. 71).

Still, the number of variables that could be tested at the end of the 1970s was limited. With the onset of the molecular genomics era the number of variables increased dramatically. In, 1992 Bonné-Tamir's group studied 12 molecular restriction fragment length polymorphism (RFLP) variables. Yet the technique was rapidly supplanted by a variety of techniques based on the polymerase chain reaction (PCR), and researchers increasingly pointed to the difficulties of using a small number of variables. For example, a sample of Roman Jews turned out to be most like Amazon Basin Indians with respect to several markers, and like Cambodians with respect to others. Another sample of markers suggested a difference between Yemenites and some native Africans, even though similarities to the Yemenites to populations further eastwards (Kidd et al. 1992).

As Bonné-Tamir and colleagues extended their studies to include more molecular markers, they were able to conclude that "results with the new DNA markers do not show a consistent pattern in the genetic profiles and affinities of the Israeli communities." Immigrants from Yemen and Morocco show a closer relationship using the new markers than they showed using the old markers. The Ashkenazi, on the other hand, show a closer relationship to the Mediterranean pattern with respect to some markers and a less close relationship with respect to other markers to the extent that "the theory of a Mediterranean origin of the Ashkenazi community could neither be supported nor denied conclusively" (Bonné-Tamir et al. 1992, p. 90).

---

[28] R. C. Lewontin made the following point in the *Annual Review of Genetics, Directions in evolutionary biology* (2002): "On the conceptual side, unlike for molecular, cellular, and developmental biology, there is no basic mechanism that evolutionists are attempting to elucidate. There is no single cause of the evolutionary change in the properties of members of a species. Natural selection may be involved but so are random events, patterns of migration and interbreeding, mutational events, and horizontal transfer of genes across species boundaries. The change in each character of each species is a consequence of a particular mixture of these causal pathways."

The conflicting and non-cohesive data indicate the extent of the problem: the frequencies of the alleles of the examined genes in the examined populations were determined not only by factors such as population mixing or isolation, or random sampling from a small gene pool during crises (so called "bottleneck" effects), but also by forces of natural selection. All these factors and others shaped the genetic patterns of populations, which may be defined only to a certain degree of precision by demographic variables of culture, religion, language, education, or place of residence. Even with no common biological roots, people with certain traditions and religion, and with specific ideas and notions, do prefer to marry each other. Likewise, when there were genetic links between groups in the past, many were completely blurred, and there is little point in persisting in the search for common markers between people defined as Jews or non-Jews on the basis of social or cultural-religious indicators. Suffice it to recall that in Germany of the 1930s, in spite of the efforts of the best scientists and the application of the most sophisticated scientific methods of the time, the Nazis had to fall back on attaching the Yellow Star to the garments of the Jews in order to identify them.

Chapter 9
From DNA to Politics

> MK Zeev Boim: "What is there in Islam at all? What is there in the Palestinians specifically? Is this a cultural deprivation? Is this a genetic defect?"
>
> MK Yechiel Khazan: "It is known that for generations Arabs murder Jews. Nothing can be done: It is in the blood. This is something genetic." (February 24, 2004, memorial session in the Knesset for the victims of the terror attack on a bus on Israel's main coastal highway.)

In the previous chapters I have tried to describe the efforts of many scientists to challenge or prove the Mediterranean links between the gene pools of the Israeli ethnic groups. Whether it was blood types or skull dimensions, diseases or DNA sequences, no matter how much data was gathered, conclusions were drawn from premises that were basically assumptions. The more genetic research progressed, the clearer it became that it would be impossible to derive from genetics unequivocal conclusions concerning the historical interrelations or connections between the Jewish ethnic groups on the one hand, and between them and Palestinians who have been living in the country, some of them for many generations, on the other hand. In the end, *it is the historical data that ought to to give meaning to gene distributions and not vice versa.* Most Jewish communities were as a rule small and relatively, though not hermetically, closed and many of the presumably selectively neutral genetic markers were found to significantly affect survival, often in unpredictable directions. It became increasingly obvious that it was essential to consider a large number of variables and to stay as close as possible to the genotypic level proper in order to try to follow in the footsteps of history on the gene pools of the populations that gathered in Israel. It is in the nature of genetic analysis to deduce from the apparent effect (the phenotype) to that of the immanent (genotypic) level, the most explicit expression of which is the phenotype of the DNA molecule and its nucleotide sequences. The beginning of the twenty-first century is marked by the universal detailed sequencing of the genotypes of any individual organism, and the examination of variability at a very large number of sites and the effects of their interactions.

The more biologists realized that intra-group variability exceeds by orders of magnitude traditional inter-group variability (Lewontin and Hubby 1966), the more it was obvious that races are not natural categories (Roberts 2011). As has already been noted, subjective differentiation and grouping are essential properties for survival of living beings, and humans are arguably the most sophisticated classifiers into categories. Thus, at best races may be considered projectible entities in so far as they support inductive inferences of biological entities of classification (Gannett 2004, 2013).

With the emergence of the twenty-first century genetic analysis was completely transformed as epitomized by the successful Human Genome Project – DNA sequencing of whole individual genomes. Within a short time, detailed sequencing of the genome of any organism, as well as the synthesis of any desired sequence, has become possible. Natural genetic variability was now studied directly at the level of DNA sequences, rather than deduced from morphological, physiological, behavioral, or clinical patterns.[1] Although only a minor fraction of the cell's DNA is involved in coding protein sequences, each coding triplet of nucleotides, or *codon*, represents a specific amino-acid (and one of three stop codons), such as ATGAAAGCTCGGTGCAAGTCG coding for the sequence methionine, lysine, alanine, arginine, cysteine, lysine, serine (notice that the code is redundant; most amino-acids are represented by more than one triplet). A mutation in the sequence may change the protein: changing the first T in the above sequence by a C will change the peptide to threonine, lysine, alanine, and so on. Deleting or inserting a fragment may have more profound consequences.

A most important consequence of the redundancy of the genetic code was that upon analyses of population variability at the molecular level, it turned out that many hereditary properties that appeared to be identical in form and function varied at the level of the DNA sequence. Since the probability of a mutational change of a specific nucleotide in the DNA sequence is low,[2] it would be reasonable to expect two individuals who have precisely the same molecular change in their DNA sequence to be related, and to attribute the occurrence of the unique mutation to a common ancestor. Likewise, two individuals affected by the same disease, as far as the clinicians are concerned, are *not* necessarily relatives once it turns out that mutations at two different nucleotides (of roughly one thousand nucleotides that code an average protein) caused the similar phenotypic change; nor would they necessarily have a common ancestor (at least since the occurrence of the differentiating mutation/s). However, it must be kept in mind that a low probability of the occurrence of a second similar mutation at the same DNA site, does exist. Furthermore, in the effort to

---

[1] See footnote 6 in Chap. 8. As noted, whereas DNA sequences are composed of the four nucleotides, adenine (A), thymine (T), cytosine (C), and guanine (G), proteins, the major component of enzymes, are composed of precise sequences of twenty amino-acids, coded by sequences of DNA nucleotide triplets.

[2] The rate of spontaneous mutations (those not induced by man-made agents such as X-ray radiation, or treatment with "mutagenic" materials) is roughly one in a thousand million (10^{-9}) per nucleotide per cell generation.

establish common ancestors, probabilities are greatly enhanced when the comparisons are carried out for numerous nucleotides, or more specifically, for longer DNA sequences, operationally defined as distinct entities, which can be followed in many individuals and over many generations for longer periods. Such are the *haplotypes*, which may be conceived as entities of DNA sequences that are similar to the *genes* as the entities of functional genetics: like the genes, the haplotypes are entities in which mutations occur and, most importantly, recombination may occur between the mutational variants. Computer programs, such as *Structure*, were specifically designed to construct the most probable *diverging phylogenetic trees* that the available data on the distribution of mutations in haplotype sequences and their recombination suggest.

9.1 Similar but Different

The introduction of detailed DNA sequencing has added new dimensions that enable geneticists to identify factors that contribute to present-day variability between and within human communities. When the gene for the G6PD deficiency of Kurdistani Jews was examined at the level of the DNA sequences, the "Mediterranean" allele was found to be common among Kurdistani Jews: their allele carries the same mutation common in affected non-Jewish Mediterranean people, although these other populations may vary also at additional sites of the DNA sequence coding for the G6PD enzyme. Given that over the years there were chances for several mutation events to occur in this sequence (not necessarily only such that cause enzyme deficiency) and that environmental circumstances encouraged the survival of mutated genes that do affect the enzyme production in malaria infected areas, the widespread distribution of the same mutation throughout the Mediterranean Basin would *suggest* the antiquity of this mutation and the genealogical link of all these people, which would somehow lead to the same ancestor in whom the now prevalent mutation occurred.

But it does not indicate that all present-day carriers of that mutation stem from a Jewish root. The allele could have been, for example, of non-Jewish origin and distributed to Mediterranean peoples who then spread it further, primarily in communities suffering from malaria. Likewise, individuals of each of the Jewish communities in the East could have acquired the mutated allele independently by intermarriage, and then successfully distributed it further among their community members who stayed in areas infected by malaria where the allele had some selective advantage (Beutler and Kuhl 1992; Oppenheim et al. 1993).

β-thalassemia (clinically discernible from α-thalassemia. *Vide infra*), also known as Mediterranean Anemia, is common in some of these communities. Physiologically and clinically, most thalassemia patients are quite similar, although over 200 different mutations at the molecular level were detected in the gene's DNA sequence. In other words, mutations in the gene occurred independently many times, survived, were reasonably selected for, and became established in various members of these malaria-infected populations. Since many of the alleles are limited to specific communities, the presence of an allele may often detect with high probability even the person's village of origin. In several cases, progeny stemming from a specific village (within a range of several generations) could be pegged to their origins even when they were now living thousands of miles apart.

More refined differences could also be detected. If indeed one allele for deficient G6PD spread early on in the populations of the Near East, the present changes in the thalassemia gene must have occurred later, after the populations of common origin had already dispersed. A distinct molecular pattern in different Jewish communities indicates that the communities were effectively reproductively separate in the generations following the appearance of the mutations.

Whereas the identical change in the base sequence related to β-thalassemia in neighboring Jews and non-Jews community indicates blood relatedness between them, molecular tests showed that the β-thalassemia among Israeli Arabs, Druze, and Samaritans (and among a few affected Ashkenazi Jews) all carry mutations different from those of Jews of Middle Eastern origins (Filon et al. 1994; Rund et al. 1991). Accordingly, *if* Israeli or Palestinian Arabs are the progeny of the ancient inhabitants of the country, then the β-thalassemia that was found among them and among the Middle Eastern Jews was established after these populations were separated from each other (*or* they never were united or in contact).

The allele of the gene for thalassemia among the Jews of North Africa is, at the molecular level, the same one that is prevalent in Spain and Portugal. It is also widespread across the Mediterranean. This indicates, at least, that there was no hermetic isolation of the gene pools of the North African Jewish communities and those of the Iberian Peninsula. On the other hand, the presence of thalassemia patients of Ashkenazi-Lithuanian origin who carried an allele prevalent in the Mediterranean supports the claims of some migration, or rare cases of intermarriage of a member of an eastern Mediterranean community with a spouse of a central European community (Rund et al. 1992; Pras et al. 1992).

Thalassemia is also prevalent among Yemenite Jews. However, in addition to β-thalassemia, a clinically and physiologically distinct disease, α-thalassemia, in which another gene is affected, is found among them. Most known cases of α-thalassemia are missing a short segment of 3.7Kb in length, in their DNA. Notwithstanding, in several Yemenite α-thalassemia patients, another segment unique to them is missing (Shalmon et al. 1994). This is consistent with the Yemenite Jews being a relatively isolated community, even among the Jewish communities in the Middle East. These data on thalassemia also support the claims that the populations of the Mediterranean Basin, the Middle East and Yemen, lived under conditions in which different rare mutations that appeared in the same gene were maintained by the selective advantage they endowed to *heterozygotes* in these locations, in spite of the severity of the disease of the *homozygotes*. Therefore, the distribution of thalassemia, like that of G6PD-deficiency, may only indicate a possible pattern of the phylogenetic relationships between communities that were not hermetically separated from each other.

Evidence indicates that communities of Moroccan Jews lived in relative isolation (at least for several generations), separated from the Jews of Tunis and Libya:

9.1 Similar but Different

Several 'hereditary diseases,' which are very rare in other communities, are common among Libyan Jews,[3] while in Moroccan Jews, other "hereditary diseases" are prevalent (Levin et al. 1967).

Thus, research investigating hereditary diseases at the molecular level, while meaningfully increasing the possibility of discerning variants of the phenomena, also highlights the role played by non-phylogenetic factors in explaining the distribution patterns of the genes. It warns of the possible dependence of explanations of the basic assumptions of investigators concerning the histories of the communities involved. Although researchers usually refrained from drawing far reaching conclusions, it was not always possible to keep research and politics separate, once the findings in genetics became a subject that caught the eye of the media and the public at large. Many researchers actually considered the political and social meanings and consequences of such studies to be too urgent to be left to investigators who usually come up with several alternative interpretations.

Thus noted Uri Seligsohn in his introductory lecture to the Israel Academy of Sciences: "There exists in the country a 'gold mine' of genetic evidence that calls for the understanding of physiological processes, through which light may be shining on the history of the Jewish people" (Seligsohn 2002). Indeed, in his research Seligsohn isolated and cloned the gene that codes for Factor XI – one of the blood clotting factors – and followed the details of its molecular variability among Ashkenazi and Iraqi Jewish patients who lacked the proper functioning of this factor (Peretz et al. 1997). Seligsohn and colleagues found that whereas two different major mutations at the DNA sequence level were involved in the disease among the Ashkenazi, only one of the two was found at a high frequency among the Iraqi. They concluded that two mutations had occurred at different occasions in the same population, and calculated the best fit for the phylogeny of the mutation found in both ethnic groups. The one, found in both Ashkenazi and Iraqi Jews, had occurred in an ancestor who lived some 2,500 years ago. The second mutation, unique to the Ashkenazi, occurred in an ancestor who lived some 700 years ago. This could indicate, for example, that a person of an otherwise separate and distinct Iraqi community was married to an Ashkenazi partner – and there is evidence for such events of a mutation that wandered from one ethnic group to another. But Seligsohn ignored this possibility and asserted unequivocally: "The genetic identity between Iraqi Jews who were living in the Middle East from the time of the exile to Babylon and Ashkenazi Jews from Eastern and Central Europe completely shatters the assumption that the Ashkenazim are progeny of the Khazars" (Seligsohn 2002).[4]

A somewhat different perspective is gained from following the distribution of the hereditary factor of cystic fibrosis (CF), which is one of the most prevalent

---

[3] For the distribution of HLA blood antigens, see Bonné-Tamir et al. (1978). Familial Mediterranean Fever (FMF) has already been mentioned. For other diseases in this community, see Hsia et al. (1991); Weinberger et al. (1974). See also Rosenberg et al. (2001).

[4] Nurit Kirsh, among others, noted that the Israeli researchers emphasized every detail of similarity found among ethnic groups while noticing fewer points of disagreement (Kirsh 2003, p. 649). These researchers' explanations were not necessarily wrong; they were biased.

"hereditary diseases" common among Caucasians. It affects approximately one in every 2,500 live births. The very fact that such a disease, which was until recently lethal at a young age, reached such a high frequency raises the suspicion that the carriers of the gene in one dose (the heterozygotes who got the allele for the disease from only one parent) who are not sick may have some selective advantage over those who do not carry a "defective" CF allele. In Israel, the frequency of CF births is only one in 5,200, but there are great differences among the ethnic groups: Whereas among the Ashkenazim, the frequency is similar to that in Europe (1:3,300), among persons originating in Yemen, Morocco, Iran, and Iraq, the disease is rarer. On the other hand, the frequency is as high as, 1:2,500 among Jews originating in Libya, Georgia, Greece, and Bulgaria (B. Kerem, personal communication). There are two major clinical subgroups of CF: one clinically severe with pancreatic insufficiency and the other of patients with a milder disease with pancreatic sufficiency, although in some of those milder cases, male fertility is also affected. However, there is a wide variability in lung disease, the major morbidity and mortality caused by CF, in both subgroups.

Beyond the clinical variability, most of the variability is at the level of the mutations at the DNA sequence: there are about two thousand mutations in the relevant gene. The mutation prevalent in circa 70 percent of the chromosomes of CF patients in the world (ΔF508) is also the mutation that is frequent among CF Israeli Jews from Morocco and Libya.[5] This mutation is rare among patients of Tunisian origin and is completely absent in Jews of Georgian, Egyptian, and Yemenite origins. Among Ashkenazim in Israel, it is only one of six well-known mutations, the most frequent of which (W128X) comprise about 48 percent of CF chromosomes (Kerem et al. 1995). Such results correspond to a claim that the Jewish populations were isolated from each other to a large extent at least during recent centuries. Yet the data also suggest that the Jews were not isolated from non-Jews in their environment, and indicate that there was a transfer of genes between Jewish and non-Jewish communities. The relatively high frequencies of the European mutations among Moroccan and Libyan Jews, as well as its rarity in Tunisian Jews, reflect a probable combination of intermarriages between Jews and non-Jews (one such intermarriage is enough to introduce the allele from one population to another) with a high rate of inbreeding within the community. At the same time, it appears that a certain rate of intermarriage between communities in these countries was maintained. Of course, the possibility that the same mutation at the DNA sequence occurred again – as an independent event in different populations – cannot be rejected as even events with very low probability do occur repeatedly.

In Jewish communities where the CF disease is relatively frequent (but not among Ashkenazim), the frequency of an additional mutation, besides the European ΔF508, is relatively low, which suggests that the population can carry only a given final load of alleles for the disease. This also provides another indication that the

---

[5] The gene whose alleles may cause cystic fibrosis is defined as a DNA sequence in the human chromosome called the "CF chromosome." It is located to the long arm of chromosome 7 at position 31.2. Cytogenetic Location: 7q31.2.

frequency of CF births in the population is determined by an equilibrium of selection for carriers of the mutation (i.e., the mutation in single dose somehow is advantageous to its carriers) and selection against the patients who carry a double dose of the mutated allele. It appears that another unresolved role of the proper protein appears to be a crucial factor in the living system, which again may point to the futility of our determinist molecular adjudication.

A similar picture is obtained with respect to Gaucher's disease (see Chap. 8). Clinicians discern three types of the disease, and a different type is found in different populations. Clinically, Gaucher Type, 1 among the Ashkenazi Jews is the less severe. Isolation of the relevant DNA sequences demonstrates many different mutations at different sites along the gene's DNA sequence. Among Ashkenazi Jews, more than 70 percent of the chromosomes involved in the disease carry one mutation (N370S, also called G^{1226}), and more than, 10 percent carry another mutation (84GG, or GG^{84}). Among non-Jews who suffer from Gaucher's disease Type, 1, the same mutations are found as among Jews, but at rather different frequencies. On the other hand, the mutation L444P, which causes Gaucher in only 3 percent of Jewish patients, is responsible for 26 percent of non-Jewish Gaucher patients. There are very few known Gaucher patients among non-Ashkenazi Jews (Grabowski 1997). It appears that here too molecular analysis reveals a certain transfer between the Jewish and the non-Jewish populations, even though the Jewish populations were rather isolated from the non-Jewish populations.

The onset of the molecular age has also brought about upheavals in the perception of the distribution of the gene involved in Tay-Sachs disease. As long as diagnosis was at the level of the affected enzyme (the α-subunit of the enzyme β-hexosaminidase), it was assumed that all patients have the same mutation originating in the Ashkenazi community of northern Poland or Lithuania. Today it is clear that Tay-Sachs is also a heterogeneous disease at the molecular level, and that even among the Ashkenazi Jews, more than one mutation is present. Most mutations are rather rare, but the disease is found also among Moroccan Jews, and at least one additional mutation, besides the one that is common among Ashkenazim, prevails (Navon and Proia 1992). The picture is similar with respect to Nieman-Pick's disease, which also belongs to this group of sphingomyeloid diseases. At the molecular level, three mutations comprise more than 95 percent of the cases of the clinical dominant type among Ashkenazi Jews. The prevalence of this disease in another isolate of the French-Canadians in the district of Quebec suggests the importance of the impact of the "founder effect" in the distribution of such diseases. Following the Canadian patients (who are clinically different from the Ashkenazi Jewish patients), there are indications that all of them stem from the same family that immigrated to Nova Scotia in the seventeenth century.

Gideon Bach and colleagues in Jerusalem repeatedly emphasize the high frequency of the cluster of diseases of lipid accumulation among Ashkenazi Jews. Another disease belonging to this cluster, discovered only in recent decades, is MLIV (Mucolipoidosis Type IV). It has been found in Israel and all over the world, especially among Ashkenazi Jews (Bach et al. 1992). It is difficult to avoid speculating that the high frequency of these diseases among the Ashkenazim also involves an increased selective value of the carriers of these alleles in that ethnic group (see Chap. 8).

This short superficial overview reveals that in spite of the advances in genetic analyses of the DNA sequencing level – or maybe because of them – researchers have not succeeded in uncovering common biological roots or the pattern of an unambiguous (vertical in time scale) branching phylogenetic tree of the Jewish *eidoth*. More and more indications of the impact of intra-breeding, selection and other local factors (most of them yet unexplained) are emerging. Since many communities were relatively small, sometimes only a few individuals as a result of persecutions and pests, and since they were relatively isolated both from other Jewish and non-Jewish communities, allele frequencies of many genes in different sites were unaccountably fixed. Among the exceptional impacts were factors such as natural selection that positively affected alleles otherwise known for their deleterious effects – many alleles with very slight positive effects (in heterozygotes) have been counterbalanced by major deleterious effects of a few (in homozygotes). One way to try and overcome these difficulties would be to simultaneously follow many variables, which are not known to be involved in diseases or other mal-development. Genome Wide Association Studies emerge to provide an ultimate answer and it seems that the analyses of haplotype variability of the Y-chromosomes progresses in this direction.

9.2 The Trail of Y-Chromosome Haplotypes

As emphasized, developments in research methods, and primarily in the possibility of examining polymorphisms at the level of DNA sequences, have increasingly enabled the comparison of genetic relationships even when no morphological, physiological, or behavioral variation existed. No less significant, it has become possible to simultaneously examine polymorphism at a very large number of sites along the DNA sequences.

One of the early exciting variables that have been reexamined once DNA sequencing was introduced is that of the genetics of the Jewish priests. Traditionally, the tribe of Levi was destined for priesthood, and the male descendents of Aaron, the brother of Moses, were anointed as priests or Cohanim (pl. of Cohen). Accordingly, in our patroclinous society, all persons with this name (and its derivatives) are presumably the male progeny of Aaron (see Chap. 6). If indeed the tradition of the Cohanim was maintained, it makes sense to look for a common denominator among all these male progeny of Aaron, and their cohesion may provide another indication of the common origin of all Jews.

In 1911, Salaman "involved" the Cohanim in his attempt to identity the "unmistakably Jewish expression":

> At this point one might with advantage consider the relation which the existence of the *Kohanim* has to the question of Jewish type. [...] no Kohen, according to Jewish law, can

marry a stranger, a proselyte or the daughter of the proselyte, or a divorcée: so that we have a sect whose descent may be regarded as strictly Jewish. (Salaman 1911a, p. 279)

Salaman did not succeed. With the means at his disposal, it was impossible to establish the Jewishness of the persons examined. He had to admit: "If now we review the physiognomy of the various *Kohanim*, it will be found that they exhibit no type in any way distinct from the other Jews" (Salaman 1911a). Also Weissenberg tried at the beginning of the twentieth century to rely on the tradition of the male dynasty of "Aaronides (Kohanim) and Levites." Some of these families keep centuries-old portfolios and seek to marry only with irreproachable families. Disappointedly, Weissenberg too found that the Aaronides and Levites represent, on the whole, the same type as the common Jews. "From these results it would be fundamentally incorrect to draw the conclusion that today's East European Jews are direct descendants of the ancient Israelites" (see Efron's 1994 endnote 61, pp. 201–202).

Now, with the development of methods to follow specific DNA sequences of the human genome, interest in the Cohanim (and Levites) has gained new momentum as an instrument for proof of the common origins of the current Jewish ethnic groups in the population of the Land of Israel two thousand years ago, as narrated in the biblical story.

As noted, of the 23 chromosome pairs of humans, one pair is different in females and males: whereas females have two similar copies of this chromosome, the X-chromosome, males carry one X-chromosome, and one smaller partner, the Y-chromosome. Females normally contribute one X-chromosome (as well as one of each of the other chromosome pairs, called autosomes) to each progeny; males contribute one X-chromosome to half of their progeny and one Y-chromosome to the other half of their progeny (as well as one of each autosome pair) to all their progeny. Progeny who obtain two X-chromosomes are females; those who inherit one X-chromosome and a Y-chromosome are males. Thus, by following a marker linked to the Y-chromosome, a biological lineage leading back to an ancient common male progenitor appears to be identifiable. If indeed the priesthood has been maintained by strictly following the patrilineal tradition, then all Cohanim should carry the derivatives of Aaron's Y-chromosome – *derivative* sequences, rather than the original sequence, because no doubt rare mutations have occurred and were maintained over the millennia. Since mutations are rare, each mutation would probably be unique, and the older the sequence is, the more unique mutations will have accumulated. Furthermore, since as a rule sequences in the Y-chromosome are considered to be rather irrelevant to natural selection, the abundance of a mutation in the Y-chromosomes may be used as an indicator of its age.

Notwithstanding, this Y-chromosome, which is being faithfully transferred from one Cohen male to another, may indicate that its carrier are distinct from the rest of the Israelites, and may perhaps enable constructing a pedigree tree all the way back to Aaron the Priest.

Although the Y-chromosome is the smallest human chromosome, its DNA molecule is 57 million base-pairs long. Thus, as a rule, only a small number of segments along the chromosome are selected for study. Such segments, some of which are thousands of base-pairs long, may be concieved as *haplotypes*,[6] which provide a unique combination of polymorphisms along a Y-chromosome and may represent the whole chromosome.

> [...] we sought and found clear differences in the frequency of Y-chromosome haplotypes between Jewish priests and their lay counterparts. Remarkably, the difference is observable in both the Ashkenazi and Sephardic populations, despite the geographical separation of the two communities. [...]
> We identified six haplotypes [...]. Applying the χ^2 test to the frequencies of the Y-chromosome haplotypes distinguishes priests from the lay population. [...]
> We further identified subjects as being of Ashkenazi or Sephardic origin. [...] the same haplotype distinction can be made between priests and lay members within each population. This result is consistent with an origin for the Jewish priesthood antedating the division of world Jewry into Ashkenazi and Sephardic communities. (Skorecki, et al. 1997)

As usual, the researchers declared that they were not interested in Jewish genealogy, but rather in the examination of pure scientific parameters, namely, "the examination of the rate of evolution of the Y-chromosome and its mechanisms," though soon the researchers quite appropriately called their studies a method of "archeological genetics." This time it appeared that the efforts bore fruit: DNA markers were found that indicated common denominators that were significantly more common among the Y-chromosomes of the Cohanim than those of the Israelites. No less important, these denominators were common in Sephardi as well as in Ashkenazi Cohanim. These findings were immediately published in Israel and elsewhere. The social, political, and also religious meaning of a biological continuity, of "we are all Jews," often mentioned or implied, now attained overt corroboration:

> Researchers in genetics confirmed today something that was a holy scripture in Israel for 3,300 years.
> They examined the Y-chromosome of Jewish Kohanim and found that, indeed, they vary from those of the Jewish people. [...]
> Although, according to tradition, all 14 million Jews in the world are the children of Abraham, the molecular biologists find difficulties in reconstructing the biblical links. Two distinct Jewish populations, Ashkenazi and Sephardic, of different even though somewhat blurred genetic composition, exist [...]
> [Researchers] found that in certain respects Kohanim in the different communities vary from the rest of their respective ethnic groups and are more similar to each other. The studies confirm that their chromosomes may be calibrated as a genetic "clock" of father-to-son [...] and also supports an ancient religious tradition. The Jewish priesthood appears indeed to have been founded by a single ancestor. (*The Guardian*, January 2, 1997)

Mainly two types of polymorphisms were followed: that of *microsatellites*, a type of repetitive DNA, the number of repeats varying due to (intra-chromosomal) recombination between sequences, which may occur in up to 1/1,000 cell divisions; and *single nucleotide polymorphisms* (SNPs), due to mutations that are orders of magnitude more rare. Somewhat earlier a rather high correlation was found between

---

[6] Haplotype, a contraction of the phrase "haploid genotype," as had been mentioned repeatedly, is a set of closely linked genetic markers present on one chromosome that tend to be inherited together (not easily separated by recombination).

Y-chromosome haplotypes of Ashkenazi and Sephardi Jews that exceeded that between (non-Jewish) Mediterranean groups (Santachiara Benerecetti et al. 1992). The authors claimed that the findings corresponded to the assumption that to the extent that Sephardi and Ashkenazi Jews of common origins vary, they do so rather in morphological markers (secondary, structural discernible characteristics) that probably were of value in the specific circumstances that the different communities were exposed to, rather than in simple molecular characters, such as the Y-chromosome polymorphisms, which most probably are neutral in processes of selection.

Contrary to the data of the main body of Israelites, which indicates a considerable mixing with their non-Jewish neighbors, the variation of the Y-chromosome of the Cohanim appears to be mainly limited to derivatives (by unequal recombination and mutation) of a single prevalent haplotype – the Cohen Modal Haplotype (CMH). It is convincingly prevalent among Sephardi (61 percent) as well as Ashkenazi Cohanim (69 percent), whereas among non-Cohanim Israelites the CMH comprises only some 0.1 percent. This suggests some "gene migration" from Cohanim to Israelites (Thomas et al. 1998). "Given the relative homogeneity of Cohen Y-chromosomes in comparison with those of the Israelites, we can conclude definitively that adoption of the status has not occurred on a very large scale over a long period of time" (Goldstein 2008, p. 32). Making an educated guess as to what the ancestral chromosome was, and then calculating the distance from the current chromosomes to this imagined ancestral one – the few common progenitors date to some 3,000 years ago. These results appear to be a striking confirmation of the oral tradition). However, not all data accorded with these findings. Uzi Ritte of the Department of Genetics at The Hebrew University of Jerusalem did not find an unusual cluster among Cohanim when he used a different method to examine Y-chromosome haplotypes and compared them to that of the Israelites (Ritte et al. 1993).

The tradition of following discrete genetic markers on the one hand, and the development of methods for following a large number of variables at the level of DNA sequences on the other hand, together with the development of sophisticated computational methods for the detection of the interconnections between them, suggested renewed opportunities to examine historical claims, namely, to perform "genetic archeology". Goldstein summarizes:

> Looking at this huge mass of genetic data ... my colleagues and I were astounded to see that it was possible to predict accurately those individuals claiming Jewish ancestry on their genetic composition alone. (Goldstein 2008, p. 117)

These studies of the Cohanim support that present day Ashkenazi and Sephardi Cohanim are more genetically similar to one another than they are to either Israelites or non-Jews. Among the Cohanim we see greatly reduced diversity, and the Cohanim Y-chromosome seems to be a subset of what is seen among Israelites (Goldstein 2008). The apparent achievement of the children of the priest Aaron in maintaining their distinct status over a very long time and across very diverse socio-geographic distances is even more remarkable when juxtaposed with that of the remaining children of the tribe of Levi, the Levites (Falk 2014b).

The patroclinous inheritance of the Y-chromosome – which has long sections that are easy to follow as haplotype entities, with no threat of crossing over with a homologue chromosome, and which is largely selectively neutral – is rather analogous to that of a typical gene. Thus, in the tradition of reductionist biology, following the story of the CMH in various communities, it is very much like following the contribution of a single gene for a complex trait in a pedigree. Suffice it to remember that the inheritance of the other 22 pairs of human chromosomes is independent of the mechanism of Y-chromosome inheritance, thus allowing patterns of inheritance that may defy the strict patroclinous inheritance of the Y-chromosome. Genetic mechanics could easily produce a "Sephardic" genome with an "Ashkenazi" Y-chromosome, and vice versa.[7]

Although no haplotype frequently common to Levites was found, a cluster of haplotypes with a high degree of relatedness was found among the Ashkenazi Levites. The R-M17 (also known as R1a1) Y-chromosome haplogroup that is rare in Israelite Jewish populations (<5 percent) and generally rare or absent in populations in the Near East is prevalent in Belarusians (50 percent) and the Slavic-speaking Sorbs (66 percent). Notwithstanding, it appears that Ashkenazi Levites have received a significant male contribution of that Slavic genome. Data suggest that R-M17 chromosomes appeared among Ashkenazi Levites relatively late, somewhere between the 4th and eleventh century. According to Kevin Brook (email of January 24, 2015), the careful parsing of the branches of the haplotype revealed that the Ashkenazi variety of R1a1 comes from the Asian continental branch, the origins of which are believed to be in ancient Iran rather than in the European branch of the Slavic Belarusians Sorbs.

Considering these findings, it makes sense to assume that the strong vectors of transmission from father to son that acted to preserve the Y-chromosome, were relevant also to Israelites, though less than with respect to the Cohanim and the Levites. Indeed, among Sephardi- as well as Ashkenazi-Israelites, the similarity of Y-chromosome markers is consistent with Middle Eastern paternal origins (Behar et al. 2003; Thomas et al. 1998).

Is it sensible to draw conclusions with respect to the Ashkenazi ethnic group from the clusters of haplotypes of the Ashkenazi Levites? The data indicate that the Levite status was not kept as zealously as that of the Cohanim, and that a male or a small number of "male outsiders" joined the Levites in spite of their status. What, however, is the origin of that founder(s) of many Ashkenazi Levites? One possibility is that there was merely an intra-Levite process, of an ancient Levite sequence, the representation of which was distorted as the result of a 'bottleneck' effect: The population, including the Levites, drastically declined, and it so happened that

---

[7] My Ashkenazi daughter is married to a Sephardic Cohen. Their son, assuming that he carries on his Y- chromosome a CMH derivative, continues the patroclinous Cohanim tradition, even if he marries an Ashkenazi girl. Doing so, he would further blur the "Ashkenazi-Sephardic barrier," if such a barrier existed, and a "Sephardic" Y-chromosome will be transmitted in the future among Sephardic and Ashkenazi Cohanim, even if the biblical tradition of Aaron, the common progenitor of all Cohanim, were only a legend.

roughly half the Ashkenazi Levites of today are the progeny of a single Levite and his Y-chromosome (which may or may not contain any of his other autosomal genes). Behar and his associates do not think so, although they agree that Ashkenazi Jews passed through 'bottlenecks' due to pogroms, epidemics, immigrations, etc., followed by comprehensive population expansions.[8] They point out, however, that the Levite cluster of the R-M17 haplotype is very common in non-Jewish populations of North Eastern Europe. It is reasonable to assume that the origin of the Jewish haplotypes is in non-Jewish Europeans, some of whose male progeny acquired the name (and status) of Levites. Thus, although there is no wide-spread enthusiasm for the assumption that a Khazari or some other foreign European was involved in the origin of the Ashkenazi, even Behar and associates cannot exclude a significant input of one or more foreign founders to present-day Ashkenazi Levites (Behar et al. 2003, p. 777).

Behar and his associates accept with satisfaction the fact that the research of the biological foundation of the Jews and their interrelationships became an instrument for clarifying Jewish distinctiveness in the process of the ingathering of exiles:

> The comparative study of patterns of NRY [Non-recombining Region of the Y-chromosome] variation among Ashkenazi Jews and other populations has revealed evidence for an unexpected and unusual historical event, which was not appreciated using other, more conventional historical approaches. This finding may motivate historians and social scientists to seek further information regarding the possibility of such an event and, more generally, to include information gleaned from studies of DNA variation in the repertoire of tools used to uncover historical events of interest. (Behar et al. 2003, p. 778)

David Goldstein is more daring:

> Could Khazaria, I wonder to this day, be the source of Ashkenazi Levite R-M17 Y chromosome?
> As with much else of genetic history, there is no way to be sure. [...]
> I was initially quite dismissive of Koestler's identification of the Khazars as the "thirteenth tribe" and the origin of the Ashkenazi Jewry. Was this not just another self-aggrandizing Lost Tribe narrative bereft of evidence?
> I am no longer so sure. The Khazari connection seems no more far-fetched than the spectacular continuity of the Cohen line or the apparent presence of Jewish genetic signatures in a South African Bantu people. [...] I cannot claim the evidence proves a Khazari connection. But it does raise the possibility, and I confess that, although I cannot prove it yet, the idea does now seem to me plausible, if not likely. (Goldstein 2008, pp. 73–74)

Modern advance in molecular methods that allow large scale whole genome screening, say of single nuclear polymorphisms (SNPs), and access to the local populations of Southern Russia and the Caucasus and Caspian Sea states, stimulated new search for the imprints of Khazari history.

Whole-genome DNA sequencing of modern Caucasus populations allowed Eran Elhaik to revisit the 'Khazarian Hypothesis' that suggests that Eastern European Jews descended from the Khazars, and to compare it with the 'Rhineland Hypothesis'

---

[8] According to Behar et al. (2004a), the differences in mutation rates and elimination rates by random drift of SNPs and of microsatellites may explain many of the apparently conflicting findings concerning the relationship among *eidoth* of common origin.

that depicts Eastern European Jews as a "population isolate," which arrived in Eastern Europe roughly at the thirteenth and fifteenth centuries and emerged from a small group of German Jews who migrated eastward and expanded rapidly (Elhaik 2013).

Elhaik's complete data set contained 1,287 unrelated individuals of 8 Jewish and 74 non-Jewish populations, and genotyped over 531,315 autosomal single nucleotide polymorphisms (SNPs). The author and his colleagues applied a wide range of population genetic analyses to compare the two hypotheses. According to them a sole Judean ancestry cannot account for the vast population of Eastern European Jews in the beginning of the twentieth century without the major contribution of Judaized Khazars.

These findings portray the European Jewish genome as a mosaic of Caucasus, European, and Semitic ancestries, thereby consolidating previous contradictory reports of Jewish ancestry.

> We conclude that the genome of European Jews is a tapestry of ancient populations including Judaized Khazars, Greco-Romans Jews, Mesopotamian Jews, and Judeans and that their population structure was formed in the Caucasus and the banks of the Volga with roots stretching to Canaan and the banks of the Jordan. (Elhaik 2013)

One of the major difficulties in deciding between the two hypotheses is the unknown geographical origin of Yiddish speaking Ashkenazic Jews. In a later paper Elhaik and coleagues analysed the genomes of Ashkenazic, Iranian, and mountain Jews and non-Jews. They demonstrated that Greeks, Romans, Iranians, and Turks exhibit the highest genetic similarity with Ashkenazic Jews and localized most of them along major primeval trade routes in northeastern Turkey adjacent to primeval villages with names that may be derived from "Ashkenaz." Their results suggest that Ashkenazi Jews originated from a Slavo-Iranian confederation, which the Jews call "Ashkenazic" (i.e., "Scythian"). This is compatible with linguistic evidence suggesting that Yiddish is a Slavic language created by Irano-Turko-Slavic Jewish merchants along the Silk Roads as a cryptic trade language. Later, in the ninth century, Yiddish adopted a new vocabulary that consists of a majority of newly coined Germanoid and Hebroid elements that replaced most of the original Eastern Slavic and Sorbian vocabularies (Das et al. 2016).

9.3 Towards Genome-Wide Association Studies

As noted, the Y-chromosome, one of the smallest in the human genome, is normally transferred from father to all his sons, but not to his daughters. Most of the DNA on this chromosome has no direct function in development: there are practically no genes, defined as units of function, on the Y-chromosome, except the specific *SRY* sequence, which is crucial for male-development.

Individuals with only one X-chromosome and no Y-chromosome develop as sterile females, exhibiting the Turner syndrome. Individuals with two X-chromosomes and an additional Y-chromosome develop as sterile males, showing the Klinefelter syndrome. Following a combination of DNA sequences on the chromosome that are transmitted together as closely linked genetic markers – haplotypes of a couple of thousand DNA base-pairs – one may detect in them rare unique mutations (SNPs), the frequency of which, when neutral, is reasonably *proportional* to their age. On the other hand, collections of short tandem repeats (STR or microsatellites) of the haplotypes undergo relatively frequent recombination events that increase or decrease the number of repeats, rather *independently* of the age of the sequences. Thus, relations between SNP mutational variability and tandem repeat recombination of Y-chromosome haplotype variability seemed to offer an effective tool for a genome-wide retracing of the history of populations and for reconstructing interrelationships between populations.

Ariella Oppenheim and her associates examined variability at specific SNP binary sites and in microsatellites in Y-chromosome haplotypes in Israel, in Palestinian Moslem Arabs and in Jews of various ethnic groups (Nebel et al. 2000, 2001), and concluded that "the Arabs are more closely related to Jews than to Welsh" (Nebel et al. 2000, p. 636). In spite of the large extent of Y-chromosome and haplotype sharing, the Arabs' haplotypes also reflect a certain degree of drift and/or founder effect due to regional isolation. Still, Arab and Jewish haplotypes revealed a common pool for a large portion of Y-chromosomes, although the two Arab modal haplotypes are found at only very low frequency among Jews, and haplotypes common among the Jews differ from those common among Arabs.[9] Ashkenazi differences, they suggest, may be a result of local intermarriages. The data further suggest that in Sephardic communities, genetic drift acted, and their haplotype pattern was further determined by dilution with non-Jewish populations (Nebel et al. 2000). The data also suggested that Jews of the Orient are more closely related to people of the Northern Fertile Crescent (Kurds, Turk, Armenians) than to their Arab neighbors, whereas Palestinian Arabs and the Bedouins carry a "high frequency of the Eu10 haplotype, not found in the non-Arab groups." As the authors make the point, "These chromosomes might have been introduced through migrations from the Arabian Peninsula during the last two millennia." By placing Sephardi Jews in relation to the Ashkenazi Jews, on the one hand, and in relation to the Arabs, on the other hand, but at the same time discerning between the northeastern peoples and those imported from Arabia, they believe their work "contributed to clarifying the complex demographic history that produced the present genetic landscape of the region" (Nebel et al. 2001, p., 1095).

Michael Hammer and colleagues also built on the genetic variability of the Y-chromosome haplotypes, frontally engaging the racial issue: "Given the complex history of migration, can Jews be traced to a single Middle Eastern ancestry, or are present-day Jewish communities more closely related to non-Jewish populations from the same geographic area?" (Hammer et al. 2000, p. 6769). Examining the

---

[9] In this context, it is important to consider the short paper by Baum et al. (2005), on the meaning of the Darwinian metaphor with respect to "phylogenetic trees" and the proper way to read phylogenetic schemes.

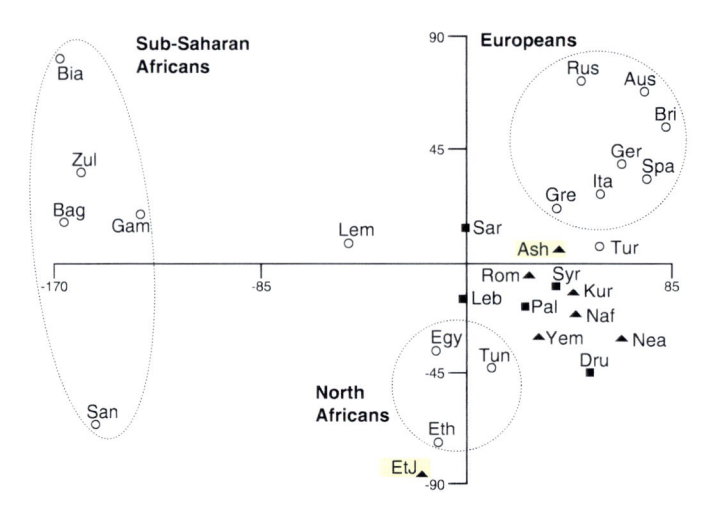

Fig. 9.1 Multivariate analysis of genetic variants of various populations based on Y-chromosome hapolotype data (after Hammer et al. 2000). *Solid triangles* represent Jewish populations; *solid squares* represent Middle Eastern populations

variability of sequences of the Y-chromosome of Jews from seven populations (Ashkenazi, Roman, North African, Kurdish, Near Eastern, Yemenite, and Ethiopian), and comparing these to non-Jewish populations from similar geographic locations, they endeavored to address the question of "whether modern Jewish Y-chromosome diversity derives mainly from a common Middle Eastern source population or from admixture with neighboring non-Jewish populations" (Hammer et al. 2000). Their conclusions are that in spite of the long period of time that they lived in various countries and in relative isolation from one another, most Jewish populations do not vary significantly from one another at the genetic level. Furthermore, in a two-dimensional projection of a multidimensional scaling plot (MDS, see Cavalli-Sforza 2000, pp. 86–91, and Cavalli-Sforza and Feldman 2003), six of the seven Jewish populations form a relatively tight cluster, interspersed with Middle Eastern non-Jewish populations, including Syrians and Palestinians (Fig. 9.1). The Ethiopian Jews are at the margins of the Near Eastern cluster. The Jewish cluster is clearly distinct and separate from the "European" cluster as well as from the "Africans South of the Sahara" cluster (Remarkably, the Lemba tribe is located halfway between the Middle Eastern cluster and the African cluster). While the interrelations between Jewish communities, including those between *eidoth*, such as Sephardi and Ashkenazi, may be considered indications of a constant trickle of intercommunity gene flow, the conspicuous overlap of the Jewish cluster with the Middle Eastern cluster strongly suggests common phylogenetic roots rather than merely culture-dependent horizontal connections.

Harry Ostrer is more explicit in his conclusions, which are based on the biblical story and histories: The Jewish people originated in the Bronze Age. Being "migratory people," they established communities throughout the Middle East and the

Mediterranean Basin. Some of these communities retained their continuity over long periods through language, religion, customs, and marital contacts. According to him, "entry into the community was possible through religious conversion, but was probably a rare event" (Ostrer 2001). He insists that studies on paternal inheritance of the Y-chromosome, as well as those of the maternal inherited mitochondrial DNA, anchor the origins of the Jews in the Middle East. Ostrer concluded by stating: "Jewishness is not determined by genetics. Nonetheless, genetic threads run through Jewish populations that provide them with a group identity" (Ostrer 2001, p. 897). In a later paper, he engaged in a whole-genome sequence analysis to prove this.[10]

Taking the finding that the Jewish populations of today and Middle East Arabs have in common 13 Y-chromosome haplotypes, Ostrer concluded that the source of the original Jews is not one but rather several local peoples, and that there was also considerable gene flow during the later history of the Jewish people. The Sephardi and Ashkenazi Cohanim Modal Haplotype (CMH), estimated to be 2000–3000 years old, is found also in 8.8 percent of the males of the Buba clan of the Lemba tribe; Ethiopian Jews share Y-chromosome haplotypes with non-Jewish Ethiopians, and Cochin Jews share serum-markers with non-Jewish India persons. "These findings indicate that these groups might have had significant admixture, or that the presence of Jewish groups in these regions results from the religious conversion of local people" (Ostrer 2001, p. 893). Ostrer denies similar mixing in other communities, such as those of European Jews.

Considering that at least forty diseases with a "Mendelian pattern" of inheritance had been identified in different Jewish communities, and that the molecular state of most of them had been clarified, Ostrer suggested how to classify their relationship. For those cases in which one or two prevalent founding mutations were detected, one could discern several possible fates for the distribution of mutations, *assuming* dispersive vertical phylogenesis (rather than horizontal admixture). One pattern of mutation distribution was shared by Jews and non-Jews alike in the ancient world and in the Mediterranean Basin even before Jews became organized as a nation. This concerns, for example, the mutation delT167 in the *GJB2* gene, involved in hearing loss; it is common among Ashkenazi Jews and is found also among the Jews of Palestine, Italy, especially Rome, Spain, and Greece, who have a similar genetic background.

A second pattern of mutations stems from Palestine before the Exile. These are shared (mainly) by diverse Jewish communities. These include the mutation affecting the gene for Factor XI for blood clotting found in both Iraqi and Ashkenazi Jews. Ostrer claims that the mutation in the gene for breast cancer susceptibility *BRCA1*, known as delAG18 (*BRCA1*:c.68_69delAG), which is prevalent in Ashkenazi, Iraqi, and Moroccan Jews, also stems from an event that occurred before the scattering of the Jews to different Diasporas (he explains its absence from some communities as the consequence of random drift at the time of the establishment of these communities, although emigration from one community to the other cannot be

---

[10] See further on, Atzmon et al. (2010).

excluded). Furthermore, the new dimensions obtained with the introduction of DNA-sequence-analyses allowed Hamel and coauthors (Hamel et al. 2011), who genotyped DNA samples from BRCA-carrier families, to state that "all mutation carriers share a common haplotype from a single founder individual." They "estimated that the mutation arose some 1800 years ago in either Scandinavia or what is now northern Russia and subsequently spread to the various populations, […] including the AJ [Ashkenazi Jewish] population." Hamel and coworkers further estimated "that c.5266dupC likely entered the AJ gene pool in Poland approximately 400–500 years ago […] from a single common ancestor and was a common European mutation long before becoming an AJ founder mutation." As for claims of BRCA1 being a "Jewish mutation," they claim that BRCA mutations too spread in a trellis-like pattern, rather than that of a closed diverging tree-like vertical pattern.

The third group comprises mutations involved in diseases that are confined to specific communities that were apparently founded later on in the Diaspora: Familial Dysautonomy is found exclusively among Ashkenazi Jews. The establishment of the mutation involving the *OCMD* (*oculopharyngeal muscular dystrophy*) gene in Bukhara Jews dates back to approximately 1243, the year Iraqi Jews began to migrate to Samarkand. Likewise, the mutation involved in the Kreuzfeld-Jacob disease in Libyan Jews is estimated to coincide with the 1492 expulsion from Spain. This mutation was observed in non-Jewish patients of similar backgrounds in Spain and in Chile, and it was probably acquired from Jews by way of intermarriage or forced conversion to Christianity (Marranos). Exciting as Ostrer's classification may be, it clearly adopts speculative assumptions that at once suggest political consequences.[11]

Others, like Amar and colleagues in Israel and overseas, deal with genes involved in tissue transplantation rejection (the HLA histocompatibility complex), whose frequencies may obviously depend on their selective impact. These authors are less cautious about making politically loaded insinuations. They opened their discussion by declaring: "The genetic makeup of today's Jewish populations is the product of the common ancestral gene pool and the introduction from people among whom, over the ages, the Jews lived" (Amar, et al. 1999, p. 726). The authors seem completely unaware of the possible political repercussions of such a naïve declaration. Furthermore, they claimed to confirm the axiom according to which "the Jews share common features, a fact that points to a common ancestry," even though "a certain degree of admixture with their pre-immigration neighbors exists despite the cultural and religious constraints against intermarriages" (Amar et al. 1999, p. 723).

As we shall see, Amar et al.'s paper was challenged for its presumed explicit political statements. It is important to keep in mind that these studies all assume that the present polymorphisms reflect repeated events of a tree-like branching from common origins of human populations, which occurred at different and successive

---

[11] In the December 2011 issue of *Discover*, Jeff Wheelwright tells the story of Shonnie Medina, a Hispanic young woman in Colorado found to be affected by "The Lethal Gene that Emerged in Ancient Palestine and Spread Around the Globe," the BRCA1 gene *185delAG*, which causes aggressive breast cancer. "Its discovery in the Hispano community confirmed events of half a millennium before in Spain that are echoing still. Most likely the mutation arrived by way of Sephardic Jews who converted to Catholicism under pressure from the Spanish Inquisition. From Spain they traveled to the New World."

occasions. Allan Templeton reminds us that such trees are nowadays accessible in computer programs, *designed* to provide the best possible vertical trees that genetic data offer (Templeton 1998). But Templeton presents the alternative trellis model, according to which there was a constant horizontal flow of genes in human populations intertwined with that of vertical evolution. Other investigators also support the pattern according to which human populations are entangled more like a woven cloth than an ordered mosaic pattern. Reducing a situation to the involvement of a finite number of discrete non-interacting factors is certainly more practical to pursue than one of a multifactor interacting system. As emphasized by Lisa Gannett, "DNA forensics research also demonstrates ways in which race is reified by scientists by the representation of what is cultural or social as natural or biological, and what is dynamic, relative, and continuous as static, absolute, and discrete" (Gannett 2004). Theories are by nature underdetermined, and it would be impossible to exclude one or the other also in the future.

9.4 DNA Sequence Analyses

The dramatic developments of whole genome DNA sequence analyses were in a way only qualitative changes at the level of ancestor analyses: it appeared that many assumptions and ad-hoc interpretations of the past could be laid aside once the evidence was straight at the genomic level. This was, of course, exaggerated. As Ellen Levy-Coffman (2005) stressed, the word "Jew" has a mosaic of meaning: culture, religion, ethnicity, and a way of life. DNA evidence did not indicate more accurately than earlier markers whether Jews were simply a people who came into being in Europe during the Diaspora years, or whether the DNA of Ashkenazi Jews reflects ancestry from ancient Central Asia and Russia tribal people.

Arguably the most dramatic change in genetic analysis was not the move to the "DNA phenotype" but rather the move from the classic bottom-up reductionist genetic analysis to top down genome-wide system analysis. Mendel's reductionist *Faktoren* analysis was a tremendously effective approach in the life sciences, just as reductionism had been earlier for physics and chemistry. Yet life and living organisms are in essence complex *interactive systems*. It is only in recent decades that conceptual and methodological tools were developed to analyze the "unsimple truths" of compound systems, essentially by using modern-day reductive tools (see e.g., Mitchell 2003, 2009). One of the most powerful tools of such modern analysis was the introduction of Genome-Wide Association Studies (GWAS): scanning markers across the whole set of DNA or genome of many individuals, together with the development of sophisticated computational methods for the detection of the interconnections between them, provided a genome-wide perspective of the variation associated with a particular phenotype, such as a disease.[12]

---

[12] Actually, the involvement of many unspecified genetic factors in gene-related traits was always in the back of the minds of experimental Mendelian geneticists. In reductive Mendelian genetics, these were usually dismissed as "modifiers." But see Rieger et al. (1991), pp. 332–3.

Fig. 9.2 Model of the reconstruction of the Ashkenazi Jewish (AJ) and European (FL) demographic history. The *wide arrow* represents an admixture event (Carmi et al. 2014)

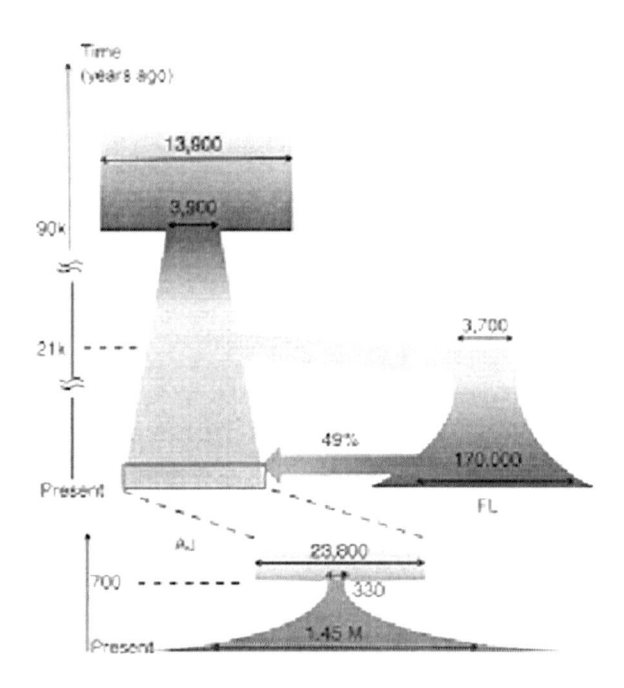

Mitochondria are extra nuclear organelles known as the powerhouse of eukaryotic cells, which (in mammals) carry a 15,000–17,000 base-pairs circular DNA molecule and less than forty genes. Recent studies of the distribution of mitochondrial DNA (mtDNA), which, like Judaism, is passed along the maternal lineage, indicate that Ashkenazi mtDNA is highly distinctive. Some 40 percent of Ashkenazi mtDNA variation has ancestry in prehistoric Europe, rather than the Near East or Caucasus. Thus, the great majority of Ashkenazi maternal lineages are assimilated within Europe (Costa et al. 2013). These results point to a significant role for horizontal phylogeneses due to the conversion of women in the formation of Ashkenazi communities in prehistoric Europe. Costa et al. concluded:

> Whereas on the male side there may have been a significant Near Eastern (and possibly east European/Caucasian) components in Ashkenazi ancestry, the maternal lineage mainly trace back to prehistoric Western Europe. [...] Overall, it seems that at least 80% of Ashkenazi maternal ancestry is due to assimilation of mtDNA indigenous to Europe, most likely through conversion. (Costa et al. 2013, pp. 2 & 8)

Advances in laboratory techniques together with sophisticated computational analyses allowed high-depth sequencing of 128 complete genomes of Ashkenazi Jews (AJ), compared with European (FL) samples of nuclear SNP arrays. Researchers now reconstructed a two dimensional picture of the "demographic history" of the AJ (Fig. 9.2) (Carmi et al. 2014). By applying the most advanced computation methods, Shai Carmi and colleagues integrated the vertical generation changes, as well as the impact of horizontal factors on the evolution of genomes, such as vectors of population-size bottlenecks and periods of intensive transpopulation admixture. Ashkenazi Jews appear to compose a distinct yet quite integral branch of European genomic tapestry.

Thus, modern detailed DNA sequencing supported previous patterns of Jewish population dynamics, but allowed more space for horizontal evolution than the strict historians' extension of vertical evolution.

9.5 Politics *versus* Science

So far the accepted convention of research, which endeavors to use data that is objective and confines opinions to the discussion section, has largely been maintained. No serious scientist will consider statements such as that of Hafez Assad, the former president of Syria, that the Jews are progeny of the Khazars and, thus, not the successors of the residents of the Land of Israel of biblical times. Also, no honest reviewer will consider other than sheer phanatic racism a declaration such as that of an Israeli lawyer that "leftism is not an ideology, but rather a genetic mental illness" (Lawyer Aharon Pappo in *Makor Rishon* [in Hebrew], September 6, 1998, quoted in *Haaretz Supplementary*, November 11, 1998). But once Israel's Prime Minister's Chief Science Advisor claimed that the Arabs in Gaza are foreign immigrants in this country since an allele of a gene originating in northern Syria is prevalent in that population, this becomes the application of scientific arguments in an explicitly political context.[13] No scientist involved in genetic research publicly responded to such claims concerning the relationship of the Palestinian Arabs to the Jews and their ethnic groups. Politics, however, could not be kept out of science.

In 2001 a Spanish-Palestinian research team published a study on "The Origin of Palestinians and their Genetic Relatedness with other Mediterranean Populations"

---

[13] See Joel Marcus, "Four Comments on Folders in the Dark," *Haaretz*, December 6, 1996, p., 1B: "A professional geneticist discovered that the Palestinians are actually Syrians." Also, Uzi Benziman, "A Letter to the Prime-Minister," *Haaretz*, December 20, 1996, p. 3B: "The new Prime Minister's Chief Science Advisor spoke in a language reminding the world of concepts of a pure race theory." The Forum of the Deans of the Medical Schools responded with a declaration saying that "it is forbidden that somewhere on the surface of the Earth, certainly not in the Jewish State, the contents of the genetic materials would provide an excuse for any kind of discrimination of any kind whatsoever and a base for political discussion." On that occasion, the Deans' Forum described eugenics as an issue that only scientists at the fringe of the scientific system dealt with. As has been repeatedly stressed in this book, eugenics served as an important branch of scientific thought. Scientists at the forefront of research and theory were leading eugenicists (see, e.g., Paul 1984). In a telephone call to Prof. Chanakuglu (October 31, 2005), who was Prime Minister Benjamin Netanyahu's Advisor, he told me that he never said what he was allegedly quoted as saying. All he did was refer to the study of Ariel Rösler (1992), who located a mutation in the gene for HSD (17-β-hydroxysteroid) in a Palestinian family in Gaza, which could be followed for eight generations to its origins in Syria.

(Arnaiz-Villena et al. 2001). Their records corresponded largely to those of similar studies performed by or in association with Israeli researchers on Arabs and Jews, and they concluded that "Palestinians are genetically very close to the Jews and other Middle Eastern populations [...]. Thus, Palestinian-Jewish rivalry is based on cultural and religious, but not on genetic, differences." Although the authors dedicate the paper "to all Palestinians and Jews who are suffering war," they emphasize – contrary to their Israeli colleagues – the contribution of the Canaanites and Philistines to the gene pool of the ancient inhabitants of the country.

The paper, however, soon became overtly political. The authors did show empathy with the Jews, "who had been several times led to Diaspora, expelled, deported, and massacred by ancient Iranians and Romans, most Western European countries, and finally Hitler," but they could not refrain from statements such as, "Israel self-proclaimed independence in 1948 and started a war against Muslim Palestinians and other Muslim neighboring countries. After several regional wars, Israel has taken more space and seized up Jerusalem" (Arnaiz-Villena et al. 2001, p. 891). Furthermore, the authors point out that the Palestinians in the Gaza-Strip "have to live mixed with Jewish colonialists," whereas the Palestinian refugees "live either in concentration camps or are scattered" in the Arab Diaspora (Arnaiz-Villena et al. 2001, p. 892).

Concise histories of the people involved in the studies are necessary factors of many of the works discussed. They obviously present the authors' perspective of history. However, such explicit "political writing" in a scientific paper resulted in pressure on the editor of the journal that published it. The paper was taken off the internet edition and regular readers were advised to extricate the pages of the paper from the printed edition.[14] Twisted and convoluted as this incident may appear to readers of the professional literature, it exposed much of the biological essence of the Palestinian-Zionist conflict that many of the participants succeeded in veiling under the cover of the presumed image of scientific objectivity. Dr. Mazin Qumsiyeh, a Palestinian-American scientist, explicitly responded to this issue and the paper's findings:

> The data provided by the paper is ironically consistent with data published in the same journal by Israeli scientists [...]. Amar et al. showed that "Israeli Arabs" (Palestinians who are Israeli citizens) are closer to Sephardic Jews than either is to Ashkenazi Jews. [...] Yet, Amar et al. incredibly concluded that "We have shown that Jews share common features, a fact that points to a common ancestry." [...] Many worked feverishly to establish links (however tenuous) between Ashkenazi Jews and the ancient Israelites [...]. But Ashkenazim are also clearly closer to Turkic/Slavic than to either Sephardim or Arab populations. (*The Ambassadors* 5(1), January 2002. http://ambassadorsd.net/archives)

According to Qumsiyeh, there is no basis for the claim that Jews and their ethnic groups are the progeny of the ancient Jews: Whereas the Middle Eastern Jews are

---

[14] See Klarreich (2001). The Jewish British geneticist, Walter Bodmer, did not understand what the fuss was about: "If the journal did not like the paper, they shouldn't have published it in the first place. Why wait until it has appeared before acting like this?" The paper appeared in an issue of the journal, for which the first author, Arnaiz-Vellena, was guest editor. Other researchers went so far as to argue that the paper was "scientifically worthless" (Risch et al. 2002).

the brethren of the Arabs, the Ashkenazi are the progeny of the Khazars, lacking any blood relationship to the ancient inhabitants of the country. Considering the extreme, often irrational responses of some Israeli researchers to claims such as Khazari contributions to the Ashkenazim, it is obvious that many researchers still have difficulty observing the borders between politics and science.

The Israeli journalist, Boaz Evron, who waited in vain for some clarification from the scientific community to claims of the genetic relationship between Jews and Palestinians, remarked:

> This loud silence is not surprising. It indicates that in the framework of the historic conceptions common in Israel there is no knowledge of how to "digest" such data.
>
> [...] Instead of searching the "remnants of the Ten Lost Tribes" in Central America, India or New Zealand, here we have got scientific evidence that the Palestinians, the inhabitants of the land, are bone of our bones, pure bred Jews [...] all that is left to do is to make them aware of it and regain them to the religion of their forefathers [...]
>
> But obviously this will not happen, because the real stake in the country has nothing to do with the Jewish origins. (Evron, "Our Jewish Palestinian Brothers," *Haaretz*, June, 18, 2000, p. 2B. TRF)

Evron reminds us that Ben Gurion and Ben-Zvi (who eventually became the first Prime Minister and the second President of Israel, respectively) had claimed that "it makes sense to assume that the majority of the Arabs in the country are actually the progeny of the inhabitants of the country who converted to Christianity when it became part of the Roman empire, and further became Moslem when the country was taken over by Arab conquerors" (Evron *vide infra*; Ben Gurion 1917).[15] Archeologist Magen Broshi noted that during the Frankian Crusader period (1099–1296), there remained in the country only about a dozen rural Jewish communities that did not go into exile (of which only one family remained in the twentieth century in Pekiin). "We do not know what became of them, but it may be guessed that they too were converted to Moslems and thus disappeared" (Broshi 2004, p. 31). Even today in several Palestinian settlements, what appear to be obvious Jewish traditions are maintained: in Yata, Hanukah-like chandeliers are put in the windows; in other places, Stars of David are found on tomb stones (sometimes later overlaid with a five-corner star).

9.6 Common Origins or Common Network?

> To claim someone has 'Viking ancestors' is no better than astrology. http://www.theguardian.com/science/blog/2013/feb/25/viking-ancestors-astrology?INTCMP=SRCH
>
> Phylogeographers compare the genealogical relationships among genetic lineages with their geographical source, to try to work out when lineages moved from one place to

---

[15] See also Chapter 7 and Ornan (1969). The subject does not disappear from the agenda. In *Haaretz Supplementary* of March 31, 2006, pp. 52–58, a long story by Aviva Lori tells of a group of Israelis who examine a theory of the origin of most Palestinians as progeny of "Jewish farmers who stuck to their land and converted to Islam."

another. http://www.theguardian.com/science/blog/2013/apr/08/unfair-genetic-ancestry-testing-astrology

Even if Chaim Sheba's idea to use the distribution of markers such as "hereditary diseases" of human communities as a biological marker for the common origins in ancient times and their fate in various Diasporas did not turn out to be a success at the molecular level (see Ostrer 2001 and the discussion above), it was a productive and fertile concept. Jews do have traditional links, whether religious or national, which shaped and dictated their lifestyles in various communities, including systems of marital links, which did affect the distribution of hereditary factors. But may the Jews of today claim a common biological root, distinct from that of the people among whom they lived, as claimed by the anti-Semites at the end of the nineteenth century, or the Zionists at the beginning of the twentieth century? Is it possible that the blood relations between contemporary Jews are the secondary consequences of the living conditions imposed on them and that they imposed on themselves throughout their Diasporas? In other words, is it possible that communities that were isolated from the people in their immediate vicinity, but linked to others (far away) by their culture, might have *induced* a fabric of blood links that we may discover today? Even though there is no doubt that fundamentalists and nationalists of a different ilk will insist on their opinions whatever the evidence shows as to the biological relationships between Jews, the challenge for students of the dynamics of human populations remains great, and more research may provide important insights regarding the input of various factors.

As research progresses, it demonstrates that relationships between members of Jewish communities and their connections with the non-Jews among whom they lived were always a complex two-way (horizontal) exchange, rather than an ordered sequence of (vertically) splitting branches from a common root. It was rather the social and cultural relationships between Jewish communities and ethnic groups that shaped the gene pool (s) of the Jews of today. Since Jews were a separate sociocultural entity, biological-genetic relationships were established between the isolates, irrespective of possible common biological roots. On the other hand, the findings of the last decades show that when the multidimensional analyses of the variation of hereditary factors are examined, not only do the Jewish communities cluster, but moreover, the cluster of Palestinians and other Middle Eastern populations overlaps that of the Jewish communities, and both are clearly distinct from Europeans and Africans alike, strongly suggesting common phylogenetic roots (see Fig. 9.1).

Modern methods of molecular genetics indicate that the gene pool of Ashkenazi Jewish populations was determined to a large extent during times of extreme decrease in the size of the populations, followed by later expansion (as expected in by the 'bottleneck' model) (Behar et al. 2004b; Nebel et al. 2005). It appears that on such occasions, alleles of Middle Eastern origins were established (Founder Effects) (Rund et al. 2004). And the scientists involved are even willing to consider contributions of the remnants of the "mysterious Khazars" (Nebel, Filon, Faerman, Soodyall, and Oppenheim 2005. See also Goldstein 2008, pp. 71–74). Moreover, there is

9.6 Common Origins or Common Network?

today more readiness to accept the historic narrative with respect to the genetic findings of other populations, such as the Libyan, Yemenite and Ethiopian Jews (Rosenberg et al. 2001).

Finally, in a comprehensive paper involving the distribution of SNPs all over the human genome, Ostrer and his associates (Atzmon et al. 2010) came out with their study's declarative title: "Abraham's Children in the Genome Era: Major Jewish Diaspora Populations Comprise Distinct Genetic Clusters with Shared Middle Eastern Ancestry." After stating that Jews "have maintained continuous genetic, cultural, and religious traditions" since the second millennium B.C.E., the authors suggest that admixture with surrounding populations had a role in the shaping of early world Jewry, but that during the last 2,000 years, such admixture may have been limited by religious law and consequently Judaism evolved from a proselytizing to an inward-looking religion.

> Genome-wide analysis of seven Jewish groups (Iranian, Iraqi, Syrian, Italian, Turkish, Greek, and Ashkenazi) and comparison with non-Jewish groups demonstrated distinctive Jewish population clusters, each with shared Middle Eastern ancestry, proximity to contemporary Middle Eastern populations, and variable degrees of European and North African admixture. (Atzmon et al. 2010, p. 850)[16]

The authors discern three "major Diaspora groups – Ashkenazi, Sephardic, and Mizrahi," that may be reduced to two major groups or clusters, Middle Eastern and European/Syrian, who are identified by the principal components of Phylogeny, and Identity-by-Descent (IBD).[17] Both Middle Eastern Jews and European/Syrian Jews formed each their own cluster as part of the larger Jewish group. Each cluster demonstrated Middle Eastern ancestry and a variable mixture with European, non-Jewish populations. However, a major difference among the two clusters was the high degree of European admixture (30–60%) among the Ashkenazi, Sephardi, Italian, and Syrian Jews, and the genetic proximity of these populations to each other relative to their low proximity to Iranian and Iraqi Jews. The calculated time of the split between Middle Eastern Jews and European/Syrian Jews is compatible with a historical divide that took place more than 2,500 years ago. Note that the genetic proximity of the European/Syrian Jews to each other and to non-Semitic southern Europeans is incompatible with theories that Ashkenazi Jews are for the most part the direct lineal descendants of converted Khazars or Slavs. Still, some admixture with local populations, including Khazars and Slavs, may have occurred: Notably, up to 50 percent of Ashkenazi Jewish Y-chromosomal haplotypes are of Middle Eastern origin, whereas the other prevalent haplotypes may be representative of early European admixture, including some Khazari or Slavic origins. The

---

[16] Note that whereas Iranian, Turkish, Syrians, etc. are specified, all Ashkenazi from the Atlantic coast to deep into Russia are pooled into one entity.

[17] A computer algorithm seeks out short, exact *pairwise* matches between individuals, and then extends from these seeds to long, inexact matches that are indicative of identity by descent (IBD). Theoretical analysis suggests that the number of IBD segments of a particular length L resulting from a shared ancestor k generations ago, decreases as a function of L. See Atzmon et al. (2010, p. 852).

major distinguishing feature between Ashkenazi and Middle Eastern Jewish communities was the absence of European Y-chromosome haplotypes in Middle Eastern Jewish populations. In conclusion, Atzmon and coauthors suggest that the data point to a common Jewish population in which local *founder* effects with subsequent genetic *drift* caused the present-day genetic differentiation.

In another study, Doron Behar and twenty associates genotyped individuals from 14 Diaspora communities and compared them with those from 69 Old World non-Jewish populations, as well as with non-Jewish populations from the Middle East and North Africa (Behar et al. 2010). Principal component analysis and structure-like analysis trace the origins of most Jewish Diaspora communities to the Levant: "Most Jewish samples form a remarkably tight sub-cluster that overlies Druze and Cypriot samples but not samples from other Levantine populations or paired Diaspora host populations. In contrast, Ethiopian Jews (Beta Israel) and Indian Jews (Bene Israel and Cochini) cluster with neighbouring autochthonous populations in Ethiopia and western India, respectively" (Behar et al. 2010, p. 238). The authors are aware of the possibility that "membership in several genetic components can imply either a shared genetic ancestry or a recent admixture of sampled individuals," but they conclude in favor of common ancestry over recent admixture due to "the fact that our sample contains individuals that are known not to be admixed in the most recent one or two generations" (Behar et al. 2010, p. 240).

In May 2012 Jon Entine reviewed Harry Ostrer's just published *Legacy: A Genetic History of the Jewish People*, in the *Jewish Daily Forward*: Ostrer had explained in detail the research efforts to expose the genetic relations between various Jewish communities to each other and to non-Jewish communities, and had concluded "that Jews are different," exhibiting "a distinctive genetic signature." However, according to Entine, Ostrer goes further by maintaining that Jews are a homogeneous group with all the scientific trappings of what we used to call a 'race', namely a group with a characteristic physical appearance, even if "[t]he preferred terms in the twenty-first century might be 'continental groups' or 'ethnic groups'" (Ostrer 2012, pp. xvi–xvii). "Few concepts have as tarnished and contentious a history as 'race' [although one] of the problems with using 'race' as an identifier is the lack of a clear definition of race" (Ostrer 2012, pp. 16–17).

Yet, when it comes to the heart of the problem, Ostrer (and his reviewer) makes explicit biological applications of this, at best, socio-cultural term. In the final chapter of Ostrer he reveals *his* predicament of defining and fostering his Jewish identity in America of today. Quoting Einstein, Ostrer specified the "features that foster Jewish identity – nationality (or race or group membership), the culture emanating from group membership, and shared religious belief" (Ostrer 2012, p. 199). Ostrer notes that "In Israel, being Jewish [… is] a nationality – Israeli is a citizenship." And Israel's Declaration of Independence stated "that Israel would be a Jewish, rather than a secular state" (Ostrer 2012, p. 208): "This belief that the Jews constitute a religious, rather than ethnic or racial, group is widespread in the United States and other Western countries.

In the United States, the 1997 […] *Revisions to the Standards for the Classification of Federal Data on Race and Ethnicity* […] does not include *Jewish* as a category in the U.S. Census. Yet the genetic studies would seem to refute this." As Ostrer

9.6 Common Origins or Common Network?

believes, he has shown that "[t]he evidence for biological Jewishness has become incontrovertible" (Ostrer 2012, p. 217). "So Jewishness at a genetic level can be characterized as a tapestry with the threads represented as shared segments of DNA and no single thread required for composition of the tapestry" (Ostrer 2012, p. 218). Nicholas Wade in his 2014 book, *A Troublesome Inheritance: Genes, Race and Human Heredity*, insists that the reintroduction of "race" appears to resolve for some the predicament of a non-religious, non-nationalist Jewish person of today. Yet, Dorothy Roberts, in a wide range discussion of legislative, biomedical and genomic studies in her *Fatal Invention: How Science, Politics, and Big Business Re-Create Race in the Twenty-First Century* (Roberts 2011), takes the unequivocal position that, rather than being a biological construct, race has always been a political and social category (Menon 2016).

Today's Jews, however, are not the only ones who claim phylogenetic relatedness to the ancient inhabitants of the Land of Israel.

> Besides Southern European groups, the closest genetic neighbors to most Jewish populations are the Palestinians, Bedouins, and Druze. [...] their genetic proximity to one another and to European and Syrian Jews suggests a shared genetic history of related Middle Eastern and non-Semitic Mediterranean ancestors who chose different religious and tribal affiliation. These observations are supported by the significant overlap of Y chromosomal haplogroups between Israeli and Palestinian Arabs with Ashkenazi and non-Ashkenazi Jewish populations that had been described previously. [...]
>
> This study demonstrates that the studied Jewish populations represent a series of geographical isolates or clusters with genetic threads that weave them together. [...] Over the past 3000 years, both the flow of genes and the flow of religious and cultural ideas have contributed to Jewishness. (Atzmon et al. 2010, p. 858)

Even though not a race in a biological sense, political Zionism, after a century of attempts to prove contemporary Jews' material, biological relationships – not merely their spiritual, cultural ones – to the ancient people of the biblical stories, in spite of widespread interspersing with local communities, finally has succeeded. It is tragic that Zionism, as well as Arab Nationalism, have failed to recognize the Palestinians, many of whom similarly appear to share phylogenetic relations to the historic inhabitants of the country, as equal partners.

Susan Martha Kahn (2013) notes the very high stakes and the subjective perspectives adopted even by experienced and essentially objective researchers, when confronted with the issue investigating whether there is a biological component to Jewishness. She compared Harry Ostrer's *Legacy* (2012) with that of Nadja Abu El-Haj's *The Genealogical Science* (2012). Both published at the same time and both, "referencing the same sets of data," arrive at entirely different answers to the age-old question: who are the Jews?" (Kahn 2013, p. 919)

> For Ostrer, these data not only confirm traditional narratives of Jewish history [...] but also provide sufficient evidence for establishing a biological basis for Jewishness. For Abu El-Haj, these studies are profoundly problematic [...] because they rely on a style of reasoning in which the notion of a biological basis for Jewishness is reinforced and legitimated through scientific discourse. In short, their first disagreement centers on the underlying hypothesis that *there is a "population" – a race, a people – of "Jews"* that traces its roots to ancient Palestine. Ostrer accepts this hypothesis; Abu El-Haj contests it. (Kahn 2013, p. 920)

According to Kahn, Ostrer's goal is to explain how this new genetic ancestry tracing "reliably confirms oral tradition and biblical stories" and "conclusively proves that contemporary Jews are overwhelmingly the direct descendants of an ancient people who originated in the Levant." Ostrer, not being the first modern Jewish scientist to embrace the notion of Jewish biological uniqueness, further contends that genetic studies "confirm a Jewish biological distinctiveness. [...] Jewishness can be characterized at the genetic level as a tapestry in which the threads are represented as shared segments of DNA" (Kahn 2013, p. 920–921). As for Abu El-Haj – the daughter of an ex-Palestinian Muslim father, "an established critic of the modern Zionist project" – Kahn notes, the whole project of genetic ancestry tracing is methodologically suspect. She identifies with an interdisciplinary group of social scientists who challenge the ways they believe the new genomics are being used to reinscribe race as a meaningful social category. Abu El-Haj concludes that most of the data "can easily be read to support the opposite assumption: that contemporary Jews have no common origins but are a miscellaneous group of people from Europe, the Middle East, and Central Asia who converted to Judaism at various points in the past" (Kahn 2013, pp. 921–922).

As Kahn stressed, "Abu El-Haj speaks to an audience different from Ostrer's – her audience comprises not only like-minded social scientists but also those academics increasingly eager to delegitimize the state of Israel and founding myths of Zionism" (Kahn 2013). Still, Kahn, like Elisabeth Goldschmidt fifty years earlier, believes that "[w]hat unites these interpretations is the sincere effort to understand and explain new genetic evidence derived from contemporary Jewish populations" (Kahn 2013, p. 923), irrespective of political biases. Let me join Susan Martha Kahn in concluding:

> Perhaps any scientific data that suggest a biological component to Jewish identity will be the subject of heated and multivocal debate. New techniques in genetic ancestry tracing may have the potential to create more consensus than discord about the nature of Jewish peoplehood [...]. Only then can we hope to find some kind of shared understanding about "who are the Jews." (Kahn 2013, pp. 923–924)

Chapter 10
Coda: Zionism and the Biology of the Jews Tomorrow

> Israel is a semi-Western country [...] but, it would be difficult to transform Israel into a Western country as long as Zionism as an ethnic ideology dictates the order of life in the country (Smocha 1999, p. 253).
>
> For me Zionism died (Ruth Dayan, the ninety-seven year old divorced widow of Moshe Dayan, on an Israeli TV program, December 24, 2014).

Although Israel is the realization of political-Zionist longing and is considered a modern Western country, its demographic future is notably directed by far-reaching, traditional, conservative policies. Childbirth, which has been encouraged since early on by its leaders and is reinforced by various state regulations,[1] reflects the strong impact of the traditional, religious, even orthodox Jewish lifestyle on the heterodox, humanist-liberal notions of the early Western Zionists. As noted by Susan Martha Kahn, there were many justifications for Israelis to desire large families in addition to honoring the commandment to "be fruitful and multiply" (Genesis I: 28). For Israeli Jews, the imperative to reproduce has deep political and historical roots. Some feel they must have children to counterbalance what they believe to be a demographic threat represented by Palestinian and Arab birthrates. Others believe they must produce soldiers to defend the fledgling state. Some feel pressure to have children in order to 'replace' the six million Jews murdered in the Holocaust. Many Jews simply have traditional notions of family life that are very child-centered (Kahn 2000, p. 3).

As has been mentioned earlier, thinking of the genetic future and planning for it was, toward the end of the nineteenth century, Francis Galton's rationale when he introduced the notion of' eugenics. And eugenics became part of the Zionist settlement effort in the Land of Israel since its establishment (see Chap. 7 and Falk 2010). However, usurped and interpreted by Nazi ideology, eugenics was rejected after

---

[1] In the 1950s, Ben-Gurion gave monetary awards to families with ten or more children. In 1967, the Israeli demographic center was established "to act systematically to realize a demographic policy directed at creating an atmosphere and conditions that encourage a high birth rate, which is so vital to the future of the Jewish people" (see Kahn 2000, p. 4).

WWII, although the notion never disappeared (see Bashford 2010). In the early years of the State of Israel, during the mass immigration, an enormous effort was directed at eliminating diseases that people brought with them from their various lands of origin. From today's perspective, it is clear that this program was often carried out ineptly, without due respect to the needs and sensitivities of individuals. It often reflected elements of the beliefs of the eugenic movement, notably giving precedence to the interests of the community rather than to those of the individuals involved (see Chap. 7).

Some of the diseases of many of the new immigrants have been treated by administering medicines, improving hygienic conditions, and offering supplementary diets for nutritional deficiencies. But the treatment or prevention of 'hereditary diseases' that were common in some communities offered a different challenge: Because the modern medical services provided by the state addressed the factors responsible for some diseases such as malaria or tuberculosis, it was hoped that hereditary factors related to diseases would similarly decrease at a slow but constant rate. However, because there were suggestions about possible selective advantage of heterozygous carriers of some genetic diseases, it was not clear what other forces might shape the frequencies of these genes under the modern Israeli conditions. Moreover, to the extent that there was agreement concerning the *biological* sources of this or that disease, there were still questions regarding how far one may interfere with the *social* aspects of the marriage patterns of the respective *eidah*. Even when the determinist eugenic attitudes of the first half of the twentieth century were no longer in vogue, questions arose about planning for the 'good of the nation.'

How far did the eugenic and the Zionist notions change concerning family planning at the age of the ingathering of exiles? There are indications that eugenic thinking, even that of racial hygiene for the benefit of each individual, is quite prevalent today as it was a century ago (Paul 2002).

The slow breakdown of the boundaries of the communities' or the *eidoth*'s marital traditions and the increasing rate of intermarriages between the communities are undoubtedly important factors in decreasing the number of persons affected by hereditary diseases. The 'dilution' of the frequency of an allele for a recessive disease in the Jewish population of Israel *per se*, decreases the probability of two heterozygotes marrying and producing affected children. On the other hand, some of the persons who under other conditions would have developed a severe disease now might develop only a mild case (or no disease at all) and may yet produce progeny who carry the undesirable allele. In such cases, it is necessary to consider the chance that eventually the disease might appear with a vengeance in future progeny. Such eugenic considerations have been sometimes applied, but rarely discussed.

In the 1960s, it was found that a simple blood test of newborns could detect children who are homozygous for the allele defective in the enzyme for the breakdown of the amino-acid phenylalanine, which results in severe mental retardation known as phenylketonuria. By providing a strict diet for many years to these individuals when young, they were spared most of the symptoms of the disease (see Paul and Brosco 2013). Consequently, since the 1960s, by law, all newborns in Israel are checked for blood phenylalanine levels (heel-stick test). Notwithstanding, the special diet imposes a very heavy burden on the affected individuals and their families (including a huge financial burden). Furthermore, it was found that most individuals born in Israel with abnormal enzyme function were of a variant that was not that of potential phenylketonuria patients. As a matter of fact, the special phenylalanie-deficient diet may sometimes even be harmful to these individuals (Paul 2002). Moreover, female patients who overcome the deficiency by following the strict diet until maturity must return to it when pregnant to prevent the development of this defect in their (heterozygous) embryo. All these early-detection procedures would in the long run increase the frequencies of carriers of the unwanted allele in the population and the numbers of persons in need of the strict phenylalanine-less diet. Thus the notion of the eugenic benefit of checking newborns for increased levels of blood-phenylanaine may be contested as being, like most medical treatments, palliative rather than preventive.

Other humane programs may have better eugenic consequences. In Israel's neighboring Cyprus, where thalassemia is prevalent, a policy was introduced that makes it obligatory for candidates for marriage to be *tested* for the allele, which if inherited from both parents may cause the severe hemolytic disease. Although nothing else is demanded by law, the *awareness* of the population to the risks of the birth of children who are homozygous for the thalassemia allele has increased and the number of thalassemia births (though not the unwanted allele's frequency) has decreased dramatically (Holtzman 1998; Chadwick 1998).

There are more fertility clinics per capita in Israel than in any other country, and Israeli fertility specialists have emerged as global leaders in the research and development of these technologies. Furthermore, all the new reproductive technologies, including artificial insemination, ovum donation, and in-vitro fertilization, are subsidized by the Israeli national health insurance. In, 1996, Israeli legislators passed the Embryo Carrying Agreement Law, making Israel the first country in the world to legalize surrogate mother agreements. This extraordinary state support for reproductive technology must be contrasted, however, with the striking degree to which treatments that limit family size remain unsubsidized (Gilbert 1997).

A more direct eugenic project has been established by the closed ultra-orthodox Ashkenazi (professedly non-Zionist) community for the detection of carriers of genes for hereditary diseases and their prevention. As noted, the "Dor Yesharim" (see Chap. 8 for the Yiddish version: "Dor Yeshorim") society carries out discreet blood tests for young men and women in preparation for marriage. The tests and results are known only to the matchmakers, active in these communities, so that they can regulate and prevent marriages that may produce affected children. To date, more than a dozen tests are performed with respect to different hereditary diseases found in these communities (see Chap. 8 and below; see also Sagi 1998). As directive, coercive, and compromising of personal autonomy the features of this program are, it has proved to be an "efficient and cultural fit in the ultra-orthodox Jewish community" (Raz and Vizner 2008). As noted, this approach is contrary to the Cyprus case, where the appeal is to the individual, who is responsible for his or her family. In the Jewish ultra-orthodox community, however, the unit of reference is the family – not the individual – which comprises the basic socio-economic unit responsible for its future (Yaniv et al. 2004). Raz and Vizner (2008, p., 1367) claim that "[t]he major finding of [their] study is that the actual meaning and practice of carrier matching [...] hinge on misunderstandings regarding the genetic basis of carrier matching. [...] Dor Yeshorim also inadvertently reinforces the message that being a carrier is something which one is better off *not* knowing, and hence that being a carrier is *bad*."

Noteworthily, several births of children with severe 'hereditary diseases' have been prevented in the ultra-orthodox community, yet it is doubtful whether these procedures could be applied outside this community (in which abortions are strictly excluded).[2]

Awareness of hereditary disease and the possibilities of detailed DNA molecular sequencing have brought an increase in pre-birth checks in the public at large and in abortions of embryos suspected of being potentially malformed. This has already been followed by statements in the press, such as "Tay-Sachs disease has been eliminated in the Jewish community."[3] Although, as a rule, the Israeli government generously supports procedures that improve childbirth by welcoming modern fertility technologies, it is noteworthy that the government does not provide any legal or material support for birth control.

This implicit eugenic policy is also exercised in the social and medical support of providing or, in most cases, *not* providing social assistance and welfare services to affected individuals. Gaucher's disease may be viewed as such a challenge, because it is limited nearly exclusively to Ashkenazim and puts a heavy burden on the affected individuals, although it is not lethal and the affected individuals survive to old age, especially since medical treatment (providing the missing enzyme) alleviates the suffering considerably. In 1999, 351 patients in Israel received therapeutic treatment at the annual cost of 21 million dollars (roughly $60,000 per patient), which must be maintained for the rest of their life (Gross 2002). Must a state that has limited resources be compelled to provide such sums to a relatively small group of citizens (who happen to be Ashkenazi) while such funds might be used to provide relief to many more patients whose suffering is more "treacherous"? These ethical issues that have been with us for many years are beyond the scope of this study.[4]

---

[2] See Jeff Wheelwright's story of *Dor Yesharim* in "Cancer's Wandering Gene" (*Discover*, Vol. 32, December, 10, 2011).

[3] "Tay-Sachs disease has been eliminated among Jews – is cystic-fibrosis next in line?" Tamara Trautman, *Haaretz*, January, 18, 2005. Among the respondents to this news, several made the point that not the disease but rather the frequency of affected newborns was suppressed. The prevalence of the factor for the disease may even increase; suppose that a family that might have produced an affected child over-compensates for the aborted child by having more healthy children, some of whom may be heterozygous for the relevant allele, thus, inadvertently increasing the frequency of that allele in the population.

[4] The issue of the costs to the State for treatment of handicapped persons was used extensively by the Nazi proponents of eugenics, or as they called it, *Rassenhygiene*. This further emphasizes that it is not the medical problem that should be suppressed, but the methods taken to solve it, or as expressed by Francis Galton, "eugenics is not immoral but unmoral" (see Chap. 3).

Many Jewish communities experienced the consequences of one or more "founder effects" following intensive inbreeding. In the past, these phenomena emphasized the Jewish "hereditary-racial" element of diseases; in today's age of molecular sequencing, they may facilitate the detection of the carriers of alleles for diseases if the ethnic group (race?) of the persons involved is known. There is tremendous molecular DNA variability in genes or haplotypes that are involved in known diseases, not all of which are relevant to the effect in question. Often it is not possible to predict the clinical impact of a molecular change at the DNA sequence. Regarding the genes involved in CF, in Gaucher, or in BRCA1 and BRCA2, it is difficult and costly to determine whether a change in any of the many nucleotides is relevant to the disease or not. But in Jewish communities that have gone through a bottleneck, the founder effect established a small number of relevant DNA sequences that should be tested if it is known that the person belongs to a Jewish Sephardi or Ashkenazi family. In the Ashkenazi population, it would be enough to check three sites to detect 94 percent of the chromosomes involved in Tay-Sachs; five sites would suffice to detect 95 percent of chromosomes involved in Gaucher; six sites to detect 97 percent of chromosomes involved in CF, and three sites to detect most chromosomes involved in Canavan disease, Niemann-Pick, familial dysautonomia, and in BRCA (Gilbert 1997).

Gabai-Kapara and her associates (Gabai-Kapara et al. 2014) found that in the Ashkenazi Jewish population of Israel, many breast cancer and ovarian cancer carriers are due to three inherited founder mutations in the cancer predisposition genes *BRCA1* or *BRCA2*. High cancer risk in carriers of these mutations identified through healthy males may provide an indication for initiating a general screening program in the Ashkenazi-Jewish population that would identify many carriers who are not evaluated by genetic testing based on family history criteria. Of course, concentrating on a small number of sites increases the risk of missing relevant rare or new mutations.

In other words, even though it is possible that the diverse Jewish communities share some common genetic elements, most Jewish communities are largely indistinguishable, in terms of most biological parameters, from the populations among whom they lived. It is uncertain to what extent this is due to socio-culturally motivated interchanges (including rape and other extramarital sex) or to common phylogenetic origins.

10.1 A Jewish State or a State for the Jews?

The Zionist establishment accepted the basic tenet that Jews had retained their blood ties over generations of life in the Diaspora. Borrowing from the romantic European nationalist movements, political Zionism grounded the Jewish Diaspora socio-cultural, religious concept not on the traditional "Blood and God" – which deferred the return to the homeland indefinitely until the coming of the Messiah – but rather on the European notion of "Blood and Soil." Ironically, Zionists thus accepted an ancient anti-Semitic notion that the 'anomaly' of the Jews is inherent in the dissociation of their "Blood" from their "Soil." [5]

World War II and the Holocaust threw many idealistic notions into disarray. Nazi dogma further tainted the motif of "Blood" with explicit racial connotations. At the same time any attempt to define Jews merely on the basis of socio-cultural criteria was perceived by the guardians of the religious concept of Judaism as an existential threat. After the 1967-war it was the "Soil," the land, rather than the "Blood" that for many became the focus of Israeli existence. Elsewhere, meanwhile, the collapse of colonialism and the shift of nationalism from Europe to the former Colonies meant that Zionism found itself on the wrong side of history. For many Zionism changed from representing the aspirations of a deprived nation to embodying the injustice of a depriving nation. The very fulfillment of the Jewish nationalist dream and the establishment of the State of Israel came to be conceived by many in the twenty-first century as a holdover from Western European colonialism.

***

At the beginning of this book, three key questions were posed: What is special or unique to the Jews? Who were the genuine Jews? And how can one today identify Jews? We may conclude that there is no unique and unifying "biology of Jews." In other words, although the social and cultural relations among Jewish communities (and the very different relations between each of them and its Gentile neighbors) have resulted in the formation of a loosely linked cluster of reproductive isolates, any general biological definition of Jews is meaningless. Despite the persistence of intra-Jewish, socio-cultural relatedness, coupled with the exclusion of Jews by Gentile society, Jewish communities have never been reproductively isolated from their neighbors. Although there is widespread evidence of certain Middle Eastern genetic components in numerous Jewish communities, there is no proof of a typical Jewish prototype. Biology alone cannot provide proof to identify Jews as such: There are no 'Jewish genes,' even though there are plenty of mutations that are pretty much restricted to a certain group of Jews. It follows that there can be no clinching biological answer to the question of identifying the original Jews, nor to any question about the shared heritage of all Jews qua Jews.

Once the State of Israel was established as the fulfillment of the Zionist dream, it no longer required biology, history, or any other justification for its national identity (see, e.g., Dagan 1999). Arguments from biology could now be diverted to domestic matters, namely, to establishing shared links among immigrants from seventy different Diasporas. At the same time, many in the Israeli political, national, and religious establishment invested the power to uphold the biological uniqueness of the Jews in opposition to the Arabs of the region (see, e.g., Yiftachel 1999).

Barring a major anti-Semitic catastrophe that would spur Jews in large numbers to seek refuge in Israel, the country's future demography no longer will be

---

[5] Marxists accuse Jews even today of the sin of accepting anti-Semitism as a fact in a world of nationalism, instead of fighting both. See, e.g., Brenner (1983).

determined by Aliya – immigration of Jews to their homeland. The irony of the situation is that Israel's current long-range demographic policy (or lack thereof) selectively encourages differential increase of three sectors of Israel's population. These most radically oppose each other and vehemently reject the secular social-liberal, national emancipation notions that led to the birth of Zionism over a century ago. These are broadly speaking, the nationalistic, often orthodox circles, who – inspired by the biblical notions of "Blood and Soil" – consider themselves the present-day Zionists; the ultra-orthodox Jewish congregations, who declare that they are immanently anti-Zionist; and the Palestinian population (mostly Moslem, but also Christian and other minorities), who are inherently anti-Zionist. How are these mutually antagonist groups going to shape the future of the State of Israel and the distinctiveness of the Jewish people?

In fact, the only one among democratic nations, the State of Israel is nowadays still debating the definition of its national identity. In pluralistic democracies throughout the Western world, this issue is deemed largely immaterial. To the extent that Jews wish to define themselves as a community they form a distinctive religious and traditional minority, whereas, anti-Semites, in any case, proceed to pin the label "Jew" on whomever they wish and by whatever criteria they adopt. It is only in the State of Israel that the struggle to define the nation by the tripartite criteria of religion, society, and biology continues to occupy the populace at large, as well as the state authorities.[6] Israel's Law of Return which guarantees the right of Aliya to "Jewish" communities and individuals, unfortunately also imprints the issue with explicit racial dimensions. Former Member of Knesset (and former head of the Jewish Agency), Avraham Burg, points out the thinly disguised "faith in the superiority of the genes" that underlies the claims by the country's national-religious establishment to anchor Jewish identity in "an affinity of spirit":

> How much abuse is heaped on the Jewish religion! What arrogance and racism are concealed beneath the words "Jews never expel other Jews" [a slogan used by many in Israel to dissuade the State from any perceived intent to dismantle settlements in the Occupied Territories]; herein lays a belief in the superiority of the genes, in the mastery of the master race, all in the name of God. […] To which I respond: Nothing is innate, nothing is automatic; even God's choice of the People of Israel is not guaranteed without [their] moral commitment, and without [their] constant toiling for self-improvement and for better human behavior. These have all been shunted aside in recent years in favor of an unholy trinity: a racist Judaism, reliant on violent settlement activity and defended by a distorted security doctrine. (Avraham Burg, "Sharon's deception may still turn out to be beneficial" [in Hebrew], *Haaretz*. August 5, 2005, p. 3B. TRF)

Israel is currently, in the words of the author Sami Smocha, an ethnic democracy: "This is a regime which distinguishes definitively between the nation and the civil

---

[6] On an Israeli news broadcast on May 5, 2007, a couple of days after he won the presidential election in France, Nicolas Sárközy was presented as Jewish, like two previous French prime ministers, Leon Bloom and Pierre Mendès-France. At once, Prof. Joshua Ben-Porat, of The Hebrew University of Jerusalem responded: Sárközy is not Jewish; only one of four of his grandparents was Jewish – as the racist Le Pen was happy to recite. Sárközy is not Jewish in his blood, or his culture, or in his identity.

population" (Smocha 1999, p. 256). Smocha proposes that the time has come to recognize that Judaism is not an ethnic-biological essence but a socio-cultural conjuncture; doing so would set Israel on the path to rejoin the Western world as a state in which all citizens enjoy equal rights:

> Over the past fifty years, Israel has taken some steps towards the model of a multi-cultural democracy. But as long as the Zionist project of constructing nation, society, and state according to a single uniform pattern endures, it will be difficult for Israel to develop into a real multi-cultural democracy. (Smocha 1999 p. 257. TRF)

Smocha argues for the emancipation of the Jewish nation from inherited notions of alleged biological unity. Shouldn't genetic research likewise shake itself loose of the effort to anchor Zionism in the supposedly shared biological origins of the Jews? Describing populations as genetic lattices, where social and cultural links are made manifest in hereditary links (among other parameters), would give a more accurate picture of the material interdependence between science and society. Doing so would also allow the expression of the aspiration to pluralism among Israel's diverse populations, a pluralism in which the Jewish nation plays a key role. Shouldn't Israel finally mature as a State for the Jews rather than merely aspire to be a Jewish State?

Bibliography

Abu El-Haj, N. (2012). *The genealogical science: The search for Jewish origins and the politics of epistemology*. Chicago/London: The University of Chicago Press.

Achad Hamorim. (1927). Question and request (in Hebrew). *Hachinuch, 10*, 268.

Adam, A. (1973). Genetic diseases among Jews. In B. Ramot (Ed.), *Genetic polymorphisms and diseases in man* (pp. 1383–1392). New York: Academic.

Adam, A., & Bonné-Tamir, B. (1997). Genetic disorders among Jews from Arab countries. In A. Teebi & T. Farag (Eds.), *Genetic disorders among Arab populations* (pp. 432–466). New York: Oxford University Press.

Ahad Ha'am. (1930). *At the crossroads (selected essays)* (in Hebrew). Berlin: Jüdischer Verlag.

Aksentijevich, I. A., Torosyan, Y., Samuels, J., Centola, M., Pras, E., Chae, J. J., Oddoux, C., Azzaro, G. W., Pia, M., Palumbo, G., Giustolisi, R., Pras, M., Ostrer, H., & Kastner, D. L. (1999). Mutation and haplotype studies of familial Mediterranean fever reveal new ancestral relationship and evidence for a high carrier frequency with reduced penetrance in the Ashkenazi Jewish population. *American Journal of Human Genetics, 46*(4), 949–962.

Almog, S. (1991). 'Judaism as illness': Antisemitic stereotype and self-image. *History of European Ideas, 13*, 793–804.

Almog, S. (1993). Pioneering as an alternative culture (in Hebrew). *Zion, 58*(3), 329–346.

Altmann, A. (1985). Moses Mendelssohn as the Archetypal German Jew. In: J. Reinharz & W. Schatzberg (Eds.), *The Jewish response to German culture: From the enlightenment to the Second World War* (pp. 17–31). Hanover: University Press of New England.

Amar, A., Kwon, O. J., Motro, U., Witt, C. S., Bonné-Tamir, B., Gabison, R., & Brautbar, C. (1999). Molecular analysis of HLA Class II polymorphisms among different ethnic groups in Israel. *Human Immunology, 60*(8), 723–730.

Andreasen, R. O. (2004). The cladistic race concept: A defense. *Biology & Philosophy, 19*(3), 425–442.

Ankori, Z. (1979). Origins and history of Ashkenazi Jewry (8th to 18th century. In R. M. Goodman & A. Motulsky (Eds.), *Genetic diseases among Ashkenazi Jews* (pp. 19–46). New York: Raven Press.

Arnaiz-Villena, A., Elaiwa, N., Silvera, C., Rostom, A., Moscoso, J., Gómez-Casado, E., Allende, L., Varela, P., & Martínez-Laso, J. (2001). The origin of Palestinians and their genetic relatedness with other Mediterranean populations. *Human Immunology, 62*, 889–900.

Atzmon, G., Hao, L., Pe'er, I., Velez, C., Pearlman, A., Palamara, P. F., Morrow, B., Friedman, E., Oddoux, C., Burns, E., & Ostrer, H. (2010). Abraham's children in the genome era: Major Jewish diaspora populations comprise distinct genetic clusters with shared Middle Eastern ancestry. *The American Journal of Human Genetics, 86*, 850–859.

Auerbach, E. (1911). The Jewish outlook in Palestine. In I. Cohen (Ed.), *Zionist work in Palestine* (pp. 172–181). Leipzig/London: T. Fischer Unwin.

Avigad, S., Cohen, B. E., Bauer, S., Schwartz, G., Frydman, M., Woo, S. L. C., Nino, Y., & Shilo, Y. (1990). A single origin of phenylketonuria in Yemenite Jews. *Nature, 344*, 168–170.

Avineri, S. (1986). *Moses Hess – Between socialism and Zionism* (in Hebrew). Tel-Aviv: Am Oved.

Bach, G., Zlotogora, J., & Ziegler, M. (1992). Lysosomal storage disorders among Jews. In B. Bonné-Tamir & A. Adam (Eds.), *Genetic diversity among Jews* (pp. 301–304). New York/Oxford: Oxford University Press.

Barkan, E. (1992). *The retreat of scientific racism*. Cambridge: Cambridge University Press.

Bashford, A. (2010). Epilogue: Where did eugenics go? In A. Bashford & P. Levine (Eds.), *The Oxford handbook of the history of eugenics* (pp. 539–558). New York: Oxford University Press.

Bashford, A., & Levine, P. (Eds.). (2010). *The Oxford handbook of the history of eugenics*. Oxford: Oxford University Press.

Baum, D. A., Smith DeWitt, S., & Donovan, S. S. S. (2005). The tree-thinking challenge. *Science, 310*(5750), 979–980.

Baur, E., Fischer, E., & Lenz, F. (1931). *Human Heredity*. New York: Macmillan.

Behar, D. M., Thomas, M. G., Skorecki, K., Hammer, M. F., Bulygina, E., Rosengarten, D., Jones, A. L., Held, K., Moses, V., Goldstein, D. B., Bradman, N., & Weale, M. E. (2003). Multiple origins of Ashkenazi Levites: Y chromosome evidence for both Near Eastern and European ancestries. *American Journal of Human Genetics, 73*, 768–779.

Behar, D. M., Garrigan, D., Kaplan, M. E., Mobasher, Z., Rosengarten, D., Karafet, T. M., Quintana-Murci, L., Ostrer, H., Skorecki, K., & Hammer, M. F. (2004a). Contrasting patterns of Y chromosome variation in Ashkenazi Jewish and host non-Jewish European populations. *Human Genetics, 114*(4), 354–365.

Behar, D. M., Hammer, M. F., Garrigan, D., Villems, R., Bonné-Tamir, B., Richards, M., Gurwitz, D., Rosengarten, D., Kaplan, M., Della Pergola, S., Quintana-Murci, L., & Skorecki, K. (2004b). MtDNA evidence for a genetic bottleneck in the early history of the Ashkenazi Jewish population. *European Journal of Human Genetics, 12*(5), 355–364.

Behar, D. M., Yunusbayev, B., Metspalu, M., Metspalu, E., Rosset, S., Parik, J., Rootsi, S., Chaubey, G., Kutuev, I., Yudkovsky, G., Khusnutdinova, E. K., Balanovsky, O., Semino, O., Pereira, L., Comas, D., Gurwitz, D., Bonné-Tamir, B., Parfitt, T., Hammer, M. F., Skorecki, K., & Villems, R. (2010). The genome-wide structure of the Jewish people. *Nature, 466*(7303), 238–242.

Bein, A. (1934/1974). *Theodor Herzl*. Wien: Selbstverlag der Östrreichisch-Israelischen Gesellschaft.

Bein, A. (1962). *Theodor Herzl: A Biography* (M. Samuel, Trans.). Cleveland/New York: Meridian Books.

Bein, A. (1971). *Arthur Ruppin: Memoirs, Diaries, Letters*. London: Weidenfeld and Nicolson.

Belkind, I. (1919). *Eretz Israel* (in Hebrew). Tel Aviv: Hameir.

Belkind, I. (1928). *The Arabs in Palestine* (in Hebrew). In U. Ornan, (Ed.), 1969. *Israel Belkind: The Arabs in Palestine* (pp. 43–48). Jerusalem: Hermon.

Ben-Chetrit, E., & Levy, M. (1998). Familial Mediterranean fever. *Lancet, 351*, 659–664.

Ben-Gurion, D. (1917). Concerning the origin of the *fellahin* (in Hebrew). In U. Ornan (Ed.), 1969. *Israel Belkind: The Arabs in Palestine* (pp. 43–48). Hermon: Jerusalem.

Ben-Horin, M. (1956). *Max Nordau: Philosopher of human solidarity*. London: The London Jewish Society.

Benyamini, A. (1928). *Our children* (in Hebrew). Tel Aviv: Eitan Press.

Berkowitz, M. (1997). *Western Jewry and the Zionist Project, 1914–1933*. Cambridge: Cambridge University Press.

Bernstein, F. (1925). Zusammenfassende Betrachtungen über die erblichen Blutstrukturen des Menschen. *Zeitschrift für induktive Abstammungs- und Vererbungslehre, 37*, 237–270.

Bernstein, M. A. (1994). *Foregone conclusions. Against apocalyptic history*. Berkeley: University of California Press.

Beutler, E. (1993). Gaucher disease as a paradigm of current issues regarding single gene mutations of humans. *Proceedings of the National Academy of Science, Washington, 90*, 5384–5390.

Beutler, E., & Kuhl, W. (1992). Variability and population genetics of glucose-6-phosphate dehydrogenase (G6PD). In B. Bonné-Tamir & A. Adam (Eds.), *Genetic diversity among Jews* (pp. 60–69). New York: Oxford University Press.

Biale, D. (1992). *Eros and the Jews: From Biblical Israel to Contemporary America*. New York: Basic.

Biale, D. (1994). *Eros and the Jews* (in Hebrew). Tel-Aviv, Am Oved.

Bialik, C. N. (1934, May). The present hour. *The Young Zionist, 8*(4), 6–7.

Bilski Ben-Hur, R. (1993). *Every Individual a King: The Social and Political Thought of Ze'ev Vladimir Jabotinsky* (S. C. Abramson, Trans.). Washington, DC: B'nai B'rith Books.

Bloch, A. (1913). Origin and evolution of the blond Europeans. *The Smithsonian Report for, 1912*, (#2219). pp. 609–630. Washington, DC: Washington Government Printing Office.

Bloom, E. (2005). *Ruppin and the production of the Modern Hebrew Society (1908–1942)* (in Hebrew). Unpublished doctoral dissertation. Tel Aviv University, Israel.

Bloom, E. (2007). What "The Father" had in mind? Arthur Ruppin (1876–1943), cultural identity, weltanschauung and action. *History of European Ideas, 33*(3), 330–349.

Bloom, E. (2008). *Arthur Ruppin and the production of the Modern Hebrew Culture*. Ph.D. thesis submitted to the Senate of Tel-Aviv University.

Bobo, L. D. (2008). Foreword. In B. A. Koenig, S. S.-J. Lee, & S. S. Richardson (Eds.), *Revisiting race in a genomic age* (pp. ix–xii). New Brunswick: Rutgers University Press.

Bodenheimer, F. S, (1936). *The biological background of human population theory* (in Hebrew). Tel Aviv: Stibel.

Bodenheimer, F. S. (1959). *A biologist in Israel: A book of reminiscences*. Jerusalem: Biological Studies Publishers.

Bolnick, D. A. (2008). Individual ancestry inference and the reification of race as a biological phenomenon. In B. A. Koenig, S. S.-J. Lee, & S. S. Richardson (Eds.), *Revisiting race in a genomic age* (pp. 70–85). New Brunswick: Rutgers University Press.

Bonné-Tamir, B. (1980). A new perspective on the genetics of the Jews (in Hebrew). *Mada, 24*(4–5), 181–186.

Bonné-Tamir, B. (2010). *Living with genes: Fifty years of research on genetics of Israeli populations* (in Hebrew). Jerusalem: Carmel.

Bonné-Tamir, B., Ashbel, S., & Bar-Shani, S. (1978). Ethnic communities in Israel: The genetic blood markers of Moroccan Jews. *American Journal of Physical Anthropology, 49*, 465–471.

Bonné-Tamir, B., Ashbel, S., & Kenett, R. (1979a). Genetic markers: Benign and normal traits of Ashkenazi Jews. In R. M. Goodman & A. Motulsky (Eds.), *Genetic diseases among Ashkenazi Jews* (pp. 59–76). New York: Raven Press.

Bonné-Tamir, B., Karlin, S., & Kenett, R. (1979b). Analysis of genetic data on Jewish populations. I. Historical background, demographic features, and genetic markers. *American Journal of Human Genetics, 31*(3), 324–340.

Bonné-Tamir, B., Zoossmann-Diskin, A., & Ticher, A. (1992). Genetic diversity among Jews reexamined: Preliminary analyses at the DNA level. In B. Bonné-Tamir & A. Adam (Eds.), *Genetic diversity among Jews* (pp. 80–94). New York: Oxford University Press.

Bouchard, T. J. Jr., & Loehlin, J. C. (2001). Genes, evolution, and personality. *Behavior Genetics, 31*(3), 243–273.

Boyd, W. C. (1950). *Genetics and the races of man*. Oxford: Blackwell.

Brachyahu, M. (1926). The activity of the department for school hygiene (in Hebrew). *Hachinuch, 9*, 181–189.

Brautbar, C., Mizrahi, M., Cohen, M., Ziedler, A., Levene, C., Cohen, I., et al. (1992). HLA polymorphisms in Ethiopian Jews and other Jewish and non-Jewish groups. In B. Bonné-Tamir & A. Adam (Eds.), *Genetic diversity among Jews* (pp. 70–79). New York: Oxford University Press.

Brenner, L. (1983). *Zionism and the age of the dictators: A reappraisal*. Kent (GB), Croom Helm Westport, CN (USA): Lawrence Hill.

Broberg, G., & Roll-Hansen, N. (1996). *Eugenics and the welfare state: Sterilization policy in Denmark, Sweden, Norway, and Finland*. East Lansing: Michigan State University Press.

Broshi, M. (2004, Fall). The origin of the fellahs, or we and our neighbors (in Hebrew). *Keshet Hadasha, 9*, 26–35.

Bruchow, M. (1922). *Mendelismus* (in Hebrew). *Hatkufah, 16*, 326–341.

Buber, M. (1920). *Die jüdische Bewegung: Gesamelte Aufsätze und Ansprachen (1900–1914)*. Berlin: Juüdischer Verlag.

Buber, M. (1984). *TeUda VeYeud* (in Hebrew). Jerusalem: The Ziojnist Library.

Burla, J. (1927). On the Sephardic child (in Hebrew). *Hachinuch, 190*, 21–24.

Buss, L. W. (1987). *The evolution of individuality*. Princeton: Princeton University Press.

Bychowski, S. (1918). Nervous diseases and eugenics in Jews (in Hebrew). *Hatkufah, 2*, 289–307.

Carmi, R., Elbedour, K., Wietzman, D., Sheffield, V., & Shoham-Vardi, I. (1998). Lowering the burden of hereditary diseases in a traditional, inbred community: Ethical aspects of genetic research and its application. *Science in Context, 11*(3–4), 391–395.

Carmi, S., Hui, K. Y., Kochav, E., Liu, X., Xue, J., Grady, F., Guha, S., Upadhyay, K., Ben-Avraham, D., Mukherjee, S., Bowen, M., Thomas, T., Vijai, J., Cruts, M., Froyen, G., Lambrechts, D., Plaisance, S., Van Broeckhoven, C., Van Damme, P., Van Marck, H., Barzilai, N., Darvasi, A., Offit, K., Bressman, S., Ozelius, L. J., Peter, I., Cho, J. H., Ostrer, H., Atzmon, G., Clark, L. C., Lencz, T., & Pe'er, I. (2014). Sequencing an Ashkenazi reference panel supports population-targeted personal genomics and illuminates Jewish and European origins. *Nature Communications*. doi:10.1038/ncomms5835.

Cavalli-Sforza, L. L. (2000). *Genes, peoples, and languages*. New York: Farrar, Straus & Giroux.

Cavalli-Sforza, L. L., & Carmelli, D. (1979). The Ashkenazi gene pool: Interpretations. In R. M. Goodman & A. G. Motulsky (Eds.), *Genetic diseases among Ashkenazi Jews* (pp. 93–102). New York: Raven Press.

Cavalli-Sforza, L. L., & Feldman, M. W. (2003). The application of molecular genetic approaches to the study of human evolution. *Nature Genetics, 33*(Suppl), 266–275.

Chadwick, R. (1998). Can genetic counseling avoid the charge of eugenics? *Science in Context, 11*(3–4), 471–480.

Churchill, F. B. (2015). *August Weismann: Development, heredity, and evolution*. Cambridge, MA: Harvard University Press.

Cochran, G., Hardy, J., & Harpending, H. (2006). Natural history of Ashkenazi intelligence. *Journal of Biological Sciences, 38*(5), 659–693.

Cohen, A. M. (1963). The prevalence of diabetes. In E. Goldschmidt (Ed.), *The genetics of migrant and isolate populations* (pp. 324–325). New York: Williams & Wilkins.

Cohen, T., Bloch, N., Kelti, L., Wartski, S., Hurwitz, D., Barak, E., & Goldschmidt, E. (1963). The population of Kurdish Jews. In E. Goldschmidt (Ed.), *The genetics of migrant and isolate populations* (pp. 344–349). New York: Williams & Wilkins.

Cohen, T., Vardi-Saliternik, R., & Friedlander, Y. (2004). Consanguinity, intracommunity and intercommunity marriages in a population sample of Israeli Jews. *Annals of Human Genetics, 31*(1), 38–48.

Corcos, A. F. (2005). *The myth of the Jewish race: A biologist's point of view*. Bethlehem: Lehigh University Press.

Costa, M. D., Pereira, J. B., Pala, M., Fernandes, V., Olivieri, A., Achilli, A., Perego, U. A., Rychkov, S., Naumova, O., Hatina, J., Woodward, S. R., Eng, K. K., Macaulay, V., Carr, M., Soares, P., Pereira, L., & Richards, M. B. (2013). A substantial prehistoric European ancestry amongst Ashkenazi maternal lineages. *Nature Communications, 4*(2543), 1–10. doi:10.1038/Ncomms3543.

Crow, J. F., & Kimura, M. (1970). *An introduction to population genetics theory*. New York/Evanston/London: Harper & Row.

Dagan, H. (1999). The concept of homeland and and the Jewsih ethos: A chronical of dissonance (in Hebrew). *Alpayim, 18*, 9–23.

Das, R., Wexler, P., Pirooznia, M., & Elhaik, E. (2016): Localizing Ashkenazic Jews to primeval villages in the ancient Iranian lands of Ashkenaz. [Advance Access]. *Genome Biology and Evolution*. doi:10.1093/gbe/evw046.

Davies, M., & Hughes, A. G. (1927). An investigation into the comparative intelligence and attainments of Jewish and non-Jewish school children. *British Journal of Psychology (General Section), 18*(2), 134–146.

Doron, J. (1980). Rassenbewusstsein und naturwissenschaftliches Denken im deutschen Zionismus während der Wilhelminischen Ära. *Jahrbuch des Instituts für deutsche Geschichte, 9*, 389–427.

Dugdale, R. L. (1877). *The Jukes: A study of crime, pauperism and heredity*. New York: Putnam.

Dunlop, D. M. (1954). *The history of the Jewish Khazars*. Princeton: Princeton University Press.

Efron, J. M. (1993). Scientific racism and the mystique of Sephardic racial superiority. In *Leo Baeck Institute Year Book* (Vol. 38, pp. 77–96). London: Secker & Warburg.

Efron, J. M. (1994). *Defenders of the race: Jewish doctors and race science in Fin-de-Siècle Europe*. New Haven/London: Yale University Press.

Einstein, A. (1930). *About Zionism: Speeches and Letters* (L. Simon, Trans.). London: The Soncino Press.

Eisensadt, S. N. (1948). *Introduction to the study of the sociological structure of the oriental ethnic groups* (in Hebrew). Jerusalem: Mossad Szold.

Eisensadt, S. N. (1954). *The absorption of immigrants: A comparative study based mainly on the Jewish Community in Palestine and the State of Israel*. London: Routledgte & Kegan Paul.

Elhaik, E. (2013). The missing link of Jewish European ancestry: Contrasting the Rhineland and the Khazarian Hypotheses. *Genome Biology and Evolution, 5*(1), 61–74. doi:10.1093/gbe/evs119.

Elon, A. (2002). *The pity of it all: A history of Jews in Germany, 1743–1933*. New York: Henry Holt & Company.

Endelman, T. M. (1987). Conversion as a response to antisemitism in modern Jewish history. In J. Reinharz (Ed.), *Living with antisemitism: Modern Jewish responses* (pp. 59–83). Hanover/London: Brandeis University Press.

Endelman, T. M. (2004). Anglo-Jewish scientists and the science of race. *Jewish Social Studies, 11*(1), 52–92.

Eshkoli, A. Z. (1937). *The human race: Race research and race theory* (in Hebrew). Jerusalem: Reuben Mass.

Etinger, S. (1972). *Modern times* (in Hebrew). In H. H. Ben-Sasson (Ed.), *History of the Jewish People* (Vol. 3). Tel Aviv: Dvir.

Falk, R. (2003–2004). Nervous diseases and eugenics of the Jews: A view from 1918. *Korot, 17*, 23–46.

Falk, R. (2009). *Genetic analysis: A history of genetic thinking*. Cambridge: Cambridge University Press.

Falk, R. (2010). Eugenics and the Jews. In A. Bashford & P. Levine (Eds.), *The Oxford handbook of the history of eugenics* (pp. 462–476). New York: Oxford University Press.

Falk, R. (2011). Evolution as progressing complexity. In S. Gissis & E. Jablonka (Eds.), *Transformations of Lamarckism* (pp. 381–384). Cambridge, MA: The MIT Press.

Falk, R. (2013). Biology comes of age [Special issue: Topics in evolutionary biology in memory of Prof. Uzi Ritte] *Israel Journal of Ecology & Evolution, 59*(4), 186–188. doi: http://dx.doi.org/10.1080/15659801.2013.898403

Falk, R. (2014a). Logic and Philosophy of Science [Review of the book *Genetics and Philosophy: An Introduction*, by P. Griffiths & K. Stotz]. *Philosophy of Science, 81*, 470–475.

Falk, R. (2014b). Genetic markers cannot determine Jewish descent. *Frontiers in Genetics: Evolutionary and Population Genetics, 5*(Article 462), 1–10. doi: 10.3389/fgene.2014.00462

Fattal, S. (2003). *In the alleys of Bagdad* (in Hebrew). Jerusalem: Carmel.

Feldman, W. M. (1939). Ancient Jewish eugenics. *Medical Leaves, 2*, 28–37.

Filon, D., Oron, V., Krichevski, S., Shaag, A., Shaag, Y., Warren, T. C., Goldfarb, A., Sheneor, Y., Koren, A., Aker, M., Abramov, A., Rachmilewitz, E. A., Rund, D., Kazazian Jr., H. H., & Oppenheim, A. (1994). Diversity of β-globin mutations in Israeli ethnic groups reflects recent historic events. *American Journal of Human Genetics, 54*, 836–843.

Firsht, S. (1935). From the history of the movement for the regulation of births (in Hebrew). In: J. Meir, & I. Rivkai (Eds.), *Year book – Mother and Child, 1935*.

Fishberg, M. (1911). *The Jews: A study of race and environment*. London: The Walter Scott Publishing Co..

Fisher, R. A. (1930). *The genetical theory of natural selection*. Oxford: Clarendon.

Frankenstein, K. (1938). Delinquency and neglect in Jerusalem youth (in Hebrew). *Hachinuch, 11*(1), 26–35.

Frisch, A., Comombo, R., Michaelovsky, E., Karpati, M., Goldman, B., & Peleg, L. (2004). Origin and spread of the, 1278insTATC mutation causing Tay-Sachs disease in Ashkenazi Jews: Genetic drift as a robust and parsimonious hypothesis. *Human Genetics, 114*(4), 366–376.

Futuyma, D. J., & Slatkin, M. (Eds.). (1983). *Coevolution*. Sunderland: Sinauer.

Gabai-Kapara, E., Lahad, A., Kaufman, B., Friedman, E., Segev, S., Renbaum, P., Beeri, R., Gal, M., Grinshpun-Cohen, J., Djemal, K., Mandell, J. B., Lee, M. K., Beller, U., Catane, R., King, M.-C., & Levy-Lahad, E. (2014). Population-based screening for breast and ovarian cancer risk due to BRCA1 and BRCA2. *Proceedings of the National Academy of Sciences, USA, 111*(39), 14205–14210.

Galton, F. (1878). Composite portraits. *Nature, 18*, 97–100.

Galton F. (1910, July 29). Eugenics and the Jew. *The Jewish Chronicle*.

Gannett, L. (2004). The biological reification of race. *British Journal of Philosophy of Science, 55*, 323–345.

Gannett, L. (2013). Projectibility and group concepts in population genetics and genomics. *Biological Theory, 7*(2), 130–143. doi:10.1007/s13752-012-0085-8.

German, J. (1969). Blooms' syndrome: I. Genetical and clinical observations in the first twenty-seven patients. *American Journal of Human Genetics, 21*(2), 196–227.

Gilbert, F. (1997). Genetic diseases and testing in Ashkenazi Jews: Part I. *Genetic Testing, 1*(1), 3.

Gilman, S. L. (1985). *Difference and pathology: stereotypes of sexuality, race, and madness*. Ithaca: Cornell University Press.

Gilman, S. L. (2010). *Disease & diagnosis: The second age of biology*. New Brunswick: Transaction Publishers.

Gissis, S. (2008). When is 'race' a race? 1946–2003. *Studies in History and Philosophy of Biological and Biomedical Science, 39*, 437–450.

Gissis, S. B. (2011). Visualizing "race" in the eighteenth century. *Historical Studies in the Natural Sciences, 41*(1), 41–103.

Gissis, S. B., & Jablonka, E. (Eds.). (2011). *Transformations of Lamarckism: From subtle fluids to molecular biology*. Cambridge, MA: The MIT Press.

Glad, J. (2011). *Jewish eugenics*. Washington, DC: Wooden Shore.

Goldschmidt, E. (1963). *The genetics of migrant and isolate populations*. New York: Williams & Wilkins.

Goldschmidt, E., Ronen, A., & Ronen, I. (1960). Changing marriage systems in the Jewish communities of Israel. *Annals of Human Genetics, 24*, 191–204.

Goldstein, J. (1985). The Wandering Jew and the problem of psychiatric anti-semitism in Fin-de-Siècle France. *Journal of Contemporary History, 20*(4), 521–552.

Goldstein, D. B. (2008). *Jacob's legacy: A genetic view of Jewish history*. New Haven/London: Yale University Press.

Goodman, R. M., & Motulsky, A. G. (1979). *Genetic Diseases Among Ashkenazi Jews*. New York: Raven Press.

Gould, S. J. (1981). *The mismeasure of man*. New York: Norton.

Gould, S. J. (1996). *Full house: The spread of excellence from Plato to Darwin*. New York: Three Rivers Press.

Grabowski, G. A. (1997). Gaucher disease: Gene frequencies and genotype/phenotype correlations. *Genetic Testing, 1*(1), 5–12.

Griffiths, P., & Stotz, K. (2013). *Genetics and philosophy: An introduction*. New York: Cambridge University Press.

Gross, M. L. (2002). Ethics, policy, and rare genetic disorders: The case of Gaucher disease in Israel. *Theoretical Medicine and Bioethics, 23*(2), 151–170.

Gurevitch, J., & Margolis, E. (1955). Blood groups in Jews from Iraq. *Annals of Human Genetics, 19*, 257–259.

Gurevitch, J., Hermoni, D., & Polishuk, Z. (1951). Rh blood types in Jerusalem. *Annals of Eugenics, London, 16*, 129–130.

Gurevitch, J., Hermoni, D., & Margolis, E. (1953). Blood groups in Kurdistani Jews. *Annals of Eugenics, London, 18*, 94–95.

Gurevitch, J., Hasson, E., Margolis, E., & Poliakoff, C. (1954). *Blood groups in Jews from Tripolitania and Cochin, India.* Paper presented at the *Fifth International Congress on Blood Transfusion,* Paris.

Gurevitch, J., Hasson, E., Margolis, E., & Poliakoff, C. (1955a). Blood groups in Jews from Cochin, India. *Annals of Human Genetics, 19*, 254–256.

Gurevitch, J., Hasson, E., Margolis, E., & Poliakoff, C. (1955b). Blood groups in Jews from Tripolitania. *Annals of Human Genetics, 19*, 260–261.

Gurevitch, J., Hasson, E., & Margolis, E. (1956). Blood groups in Persian Jews. A comparative study with other oriental Jewish communities. *Annals of Human Genetics, 21*, 135–138.

Haeckel E. (1876). *The history of creation: Or the development of the earth and its inhabitants by the action of natural causes.* Revised translation by Prof. E. Ray Lancaster. London: King.

Haldane, J. B. S. (1949). Disease and evolution. *La Ricerca Scientifica, 19*(Suppl. A), 68–76.

Halkin, H. (2002). *Across the Sabbath River: In Search of a Lost Tribe of Israel.* Boston: Houghton Mifflin Co..

Hamel, N., Feng, B.-J., Foretova, L., Stoppa-Lyonnet, D., Narod, S. A, Imyanitov, E., Sinilnikova, O., Tihomirova, L., Lubinski, J., Grunwald, J., Gorski, B., Hansen, T. O., Nielsen, F. C., Thomassen, M., Yannoukakos, D., Konstantopoulou, I., Zajac, V., Ciernikova, S., Couch, F. J., Greenwood, C. M. T., Goldgar, D. E., & Foulkes, W. D. (2011). On the origin and diffusion of BRCA1 c.5266dupC (5382insC) in European populations. *European Journal of Human Genetics, 19*(3), 300–306. doi:10.1038/ejhg.2010.203.

Hammer, M. F., Redd, A. J., Wood, E. T., Bonner, M. R., Jarjanazi, H., Karafet, T. M., Santachiara-Benerecetti, S., Oppenheim, A., Jobling, M. A., Jenkins, T., Ostrer, H., & Bonné-Tamir, B. (2000). Jewish and Middle Easternnon-Jewish populations share a common pool of Y-chromosome biallelic haplotype. *Proceedings of the National Academy of Science, Washington, 97*, 6769–6774.

Harari, Y. N. (2015). *The history of tomorrow* (in Hebrew). Or-Yehudda: Kinneret, Zmora-Beitan, Dvir.

Harpending, H., & Cochran, G. (2002). In our genes. *Proceedings of the National Academy of Sciences, Washington, 99*(1), 10–12.

Harris, H. (1966). Enzyme polymorphism in man. *Proceedings of the Royal Society of London. Series B, 164*, 298–310.

Hart, M. (1995). Picturing Jews: Iconography and racial science. In P. Y. Medding (Ed.), *Values, interests and identity: Jews and politics in a changing world* (pp. 159–175). Oxford: Oxford University Press.

Hart, M. B. (2000). *Social science and the politics of Modern Jewish identity.* Stanford: Stanford University Press.

Herzl, T. (1902). *Altneuland.* Leipzig: Hermann Seemann Nachfolger.

Hess, J. M. (2002). Memory, history, and the Jewish question: Universal citizenship and the colonization of Jewish memory. In A. Confino & P. Fritzsche (Eds.), *The work of memory: New directions in the study of German society and culture* (pp. 40–61). Urbana/Chicago: University of Illinois Press.

Hilmar, K., & Hilmar, E. (1986). *Juden in Sulingen, 1753–1938*. Sulingen: Kreisizeitung Verlagsgesellschaft.

Holtzman, N. A. (1998). Eugenics and genetic testing. *Science in Context, 11*(3–4), 397–417.

Hsia, K., Meiner, Z., Kahan, E., Cass, C., Kahana, I., Avraham, D., Scarlato, G., Abramsky, O., Prusiner, S., & Gabison, R. (1991). Mutation of the prion protein in Libyan Jews with the Creutzfeld-Jakob disease. *The New England Journal of Medicine, 324*, 1091–1097.

Hubby, J. L., & Lewontin, R. C. (1966). A molecular approach to the study of genic heterozygosity in natural populations. I. The number of alleles at different loci in *Drosophila pseudoobscura*. *Genetics, 54*, 577–594.

Hull, D. L. (1973). *Darwin and his critics*. Chicago/London: The University of Chicago Press.

Jablonka, E., & Lamb, M. J. (1995). *Epigenetic inheritance and evolution: The Lamarckian dimension*. Oxford: Oxford University Press.

Johannsen, W. (1909). *Elemente der exakten Erblichkeitslehre*. Jena: Gustav Fischer.

Jones, S. (1997). *In the blood: God, genes and destiny*. London: Flamingo.

Kahn, S. M. (2000). *Reproducing Jews: A cultural account of assisted conception in Israel*. Durham: Duke University Press.

Kahn, S. M. (2013). Who are the Jews? New formulations of an age-old question. *Human Biology, 85*(6), 918–924.

Kaplan, S. (2003). If there are no races, how can Jews be a "race"? *Journal of Modern Jewish Studies, 2*(1), 79–96.

Katz, J. (1979). *Anti-semitism: From religious hatred to racial rejection* (Hebrew). Tel Aviv: Am-Oved.

Katz, J. (1980). *From prejudice to destruction: Anti-semitism, 1700–1933*. Cambridge, MA: Harvard University Press. http://www.jewishvirtuallibrary.org/jsource/judaica/ejud_0002_0017_0_17174.html.

Katz, J. (1986). *Jewish emancipation and self-emancipation*. Philadelphia: The Jewish Publication Society.

Katz, S. & Head, M. (1997). *The history of the Hebrew University in Jerusalem: Roots and beginnings* (in Hebrew). Jerusalem: Magnes Press.

Kautsky, K. (1926). *Are the Jews a race?* New York: International Publishers.

Keller, E. F. (2010). *The mirage of a space between nature and nurture*. Durham/London: Duke University Press.

Kerem, E., Kalman, Y. M., Yahav, Y., Shoshani, T., Abeliovich, D., Szeinberg, A., Rivlin, J., Blau, H., Tal, A., Ben-Tur, L., Springer, C., Augarten, A., Godfrey, S., Lerer, I., Branski, D., Friedman, M., & Kerem, B. (1995). Highly variable incidence of cystic fibrosis and different mutation distribution among different Jewish ethnic groups in Israel. *Human Genetics, 96*, 193–197.

Kevles, D. J. (1985). *In the name of eugenics*. New York: Alfred A. Knopf.

Kidd, K. K., Kidd, J. R., Bonné-Tamir, B., & New, M. I. (1992). Nuclear DNA polymorphisms and population relationships. In B. Bonné-Tamir & A. Adam (Eds.), *Genetic diversity among Jews* (pp. 33–44). New York/Oxford: Oxford University Press.

Kirsh, N. (2003). Population genetics in Israel in the 1950s. The unconscious internalization of ideology. *Isis, 94*(4), 631–655.

Klarreich, E. (2001). Genetics paper erased from journal over political content. *Nature, 414*(6862), 382.

Koenig, B. A., Lee, S. S.-J., & Richardson, S. S. (Eds.). (2008). *Revisiting race in a genomic age*. New Brunswick: Rutgers University Press.

Koestler, A. (1946). *Thieves in the night*. New York: The Macmillan Company.

Koestler, A. (1976). *The thirteenth tribe: The Khazar Empire and Its Heritage*. New York: Random House.

Kretschmer, E. (1921). *Körperbau und Charakter: Untersuchungen zum Konstitutions-problem und zur Lehre von den Temperramenten*. Berlin: Julius Springer.

Kretschmer, E. (1936/1970). *Physique and character. An investigation of the nature of constitution and of the theory of temperament* (2nd ed.). New York: Cooper Square Publishers, Inc.

Laland, K. N., Uller, T., Feldman, M. W., Sterelny, K., Müller, G. B., Moczek, A., et al. (2015). The extended evolutionary synthesis: its structure, assumptions and predictions. *Proceedings of the Royal Society of London B, 282*, 20151019. doi:http://dx.doi.org/10.1098/rspb.2015.1019.

Lamarck, J. B. (1809/1984). *Zoological philosophy: An exposition with regard to the natural history of animals* (H. Elliot, Trans.). Chicago/London: The University of Chicago Press.

Laqueur, W. (1972). *A History of Zionism*. New York: Schocken Books.

Laski, H. (1912). A Mendelian view of racial heredity. *Biometrika, 8*, 424–430.

Lavater, J. C. (1984). *Physiognomische Fragmenter (Eine Auswahl, herausgegeben von C. Siegrist)*. Stuttgart: Philipp Reclam.

Lazarus, M. (1900–1901). *The Ethics of Judaism* (Vols. 1–2). Philadelphia: The Jewish Publication Society of America.

Lazarus, M. (1925). *Was heist national?* Berlin: Philip Verlag.

Lederberg, J. (1999). J. B. S. Haldane (1949) on infectious disease and evolution. *Genetics, 153*(1), 1–3.

Lenoir, T. (1982). *The strategy of life*. Chicago: University of Chicago Press.

Levin, S., Moses, S. W., Chayoth, R., Jagoda, N., & Steinitz, K. (1967). Glycogen storage disease in Israel. *Israel Journal of Medical Sciences, 3*, 397–410.

Levy-Coffman, E. (2005). A mosaic of people; the Jewish story and a reassessment of the DNA evidence. *Journal of Genetic Genealogy, 1*, 12–13.

Lewontin, R. C. (1972). The apportionment of human diversity. *Evolutionary Biology, 6*, 381–398.

Lewontin, R. C. (2002). Directions in evolutionary biology. *Annual Review of Genetics, 36*, 1–18.

Lewontin, R. C., & Hubby, J. L. (1966). A molecular approach to the study of genic heterozygosity in natural populations. II. Amount of variation and degree of heterozygosity in natural populations of *Drosophila pseudoobscura*. *Genetics, 54*, 595–609.

Li, C. C. (1955). *Population genetics*. Chicago: The University of Chicago Press.

Lipphardt, V. (2012, April). Isolates and crosses in human population genetics: Or, A contextualization of German race science. *Current Anthropology, 53*(Suppl. 5), 69–82.

Livneh-Freudenthal, R. (2001). From "A nation dwelling alone" to "A nation among the nations" or: "The return to history" – Between universalism and nationalism. In R. Livneh-Freudenthal & E. Reiner (Eds.), *Streams into the sea: Studies in Jewish culture and its context* (pp. 153–177). Tel Aviv: Alma College.

Livneh-Freudenthal, R. (2005). Jewish studies – The paradigm and initial patrons. In M. Mach & Y. Jacobson (Eds.), *Historiography and the science of Judaism. Te'uda, 20*, 187–214.

Lombroso, C. (2007). *Criminal Man* (M. Gibson & N. H. Rafter, Trans.). Durham: Duke University Press. (Original work, L'uomo deliquente, published in 1876.)

Lorenz, K. (1940). Durch Domestikation verursachte Stürungen arteigene Verhaltens. *Zeitschrift für angewandte Psychologie und Charakterkunde, 59*(2), 2–81.

MacDonald, K. (1994). *A People that shall Dwell Alone: Judaism as a group evolutionary strategy*. Westport: Praeger.

Marcus, A. W., Ebel, E. R., & Friedman, D. A. (2015). Commentary: Portuguese Crypto-Jews: the genetic heritage of a complex history. *Frontiers in Genetics, 6*. doi:10.3389/fgene/2015.00012.

Matmon, A. (1933). *The improvement of the race of the human species and its value to our people* (in Hebrew). Tel Aviv: Biological-Hygienic Library.

McKusick, V. A. (1979). Nonhomogeneous distribution of recessive diseases. In R. M. Goodman & A. M. Motulsky (Eds.), *Genetic diseases among Ashkenazi Jews* (pp. 271–284). New York: Raven Press.

Meir-Glitzenstein, E. (2012). *The exodus of the Yemenite Jews: A failed operation and a formative myth* (in Hebrew). Tel Aviv: Resling.

Menon, A. V. (2016). Review: Dorothy Roberts, *Fatal intervention: How science, politics, and big business re-create race in the twenty-first century*. *Spontaneous Generation: A Journal for the History and Philosophy of Science, 8*(1), 109–111. doi:10.4245/sponge.v8i1.20388.

Mitchell, S. D. (2003). *Biological complexity and integrative pluralism*. Cambridge: Cambridge University Press.

Mitchell, S. D. (2009). *Unsimple Truths: Science, complexity, and policy*. Chicago/London: The University of Chicago Press.

Montagu, A. (1974). *Man's most dangerous myth: The fallacy of race* (5th ed., revised and enlarged). London/Oxford/New York: Oxford University Press.

Mosse, G. L. (1964). *The crisis of German ideology: Intellectual origins of the Third Reich*. New York: Grosset & Dunlap.

Mosse, G. L. (1970). *Germans and Jews*. New York: Howard Fertig.

Motulsky, A. G. (1980). Ashkenazi Jewish gene pool: Admixture, drift and selection. In A. W. Eriksson, H. Forsius, H. R. Nevanlinn, P. L. Workman & R. K. Norio (Eds.), *Population structure and genetic discoveries* (pp. 353–365). London: Academic.

Motulsky, A. G. (1995). Jewish diseases and origins. *Nature Genetics, 9*, 99–101.

Mourant, A. E., Kopeć, A. C., & Domaniewska-Sobczak, K. (1978). *The genetics of the Jews*. Oxford: Oxford University Press.

Muhsam, H. V. (1964). The genetic origin of the Jews. *Genus, 20*(1–4), 3–30.

Muller, H. J. (1922). Variation due to change in the individual gene. *The American Naturalist, 56*, 32–50.

Muller, H. J. (1933). The dominance of economics over eugenics. *Scientific Monthly, 37*, 40–47.

Murray, C., & Harnstein, R. (1994). *The Bell Curve*. New York: Free Press.

Nardi, N. (1948). The intelligence of Jewish children (in Hebrew). *Hachinuch, 21*, 257–270.

Navon, R., & Proia, R. L. (1992). Tay-Sachs disease mutations among Moroccan Jews. In B. Bonné-Tamir & A. Adam (Eds.), *Genetic diversity among Jews* (pp. 259–266). New York/Oxford: Oxford University Press.

Nebel, A., Filon, D., Weiss, D. A., Weale, M., Faerman, M., Oppenheim, A., & Thomas, M. G. (2000). High-resolution Y chromosome haplotypes of Israeli and Palestinian Arabs reveal geographic substructure and substantial overlap with haplotypes of Jews. *Human Genetics, 107*, 630–641.

Nebel, A., Filon, D., Brinkmann, B., Majumder, P. P., Faerman, M., & Oppenheim, A. (2001). The Y chromosome pool of Jews as part of the genetic landscape of the Middle East. *American Journal of Human Genetics, 69*, 1095–1112.

Nebel, A., Filon, D., Faerman, M., Soodyall, H., & Oppenheim, A. (2005). Y chromosome evidence for a founder effect in Ashkenazi Jews. *European Journal of Human Genetics, 13*(3), 388–391.

Nelkin, D., & Lindee, M. S. (1995). *The DNA Mystique. The Gene as a Cultural Icon*. New York: W. H. Freeman and Compay.

Neufeld, E. (1992). The molecular basis of Tay-Sachs disease and related $gm2$ gangliosidoses among Jews and non-Jews. In B. Bonné-Tamir & A. Adam (Eds.), *Genetic diversity among Jews* (pp. 97–103). New York/Oxford: Oxford University Press.

Nogueiro, I., Teixeira, J. C., Amorim, A., Gusmão, L., & Alvarez, L. (2015). Portuguese crypto-Jews: The genetic heritage of a complex history. *Frontiers of Genetics, 6*. doi: http://dx.doi.org/10.3389/fgene.2015.00012

Nordau, M. (1883/1895). *The conventional lies of our civilisation*. Chicago: Laird & Lee.

Nordau, M. (1895). *Degeneration* (Trans. from 2nd German ed.). London: William Heinemann.

Nordau, M. (1909). *Zionistiche Schriften*. Köln: Jüdischer Verlag.

Oppenheim, A., Jury, C. L., Rund, D., Vulliamy, T. J., & Luzzatto, L. (1993). G6PD Mediterranean accounts for the prevalence of G6PD deficiency in Kurdish Jews. *Human Genetics, 91*, 293–294.

Ornan, U. (Ed.). (1969). *Israel Belkind: The Arabs in Palestine* (in Hebrew). Jerusalem: Hermon.

Ornstein, A. (1938). The skill of criticism of our children (in Hebrew). *Hachinuch, 11*, 61–69.

Ostrer, H. (2001). A genetic profile of contemporary Jewish populations. *Nature Reviews Genetics, 2*(11), 891–898.

Ostrer, H. (2012). *Legacy: A genetic history of the Jewish people*. New York: Oxford University Press.

Oz, A., & Oz-Salzberger, F. (2014). *Jews and words*: New Haven: Yale University Press.

Paul, D. B. (1984). Eugenics and the left. *Journal of the History of Ideas, 45*, 567–590.

Paul, D. B. (1995). *Controlling human heredity., 1865 to the present*. Amherst: Humanities Press.

Paul, D. B. (2002). From reproductive responsibility to reproductive autonomy. In L. S. Parker & R. A. Ankeny (Eds.), *Mutating concepts, evolving disciplines: Genetics, medicine and society* (Vol. 75, pp. 63–85). Dordrecht: Kluwer.

Paul, D. B. (2009). Darwin, social Darwinism, and eugenics. In J. Hodge & G. Radick (Eds.), *Cambridge companion to Darwin* (2nd ed., pp. 219–245). Cambridge: Cambridge University Press.

Paul, D. B. (2016). Reflections on the historiography of American eugenics: Trends, fractures, tensions. *Journal of the History of Biology*. doi: 10.1007/s10739-016-9442-y

Paul, D. B., & Brosco, J. P. (2013). *The PKU paradox: A short history of a genetic disease*. Baltimore: Johns Hopkins University Press.

Pearson, K., & Moul, M. (1925). The problem of alien immigration into Great Britain, illustrated by an examination of Russian and Polish Jewish children. *Annals of Eugenics, 1*, 5–55.

Peled, Y. (1999). Meir Kahana (in Hebrew). In A. Ophir (Ed.), *Fifty to forty-eight* (Special issue of *Theoria Ubikoret, 12–13*), 321–327. Jerusalem: Van-Leer Institute and Hakibutz-Hameuchad.

Peretz, H., Mulai, A., Usher, S., Zivelin, A., Segal, A., Weisman, Z., Mittelman, M., Lupo, H., Lanir, N., Brenner, B., Shpilberg, O., & Seligsohn, U. (1997). The two common mutations causing factor XI deficiency in Jews stem from distinct founders: One of ancient Middle-Eastern origin and another of more recent European origin. *Blood, 90*(7), 2654–2659.

Phillips, C. (2002). The disappeared [Review of the book *Across the Sabbath River: In Search of a Lost Tribe of Israel* by Hillel Halkin]. *The New Republic*, 39-42.

Pick, D. (1989). *Faces of degeneration. A European disorder, c., 1848 – c., 1918*. Cambridge: Cambridge University Press.

Poliak, Ab. N. (1951). *Khazaria: The history of a Jewish Kingdom in Europe* (in Hebrew). (3rd ed. updated). Tel Aviv: Mossad Bialik.

Poppel, S. M. (1976). *Zionism in Germany, 1897–1933*. Philadelphia: Jewish Publication Society of America.

Post, R. H. (1973). Jews, genetics and disease. In A. Shiloh & I. C. Selavan (Eds.), *Ethnic groups of America: Their morbidity, mortality and behavior disorders* (Vol. I – The Jews. pp. 67–71). Springfield: Charles C Thomas.

Pras, M., Zemer, D., Langevitz, P., & Sohar, E. (1992). Familial Mediterranean fever: A genetic disorder prevalent in Sephardic Jews. In B. Bonné-Tamir & A. Adam (Eds.), *Genetic diversity among Jews* (pp. 223–227). New York/Oxford: Oxford University Press.

Ragins, S. (1980). *Jewish responses to anti-semitism in Germany, 1870–1914: A study in the history of ideas*. Cincinnati: Hebrew Union College Press.

Rav Tzair. (1911–1912). On religious education in school (in Hebrew). *Hachinuch, 2*, 185–193.

Raz, A. E., & Vizner, Y. (2008). Carrier matching and collective socialization in community genetics: Dor Yeshorim and the reinforcement of stigm. *Social Science & Medicines, 67*, 1361–1369.

Reinharz, J. (1985). The Zionist response to antisemitism in the Weimar Republic. In J. Reiharz & W. Schatzberg (Eds.), *The Jewish response to German culture: From the enlightenment to the Second World War* (pp. 266–293). Hanover/London: University Press of New England.

Reuter, S. Z. (2006). The Genuine Jewish Type: Racial ideology and anti-immigrationism in early medical writing about Tay-Sachs disease. *The Canadian Journal of Sociology, 31*(3), 291–323.

Richards, R. J. (2013). The relation of Spencer's evolutionary theory to Darwin's, In *Was Hitler a Darwinian? Disputed questions in the history of evolutionary theory* (pp. 116–134). Chicago: The University of Chicago Press.

Richarz, M. (1982). *Jüdisches Leben in Deutschland. Selbstzeugnisse zur Sozialgeschichte, 1918–1945*. Stuttgart: Deutsche Verlags-Anstalt.

Rieger, R., Michaelis, A., & Green, M. M. (1991). *Glossary of genetics and cytogenetics* (5th ed.). Berlin: Springer.

Risch, N., Piazza, A., & Cavalli-Sforza, L. L. (2002). Dropped genetics paper lacked scientific merit. *Nature, 415*(6868), 115.

Ritte, U., Neufeld, E., Broit, M., Shavit, D., & Motro, U. (1993). The differences among Jewish communities – maternal and paternal contributions. *Journal of Molecular Evolution, 37*, 435–440.

Roberts, D. (2011). *Fatal invention: How science, politics, and big business re-create race in the twenty-first century*. New York: The New Press.

Robinson, J. (1923). *The knowledge of our nation: Demography and nationology* (in Hebrew). Berlin: Ayanot.

Robstein, B.-Z. (1924). Critic and bibliography (in Hebrew). *Hachinuch, 7*(6), 38–42.

Roll-Hansen, N. (2000). Eugenics practice and genetic science in Scandinavia and Germany. *Scandinavian Journal of History, 26*(1), 75–82.

Ronen, A., Ronen, I., & Goldschmidt, E. (1963). Marriage systems. In E. Goldschmidt (Ed.), *The genetics of migrant and isolate populations* (pp. 340–343). New York: Williams & Wilkins.

Rosenberg, N. A., Woolf, E., Pritchard, J. K., Schaap, T., Gefel, D., Shpirer, I., Lavi, U., Bonné-Tamir, B., Hillel, J., & Feldman, M. W. (2001). Distinctive genetic signatures in the Libyan Jews. *Proceedings of the National Academy of Science, Washington, 98*(3), 858–863.

Rösler, A. (1992). Steroid, 17β-hydroxysteroid dehydrogenase deficiency in man: An inherited form of male pseudohermaphroditism. *Journal of Steroid Biochemical and Molecular Biology, 43*(8), 989–1002.

Ross, L. F., Paul, D. B. & Brosco, J. P. (2015). 50 years ago in *The Journal of Pediatrics*: Phenylketonuria in a Negro Infant. *The Journal of Pediatrics, 167*(2), 304.

Roth, J. (1950). *The Radetzky March* (J. Neugroschel, Trans.). Woodstock: The Overlook Press.

Rubin, I. (1934). The ingathering of exiles from a eugenic perspective (in Hebrew). *Moznaim, 1*(4), 89–93.

Rudi, Z. (1927). The biological foundations of education (in Hebrew). *Hachinuch, 10*, 211–219.

Rund, D., Cohen, T., Filon, D., Dowling, C. E., Warren, T. C., Barak, I., Rachmilewitz, E., Kazazian Jr., H. H., & Oppenheim, A. (1991). Evolution of a genetic disease in an ethnic isolate: β-Thalassemia in the Jews of Kurdistan. *Proceedings of the National Academy of Science, Washington, 88*, 310–314.

Rund, D., Filon, D., Rachmilewitz, E. A., Kazazian, H. H. J., & Oppenheim, A. (1992). Diversity of molecular lesions causing β-thalassemia in Israeli Jewish ethnic groups. In B. Bonné-Tamir & A. Adam (Eds.), *Genetic diversity among Jews* (pp. 228–236). New York/Oxford: Oxford University Press.

Rund, D., Filon, D., Jackson, N., Asher, N., Oren-Karni, V., Sacha, T., Czekalska, S., & Oppenheim, A. (2004). An unexpected high frequency of heterozygosity for alpha-thalassemia in Ashkenazi Jews. *Blood Cell Molecules and Diseases, 33*, 1–3.

Ruppin, A. (1903). *Darwinismus und Sozialwissenschaft*. Jena: Gustav Fischer.

Ruppin, A. (1911). *Die Juden der Gegenwart: Eine sozialwissenschaftliche Studie*. Köln und Leipzig: Jüdischer Verlag.

Ruppin, A. (1913). *The Jews of to-day* (Bentwich, Trans.) Introduction by J. Jacobs. London: G. Bell & Sons.

Ruppin, A. (1930a). *Soziologie der Juden* (Vol., 1). *Die soziale Struktur der Juuden*. Berlin: Jüdischer Verlag.

Ruppin, A. (1930b). *The sociology of the Jews* (Vols. 1–2) (in Hebrew). Berlin/Tel Aviv: Stibel.

Ruse, M. (1996). *Monad to man. The concept of progress in evolutionary biology*. Cambridge, MA: Harvard University Press.

Sachs, L., & Bat-Miriam, M. (1957). The genetics of Jewish populations: I. Finger print patterns in Jewish populations in Israel. *American Journal of Human Genetics, 9*(2), 117–126.

Sagi, M. (1998). Ethical aspects of genetic screening in Israel. *Science in Context, 11*(3–4), 419–429.

Salaman, R. N. (1911a). Heredity and the Jew. *Journal of Genetics, 1*, 273–292.
Salaman, R. N. (1911b–1912). Heredity and the Jew. *Eugenic Review, 3*, 187–200.
Salaman, R. N. (1920). *Palestine reclaimed*. London: George Routledge & Sons.
Salaman, R. N. (1923). Some notes on the Jewish problem. In C. B. Davenport, H. F. Osborn, C. Wissler, & H. H. Laughlin (Eds.), *Eugenics in race and state: Second International Congress of eugenics* (2nd ed., pp. 134–153). Baltimore: Williams & Wilkins.
Salaman, R. N. (1925). What has become of the Philistines? *Quarterly Statement of Palestine Exploration Fund, 57*, 1–17.
Salaman, R. N. (1950). Foreword. In C. Roth (Ed.), *The record of European Jewry by Cecil Roth* (pp. 5–8). London: Frederick Muller.
Sand, S. (1999). The national account (in Hebrew). In A. Ophir (Ed.), *Fifty to forty-eight* (Special issue of *Theoria Ubikoret, 12–13*) (pp. 339–345). Jerusalem: Van-Leer Institute and Hakibutz-Hameuchad.
Sand, S. (2009). *The Invention of the Jewish People* (Y. Lotan, Trans.). London, New York: Verso.
Sandler, A. (1904). *Anthropologie und Zionismus*. Brünn: Jüdischer Buch- und Kunstverlag.
Santachiara Benerecetti, A. S., Semino, O., Passarino, G., Morpurgo, P., Fellous, M., & Modiano, G. (1992). Y-chromosome DNA polymorphism in Ashkenazi and Sephardi Jews. In B. Bonné-Tamir & A. Adam (Eds.), *Genetic diversity among Jews* (pp. 45–50). New York: Oxford University Press.
Sass, J. (1929). *The hygiene of the body and soul (A guide to parents, teachers and educators)* (in Hebrew). Jerusalem: Modern Hygiene Publishers.
Schallmayer, W. (1910/1902). *Vererbung und Auslese in ihrer soziologischen und politischen Bedeutung*. Jena, Gustav Fischer.
Schiebinger, L. (1993). *Nature's body*. Boston: Beacon Press.
Schmölders, C. (1997). *Das Vorurteil im Leibe: Eine Einführung in die Physiognomik*. Berlin: Akademie Verlag.
Scriver, C. R. (1992). What are genes like that doing in a place like this? Human history and molecular prosopography. In B. Bonné-Tamir & A. Adam (Eds.), *Genetic diversity among Jews* (pp. 319–329). New York/Oxford: Oxford University Press.
Sekula, A. (1989). The body and the archive. In R. Bolton (Ed.), *The contest of meaning: Critical histories of photography* (pp. 343–389). Cambridge, MA: MIT.
Seligsohn, U. (2002, May). Genetic and historic aspects of the blood-clotting systems in Jews (in Hebrew). *Letter of the Israel National Academy of Sciences, 22*, 7–8.
Sesardic, N. (2005). *Making sense of heritability*. Cambridge: Cambridge University Press.
Sesardic, N. (2010a). Race: a social destruction of a biological concept. *Biology & Philosophy, 25*(2), 143–162.
Sesardic, N. (2010b). Nature, nurture, and politics. *Biology & Philosophy, 25*(3), 433–436.
Shalmon, L., Kirschmann, C., & Zaizov, R. (1994). A new deletional alpha-thalassemia detected in Yemenites with hemoglobin H disease. *American Journal of Hematology, 45*, 201–204.
Sheba, C. (1971). Jewish migration in its historical perspective. *Israel Journal of Medical Sciences, 7*(12), 1333–1341.
Sheldon, W. H. (1954). *Atlas of men. A guide for somatotyping the adult male at all ages*. New York: Harper & Brothers.
Shiloh, Y., Avigad, S., Kleiman, S., Weinstein, M., Schwartz, G., Woo, S. L. C., & Cohen, B. E. (1992). Molecular analysis of hyperphenylalaninemia in Israel: A study of Jewish genetic diversity. In B. Bonné-Tamir & A. Adam (Eds.), *Genetic Diversity among Jews* (pp. 237–247). New York/Oxford: Oxford University Press.
Skorecki, K., Selig, S., Blazer, S., Bradman, R., Bradman, N., Warburton, P. J., Ismajlowicz, M., & Hammer, M. F. (1997). Y-chromosomes of Jewish priests. *Nature, 385*(6611), 32.
Smilanski, M. (1936). *Hadera* (in Hebrew). Tel-Aviv: Omanut.
Smith, K. M. (1955). Redcliffe Nathan Salaman, 1874–1955. *Biographical Memoirs of Fellows of the Royal Society, 1*, 239–245.

Smocha, S. (1999). Permutations in the Israeli society – after Jubilee years (in Hebrew). *Alpayim, 17*, 239–261.

Stampfer, S. (2013). Did the Khazars convert to Judaism? *Jewish Social Studies, 19*(3), 1–72. doi:10.1353/jss.2013.0013.

Stoler-Liss, S. (1998). *Zionist baby and child care: Anthropological analysis of parents' manuals* (in Hebrew). M.A. thesis. Tel Aviv University, Israel.

Stone, D. (2004). Of peas, potatoes, and Jews: Redcliffe N. Salaman and the British debate over Jewish racial origins. *Simon Dubnow Institute Yearbook* (Vol. 3, pp. 221–240).

Stratz, C. H. (1903). *Was sind Juden? Eine ethnographisch-anthropologische Studie*. Vienna-Colonge-Gratz: F. Tempsky.

Sweet, P. R. (1993). Fichte and the Jews: A case of tension between civil fights and human rights. *German Studies Review, 16*(1), 37–48.

Szeinberg, A. (1973. Investigation of genetic polymorphic traits in Jews. In B. Ramot (Ed.), *Genetic polymorphisms and diseases in man. Israel Journal of Medical Science 9*(9–10), 1171–1180.

Szeinberg, A. (1979). Polymorphic evidence for a Mediterranean origin of the Ashkenazi community. In R. M. Goodman & A. G. Motulsky (Eds.), *Genetic diseases among Ashkenazi Jews* (pp. 77–91). New York: Raven Press.

Templeton, A. R. (1998). The complexity of the genotype-phenotype relationship and the limitations of using genetic "markers" at the individual level. *Science in Context, 11*(3–4), 373–389.

Templeton, A. R. (2008). Human races: A genetic and evolutionary perspective. *American Anthropologist, 100*(3), 632–650.

Thomas, M. G., Skorecki, K., Ben-Ami, H., Parfitt, T., Bradman, N., & Goldstein, D. B. (1998). Origins of old testament priests. *Nature, 394*(6689), 138–140.

Toulmin, S. (1972). *Human understanding. The collective use and evolution of concepts*. Princeton: Princeton University Press.

Touroff, N. (1938). Heredity and environment (in Hebrew). *Hachinuch, 11*(3), 274–292.

Tsafrir, J., & Halbrecht, I. (1972). Consanguinity and marriage systems in the Jewish community in Israel. *Annals of Human Genetics, 35*(3), 343–347.

Tschurtakover, S. (1935). The origin of the yellowish Jews (in Hebrew). *Harefuah, 14*, 106–110.

Vilnai, Z. (1980). *Ariel: Encyclopedia for the Knowledge of Eretz-Israel* (in Hebrew). Tel Aviv: Am-Oved.

Vital, D. (1982). *Zionism: The formative years*. Oxford: Clarendon Press.

Vital, D. (1987). *Zionism: The crucial phase*. Oxford: Clarendon Press.

Wade, N. (2014). *A troublesome inheritance: genes, race and human heredity*. New York: The Penguin Press.

Weinberger, A., Sperling, O., Rabinovitz, M., Brosh, S., Adam, A., & DeVries, A. (1974). High frequency of cystinuria among Jews of Libyan origin. *Human Heredity, 24*, 568–572.

Weisberg, M., & Paul. D. B. (2016). Morton, Gould, and Bias: A comment on "The Mismeasure of Science". *PLoS Biology, 14*(4), e1002444. doi:10.1371/journal.pbio.1002444.

Weismann, A. (1893). *The germ-plasm. A theory of heredity*. New York: Charles Scribner's Sons.

Weiss, M. Y. (2002). Identity and essentialism: Race, racism and the Jews at the turn of the 19th century (in Hebrew). *Theoria Ubikoret, 21*, 133–161.

Wexler, P. (1993). *The Ashkenazic Jews: A Slavo-Turkic People in search of a Jewish identity*. Columbus: Slavica Publishers.

Yaniv, I., Benador, D., & Sagi, M. (2004). On not wanting to know and not wanting to inform others: Choices regarding predictive genetic testing. *Risk Decision and Policy, 9*, 317–336.

Yerushalmi, Y. H. (1982). *Assimilation and anti-semitism: The Iberian and the German models* (Leo Baeck Memorial Lecture 26). New York: Leo Baeck Institute.

Yiftachel, O. (1999). "The Day of the Earth" (in Hebrew). In A. Ophir (Ed.), *Fifty to forty-eight* (Special issue of *Theoria Ubikoret, 12–13*) (pp. 279–289). Jerusalem: Van-Leer Institute and Hakibutz-Hameuchad.

Yuval, I. J. (2005). The myth of the exiles from the land: Jewish time and Christian time (in Hebrew). *Alpayim, 29*, 9–25.

Zeiss, H., & Pintschovius, K. (1944). *Zivilisationsschaeden am Menschen*. Muenchen/Berlin: J. F. Lehmanns Verlag.

Zeitlin, S. (1959). Who is a Jew? A halachic-historic study. *The Jewish Quarterly Review, New Series, 49*(4), 241–270.

Zimmermann, M. (1986). *Wilhelm Marr: The Patriarch of anti-semitism*. New York: Oxford University Press.

Zimmermann, M. (1988). From radicalism to antisemitism. In S. Almog (Ed.), *Antisemitism through the ages* (pp. 241–254). Oxford: Pergamon Press.

Zondak, H. (1940). On the medicine and the physicians of to days (in Hebrew). *Harefuah, 18*, 29–32.

Index

A

Abu El-Haj, Nadia (1962–), 201, 202
Acquired properties, 22, 25, 26, 38, 41, 55, 58
Adam, Avinoam (1924–2008), 155, 169
Ahad Ha'am (Asher Z. Ginsberg) 1859–1927, 53–55
  Lo Zu Haderech, 53
Allele. *See* Gene
Amino-acid, 149, 150, 176, 204
Ancestor, 25, 39, 72, 77, 80, 94, 99, 129, 130, 145, 160, 163, 169, 176, 177, 179, 184, 192, 193, 197, 199, 201
Andree, Richard (1835–1912), 27, 29
Anthropology, 12–16, 19, 20, 22, 25–27, 40, 41, 44, 60, 67, 82, 84, 145–151, 159
Anti-Semitism, xi, 4, 5, 9, 12, 13, 20, 26, 30, 31–33, 44, 49, 52, 56, 57, 61, 69, 81, 87–90, 99, 101, 122, 137, 161, 198, 208, 209
Arabs, 3, 6, 14, 44, 57, 61, 63, 90, 91, 98, 99, 100, 103, 107, 114, 119, 124, 126, 136, 138–143, 152, 156, 165, 168, 175, 178, 189, 191, 195–197, 201, 203, 208
Aristotle (384-322 BCE), 39
Armenian. *See* Jewish communities
Armenoid. *See* Races of man
Aryan. *See* Races of man
Ashkenazi, 6, 29, 45, 81, 85, 86, 93–104, 106, 112–116, 121, 122, 125, 131–135, 138, 147, 148, 154–158, 163, 164, 168–173, 178–181, 184–201, 205–207. *See also* Eidah/eidoth
Assimilation, 15, 30, 32, 35, 36, 49–52, 54, 57, 63, 65, 68, 71, 76, 81, 85, 89, 90, 104–108, 194

Auerbach, Elias (1882–1971), 29, 79, 138, 139, 142
Autosome. *See* Chromosome

B

Bacon, Francis (1561–1626), xiii
Baghdad Spring Disease. *See* Hereditary diseases
Bar-Kochba Revolt (132–135 CE), 68, 140
Barrès, Maurice (1862–1923), 33
Bateson, William (1861–1923), 82
Baur, E., 122
  Menschliche Erblichkeitslehre, 122
Behar, Doron, 103, 104, 135, 158, 159, 186, 187, 198, 200
Belkind, Israel (1861–1929), 1, 14, 140–142
  On the Arabs in Palestine, 140
Ben Gurion, David (1866–1973), 14, 197, 203
Bergman, Hugo (1883–1975), 63
Bialik, Chaim Nachman (1873–1934), 5
Bilu, 2, 14, 140
Birnbaum, Nathan (1864–1937), 50
Bismarck, Otto von (1815–1898), 31, 32
Blond, 27, 85
Blood type, 152, 153, 163, 164, 166, 172, 173, 175
Bloom, Etan (1964–), 67, 69, 81, 104, 105, 107, 108, 126
Bloom's disease. *See* Hereditary diseases
Blumenbach, Johann Friedrich (1752–1840), 24, 25
Blut und Boden, 5, 59
Boas, Franz (1858–1942), 87
Bodenheimer, Fritz Shimon (1897–1959), 137, 138, 142, 170, 235

Bodenheimer, Max (1865–1940), 56
Bonné-Tamir, Batsheva (1932–), 155, 167–169, 173, 179
Bonnet, Charles (1720–1793), 20
Bottleneck, 154, 159, 174, 186, 187, 195, 198, 207. *See also* Population dynamics
Brachicephaly. *See* Dolichocephaly
Brachyahu/Bruchov, Mordechai (182–1959), 131, 132
Branching/divergent tree, 14, 42, 72, 80, 88, 162, 182, 192. *See also* Trellis
Branching pattern, 72
BRCA. *See* Hereditary diseases
Breeding, 8, 13, 15, 39, 72, 81, 129, 145, 156, 157
    inbreeding, 31, 75–77, 154, 155, 158, 167, 180, 207
    interbreeding, 107, 144, 173
Brit Shalom, 63
Brod, Max (1884–1968), 63
Buber, Martin (1878–1965), 63–65, 67, 69
Buffon, Georges-Louis Leclerc, Comte de (1707–1788), 38
Bund, Association of Jewish Workers in Lithuania, Poland, and Russia, 51
Burla, Jehuda (1886–1969), 132
Bychowski, Shneor Zalman (1865–1934), 110, 111, 137, 156
Byzant, 100, 101, 102, 140

C

Canaanite, 104, 196
Cancer. *See* Hereditary diseases
Caucasian race, 25, 104, 115, 180, 187, 188, 194
Cavalli-Sforza, Luigi L. (1922–), 172, 190
Chamberlain, Houston Stewart (1855–1927), 33, 81
Chamberlain, Joseph (1836–1914), 116
Charcot, Jean Martin (1825–1893), 110
    Le juif errant, 110
Chromosome, 149, 180, 181, 183–186, 188, 189, 207
    X-chromosome, 147, 183, 188
    Y-chromosome, 94, 103, 104, 147, 158, 160, 182–191, 200
Cohen/Cohanim, 103, 104, 182–187, 191
Cohen, Haim (1911–2002), 9
Cohen, Tirza (1925–2013), 154, 171
Colonial/ism, xii, 51, 87, 95, 107, 116, 125, 208
Conversion, 6, 72, 98, 100, 101, 191, 192, 194

Cyprus, 143, 150, 205
Cystic fibrosis. *See* Hereditary diseases

D

Darwin, Charles (1809–1882), 3, 8, 27, 39, 40, 52, 78, 88
    The Descent of Man, 40
    The Expression of the Emotions in Man and Animals, 40
    On the Origin of Species, 3, 8, 40, 52
Darwinism, 33, 40, 48, 58, 66
    social-Darwinism, 4, 33, 57, 65, 78
Davenport, Charles B. (1866–1944), 84
Degeneration, 5, 9, 26, 38, 48, 55, 58–60, 66, 72, 77, 80, 109, 111, 119–121, 127, 130, 137
Determinism, 38, 41, 43, 46, 55
Diabetes. *See* Hereditary diseases
Diaspora, 1, 6, 9, 10, 14, 15, 48, 51, 54, 60, 62, 69, 76, 93, 94, 95, 99, 101, 105, 106, 109, 113, 120, 121, 123, 124, 127–129, 131, 135, 138, 144, 146, 159, 165, 168, 191–193, 196, 198, 199, 200, 207, 208
Disease, 6, 12, 13, 16, 17, 24, 41, 45, 66, 74, 82, 95, 97, 108–113, 115, 118, 127, 128, 130, 133–136, 145–150, 152, 154–160, 162, 167, 169, 170, 175, 176, 178–182, 191–193, 198, 204–207
Disraeli, Benjamin (1804–1881), 21, 50
Divine Order, 7
DNA, xii, 13, 14, 16, 42, 45, 94, 95, 100, 103, 104, 112, 113, 134, 136, 148–151, 153, 157, 158, 160, 171, 173, 175–202, 206, 207
    sequence and analysis, xii, 13, 14, 16, 95, 100, 104, 112, 136, 148, 151, 154–156, 158, 176, 177, 179, 180–183, 185, 189, 192–195, 206, 207
Dolichocephaly, 26, 95, 98
Domestication, 59, 120
Dor Yesharim, 157, 205, 206
Dor Yeshorim. See Dor Yesharim
Dreyfus, Alfred (1859–1935), 33
Drift (genetic), 135, 149, 154, 156–158, 187, 189, 191, 200
Drumont, Édouard (1844–1917), 33
Drunkenness, 111
Druze, 178, 200, 201
Dubin-Jones syndrome. *See* Hereditary diseases
Du Bois-Reymond, Emil (1818–1896), 53

Dugdale, Richard L. (1841–1883), 43
  The Jukes: A Study of Crime, Pauperism and Heredity, 43
Dühring, Eugen (1833–1921), 32, 33

E
Eidah/eidoth, 3, 6, 14, 16, 93–118, 129–133, 138, 139, 162, 165, 166, 168, 182, 187, 190, 204
Einstein, Albert (1879–1955), 5, 96, 97, 134, 200
Emancipation, 3, 4, 6, 10–12, 17–36, 49, 52, 53, 56, 69, 89, 209, 210
Enlightenment, Age of, 3, 4, 7, 11, 17–21, 23, 34, 50, 51, 54
Eshkoli, Aaron Ze'ev (1901–1948), 133
Ethiopia, xii, 16, 108, 170, 200
Eugenics, xii, 8, 9, 12–16, 28, 45–48, 66, 72, 76, 81–83, 105, 107, 108, 110, 111, 113, 114, 116, 117, 119–142, 152, 155, 157, 195, 203–206
Evolution, xiii, 8, 9, 27, 33, 38–41, 43, 45, 48, 58, 59, 62, 65, 66, 72, 78, 84, 85, 87, 88, 90, 91, 108, 119, 127, 145, 152, 184, 193, 195
Exile, 5, 10, 14, 16, 48, 54, 73–77, 79, 80, 85, 86, 89, 105, 109, 122, 128, 129, 130, 141, 143–174, 179, 187, 191, 197, 204
Ezra and Nehemiah, 68, 77, 113

F
Facial features, 12, 13, 18, 26, 27, 71, 77, 83, 95, 113, 131, 151
Factor XI deficiency. *See* Hereditary diseases
Faibel, Berthold (1975–1937), 57
Familial disautonomy. *See* Hereditary diseases
Favism. *See* G6PD deficiency
Feitlowitz, Jacob (1881–1955), 108
Fichte, Johann Gottlieb (1762–1814), 22
fin de siècle, 3, 55, 58, 119, 120
Finger print, 20, 95, 166
Fischer, E., 122
  Menschliche Erblichkeitslehre, 122
Fishberg, Maurice (1872–1934), 29, 71, 85, 87, 89
  The Jews: A Study of Race and Society, 71
Fisher, Ronald A. (1890–1962), 41
FMF. *See* Hereditary diseases
Folklore, 29, 102
Founder effect, 135, 149, 154, 155, 159, 181, 189, 198, 200, 207
Frankenstein, Karl (1895–1990), 133
French Revolution, 10, 11, 17, 18, 35, 69
Freud, Sigmund (1856–1939), 134
Fritsch, Theodor (1852–1933), 33

G
Galileans, 73, 141
Galton, Francis (1822–1911), 8, 28, 29, 39, 41, 45–48, 95, 203, 206
  Hereditary Genius, 46
  Kantsaywhere, 48
Gaucher's diaease. *See* Hereditary diseases
Geiger, Avraham (1810–1884), 36
Gene pool, 13, 15, 79, 84, 100, 103, 104, 109, 114, 121, 122, 132, 135, 144, 146, 148, 159, 162, 165, 168, 171–175, 178, 192, 196, 198
Genetic archeology. *See* Sheba, Chaim (1908–1971)
Genetic code, 176
Genetic diseases, 112, 118, 135, 150, 154, 155, 157, 158, 162, 167, 169, 204
Genetic polymorphism, 153, 171
Genome, xii, 13, 42, 45, 136, 153, 176, 182, 183, 186–195, 199
Genotype, 41, 78, 112, 121, 151–153, 175, 184, 188, 192, 200
Gentile, 3, 36, 52, 80, 83, 85, 86, 108, 109, 112, 113, 122, 133, 147, 154–156, 162, 163, 165, 166, 208
German, 4, 9, 12, 13, 19, 20, 22–24, 27, 29–33, 35, 36, 44, 50, 52, 53, 59, 63–66, 69, 71, 86, 87, 95, 96, 105, 122, 132, 137, 143, 157, 188
Germ line, 41
Ghetto, 25, 26, 29, 46, 49, 56, 60, 68, 102, 105, 109, 116, 117, 129, 158
Ginsberg. *See* Ahad Ha'am (Asher Z. Ginsberg) 1859–1927
Gobineau, Joseph Arthur, Comte de (1816–1882), 26, 81, 127
Goethe, Johann Wolfgang von (1749–1832), 18
Goldschmidt, Elizabeth (1912–1970), 2, 6, 154, 155, 163, 167, 202
Goldstein, David B., 110, 159, 160, 168, 169, 185, 187, 198
Gordon, Yehuda Leib (1830–1892), 21
G6PD deficiency, 112, 146–148, 150, 158, 177, 178
Grattenauer, Karl Wilhelm Friedrich (1773–1838), 22

Günther, Hans F. K. (1891–1968), 12, 13, 124, 127
Gurevich, Joseph (1898–1960), 163

H

Haber, Fritz (1868–1934), 21
Hachinuch, 119, 126, 127, 130
Hachsharat haYeshuv, 67
Haeckel, Ernst (1834–1919), 3, 66, 137, 161
Haldane, John B. S. (1892–1964), 149
Halevy, Yehudah (1075–1145), 1
Haplotype, 104, 136, 158, 177, 182–189, 191, 192, 199, 200, 207
Harefuah, 85, 119
Haskala, 21, 101
Hatkufah, 127
Health, 12, 19, 45, 59, 60, 69, 81, 82, 95, 108, 109, 111, 120, 128, 130, 147, 148, 157, 205–207
Hebrew language, 10, 50, 99, 106, 119, 126, 127, 132, 139
Hebrew University, 2, 5, 14, 75, 82, 91, 119, 130, 137, 185, 209
Hebrew work, 124–126
Hebron, 1
Hegel, Friedrich (1770–1831), 11, 35, 52
Heine, Heinrich (1797–1856), 17, 21
Hemoglobin, 115, 149, 150
Herder, Johann Gottfried (1744–1803), 4, 10, 18, 22, 35, 52, 59, 130
Hereditary diseases
 APC *(Adenomatous polyposis coli)*, 135
 Baghdad Spring Disease, 146
 Bloom syndrome, 157
 BRCA (Breast Cancer Association 1&2), 112, 135, 161, 191, 192, 207
 cancer, 112, 161, 191, 192, 206, 207
 cystic fibrosis (CF), 157, 179, 180, 181, 206, 207
 diabetes, 109
 Factor XI deficiency, 135, 179, 191
 Familial Mediterranean Fever (FMF), 112, 146, 155, 156, 179
 favism, 146
 Gaucher disease, 135, 157, 158, 181, 206, 207
 G6PD deficiency, 112, 146–148, 150, 158, 177, 178
 Klinefelter syndrome, 189
 Niemann-Pick disease, 158, 207
 pentosuria, 111, 157
 phenylketonuria (PKU), 147, 155, 204, 205
 sickle cell anemia, 149, 150
 Turner syndrome, 188
 "The Wandering Jew" *(le juif errant)*, 110, 111
Herzelia, 123, 129
Herzl, Theodor (1860-1904), 2, 47, 48, 50, 53–58, 63, 65, 68, 75, 99, 116, 137, 139
 Altneuland, 47, 48, 55–57, 99
 Der Judenstaat, 2, 53–55
Hess, Moses (1812–1875), 4, 52, 69
 Rom und Jerusalem, 4, 52
Heterozygote, 84, 136, 147, 149, 150, 155, 157–159, 178, 180, 182, 204–206. *See also* Homozygote
Hibat Zion, 50, 51
Hindu-European, 25
Hirsch, Samson Raphael (1808–1888), 36
Hitler, Adolph (1889–1945), 5, 196
Hittite, 73, 80, 85, 121, 122. *See also* Race
HLA. *See* Human Leukocyte Antigens (HLA)
Holocaust, xi, 6, 13, 16, 91, 124, 136, 143, 203, 208
Holy Land, 1, 132, 170
Homo israelensis, 146, 148. *See also* Sheba, Chaim (1908–1971)
Homo sapiens, 7, 24, 144
Homozygote, 84, 136, 149, 150, 154–159, 178, 182, 204, 205. *See also* Heterozygote
Horizontal associations, xiii. *See also* Vertical depiction
Human Leukocyte Antigens (HLA), 170, 179, 192

I

Immigration, 6, 12, 13, 30, 46, 89, 107–109, 114–118, 120, 125, 126, 136, 138, 141, 143, 162, 167, 169, 170, 187, 192, 204, 209
Inbreeding. *See* Breeding
Indigenous, xii, 82, 141, 142, 194
Industrialization, 3, 34, 55, 58
Ingathering of exiles, 16, 77, 128, 129, 144, 187, 204
Inheritance of acquired characters, 24, 40, 41, 59, 76, 84, 87, 121
Intelligence, 79, 105, 115, 117, 133–137, 159
IQ. *See* Intelligence
Iran. *See* Jewish communities
Iraq. *See* Jewish communities

Isolate, 1, 2, 35, 41, 42, 72, 79, 88, 94, 97, 109, 113, 118, 135, 144, 146, 148, 149, 151, 152, 154–156, 162, 163, 166, 167, 169, 178–182, 188, 198, 201, 208

J
Jabotinsky, Zeev Vladimir (1880–1940), 61–63, 69, 89, 99, 100, 142
Jacobs, Joseph (1854–1916), 28, 29, 113
Jerusalem, 1, 2, 4, 5, 14, 52, 67, 75, 82, 91, 96–98, 101, 126, 129, 130, 132, 133, 137, 141, 160, 163, 167, 181, 185, 196, 209
Jewish battalions, 61, 114
Jewish communities, xii, xiii, 1, 6, 7, 11, 13–16, 21, 23, 45, 51, 72, 73, 79, 82, 85, 92, 94–97, 105, 106, 108, 112, 116, 117, 123, 125, 146, 154–158, 160, 163, 170–172, 175, 177, 178, 180, 182, 189, 190, 191, 197, 198, 200, 205–209. *See also Eidah/eidoth*
  Armenian, 19, 73, 139, 146, 155, 156, 168, 189
  Eastern European, 51, 53, 56, 57, 98, 102, 104, 108, 122, 187, 188
  Ethiopian, 25, 92, 93, 108, 161, 169–171, 190, 191, 199, 200
  Georgian, 25, 131, 180
  Iranian, 146, 188, 196, 199
  Iraqi, 143, 146, 154, 169, 179, 191, 192, 199
  Kurdistan, 109, 112, 146, 148, 150, 154, 164, 177
  Libyan, 112, 146, 155, 179, 180, 192, 199
  Lithuanian, 110, 178
  Mediterranean, xii, 27, 44, 45, 73, 85, 97, 104, 135, 147, 148, 151, 154, 167, 169, 171, 172, 173, 175, 177, 178, 185, 191, 195, 201
  Moroccan, 178–181, 191
  North African, 94, 143, 178, 190, 199
  Oriental, 93
  Polish, 110, 114
  Portugal, 94, 178
  Russian, 1, 53, 98, 99, 100–104, 106, 111, 114
  Yemenites, 93, 98, 112, 114, 133, 138, 146, 169, 170, 173
Jewish type, 6, 26–28, 71, 74, 81, 84–86, 96–98, 121, 129, 138, 182
Johannsen, Wilhelm (1857–1927), 41, 78, 115, 151
Jüdische Volkpartei, 75

K
Kafka, Franz (1883–1924), 63, 108
Kahn, Susan Martha, 201–203
Kant, Emanuel (1724–1804), 22, 25
Karaite, 101, 102, 113
Kautsky, Karl (1854–1938), 73, 86–91
  Are the Jews a Race?, 86
Kerem, Bat-Sheva, 157, 180
Khazarian Hypothesis, 187
Khazars, 73, 98, 100–104, 113, 164, 179, 187, 188, 195, 197–199
  chromosome-Y DNA haplotype, 199
King David, 68, 72
Klinefelter syndrome. *See* Genetic diseases
Koestler, Arthur (1905–1983), 101–103, 143, 187
  The Thirteenth Tribe, 101, 102
Kraeplin, Emil (1856–1926), 95
Kretschmer, Ernst (1888–1964), 95, 96
Krupp, Friedrich Albert (1854–1902), 65

L
La France Juive, 33
Lamarck, Jean Baptiste de (1744–1829), 8, 38, 39
Land of Israel, xi–xiii, 2, 4, 6, 10, 48, 51, 57, 63, 68, 77, 80, 90, 93, 95, 100, 144, 183, 195, 201, 203
Language. *See* Hebrew
Laski, Harold (1893–1950), 95
Lavater, Johann Caspar (1741–1801), 18–20, 43, 95
Law of Return, xii, 2, 9, 209
Lazarus, Moritz (1824–1903), 29, 30
Lemba, 159–161, 190, 191
Lenz, F., 122
  Menschliche Erblichkeitslehre, 122
Lenz, Fritz (1887–1976), 122
Lessing, Ephraim Gotthold (1729–1781), 20
Lessing, Theodor (1872–1933), 109
Levin, Shmaryahu (1867–1935), 124
Levites, 103, 183, 185, 186, 187
  progeny of priest Aaron/Aaronide, 182, 183, 185, 186
Lilienblum, Moshe Leib (1843–1910), 52
Linnaeus, Carolus (1707–1778), 7, 19, 24, 25, 37, 38
Lombroso, Cesare (1835–1909), 43, 44, 58, 95, 96
  L'uomo delinquent, 43
Lueger, Karl (1844–1910), 33

M

Mahler, Gustav (1860–1911), 21, 134
Malaria, 75, 148–151, 154, 155, 158, 177, 178, 204
Marr, Wilhelm (1819–1904), 4, 32, 44
  Der Sieg des Judenthum über das Germanenthum, 32
Maskilim, 21
Matmon (Metmann) Avraham (1900–?), 129
Mazzini, Giuseppe (1805–1872), 4, 52
Meir, Joseph (1890–1955), 130
Melting pot, 6, 143, 162–171
Mendel, Johann Gregor (1822–1994), 40, 41, 67, 86, 156, 193
Mendelssohn, Moses (1729–1786), 18, 20, 21
Microsatellite. *See* DNA
Middle East, xi, xii, 14, 73, 79–86, 93, 94, 97–100, 105–108, 112, 113, 115, 125, 132–135, 139, 144, 146, 156, 169, 171, 172, 178, 179, 186, 189–191, 196, 198–202, 208
Migration, xi, 6, 13, 25, 54, 72, 75, 80, 88, 89, 125, 141, 147, 149, 156, 160, 166, 168, 173, 178, 185, 189
Mitochondria, 94, 191, 194
Mizrahi, 199. *See also* Jewish communities
Molecular, xii, 6, 13, 15, 16, 42, 94, 103, 104, 112, 118, 123, 136, 145, 149, 150, 155, 158, 161, 171, 173, 176–179, 181, 184, 185, 187, 191, 198, 206, 207
Morocco. *See* Jewish communities
Mosaic Belief/Faith, 15, 21, 30
Motulsky, Arno (1923–), 134, 154, 157
Motzkin, Leo (1867–1933), 57
Mourant, Arthur Ernst (1904–1994), 159, 163–165, 173
Muhsam, Helmut (1914–1996), 93, 165, 166
Muller, Herman J. (1890–1967), 13, 41
Muscle Jews. *See* Nordau, Max (1849–1923)
Mutation, 97, 109, 112, 113, 146–150, 155–161, 173, 176–181, 183–185, 187, 189, 191, 192, 195, 207, 208

N

Napoleon, Bonaparte (1769–1821), 23
Nardi, Noah, 133
Nationalism, xi, 2, 24, 49, 52, 59, 60, 64, 67, 69, 86, 99–101, 123, 124, 139, 201, 208
Natives, 3, 7, 8, 25, 27, 51, 82, 99, 100, 107, 113–118, 125, 144, 151, 167, 173

Nature *versus* nurture, 39, 45, 64, 78
Nazi, xi, xii, 2, 4–6, 9, 12, 13, 30, 33, 46, 48, 59, 61, 91, 106, 122, 124, 125, 136, 144, 174, 203, 206, 208
Nordau, Max (1849–1923), 45, 56–61, 77, 100, 109
Nordmann, Johannes (1820–1887), 31, 44
North Africa. *See* Jewish communities
Nucleotides. *See* DNA

O

Orientals, 93
Ostrer, Harry, 190–192, 198–202
Ovum, 38, 205

P

Palestine, 2–4, 12–14, 26, 30, 32, 46, 48, 49, 52–56, 60, 61, 63, 65, 67–69, 75, 76, 79, 80, 83, 85, 89–91, 94, 97–99, 103, 105–108, 110, 114–133, 136–141, 146, 159, 161, 163, 191, 192, 201, xiii
Papua, 151
Pearson, Karl (1857–1936), 95, 114, 115, 117
Pedigree, 131, 183, 186
Pentosuria. *See* Hereditary diseases
Persecution, 1, 3, 12, 29, 47, 48, 54, 59, 60, 72, 77, 79, 89, 94, 98, 100, 106, 109, 111, 113, 114, 131, 134, 170, 182, xi, xii
Phenotype, 41, 78, 112, 123, 145, 151, 152, 175, 193. *See also* Genotype
Phenylketonuria (PKU). *See* Hereditary diseases
Philistines, 73, 84–86, 121, 122, 147, 196
Phylogenesis, 42, 191
Physiognomy, 18–20, 55, 95, 96, 151, 183. *See also See also* Lavater, Johann Caspar (1741–1801)
Pinsker, Leon (1821–1891), 2, 53
  Auto-emancipation, 2, 50, 53
Pioneer, 12, 20, 21, 27, 33, 75, 90, 119–142, 145, 163, 171
Plato (428–348 BCE), 39, 48
Polymorphism, 16, 150, 153, 158, 171–173, 182, 184, 185, 187, 188, 192
Population dynamics, 6, 13, 137, 145, 148, 150, 152, 162, 163, 172, 195 198
Population genetics, 12, 16, 42, 78, 84, 117, 137, 145, 152, 167
Preformation, 38
Protein, 16, 149, 176, 181

Prototype, 24, 27, 37, 39, 75, 95–98, 105,
  156, 208
  Jewish type, 6, 26–28, 71, 74, 81, 83–86,
    96–98, 121, 129 138, 182

R
Race, 3–5, 7, 8, 9, 12, 13, 16, 24–33, 37, 40,
  42, 44, 45, 47–50, 52, 55–58, 60–64,
  66–68, 71–92, 96–102, 104, 105, 107,
  108, 115, 116, 122, 123, 127–130, 132,
  133, 137–139, 141–148, 152, 168, 169,
  176, 193, 195, 200–202, 207, 209
  race psychology, 61
  Scientific racism, 4, 8, 17–36
Races of man, 25
  Amorite, 73
  Armenoid, 73, 80
  Aryan, 25–27, 33, 52, 73, 122, 159
  Hittite, 73, 80, 85, 121, 122
  Semite, 25, 26, 52, 73, 81, 99, 108, 122,
    147 (*See also* Anti-Semitism)
Rav Tzair, 126
Refugee, 48, 54, 100, 146, 157, 196,
  208, xi, xii
Renan, Ernst (1823–1892), 25, 26
Restitution, Age of (1815–1848), 23, 34
Retzius, Andreas (1796–1860), 26
Revisionist Zionists, 61
RFLP. *See* Genetic polymorphism
Rhineland Hypothesis, 187
Rishon LeZion, 1, 140
Rothschild, Baron Edmond James
  (1845–1934), 1, 140
Rubin (Rivkai), Israel (1890–1954), 128, 129
Rudi, Zvi, 44, 127, 128
Rühs, Friedrich Christian (1781–1820), 23, 24
Ruppin, Arthur (1876–1943), 12, 13, 17, 18,
  20, 21, 27–29, 63, 65–69, 73, 79–82,
  90, 96, 97, 99, 104–108, 123–125, 139,
  140, 142, 167
  Die Juden der Gegenwart, 67
  The Palestine Land Development Company
    (*Hachsharat haYeshuv*), 57, 125
  Soziologie der Juden, 96

S
Safad, 1, 75
Salaman, Redcliffe Nathan (1874–1955),
  27, 82–86, 95, 104, 113–118, 121–124,
  170, 182, 183
Sandler, Aron (1879–1954), 75–78, 86, 91
  Anthropologie und Zionismus, 75
Sardinia, 146–148, 150, 151, 156
Sass, Jacob, 128
Schallmayer, Wilhelm (1857–1919), 66
Schutzjude, 22, 23
Scientific racism, 4, 8, 17–36
Sculptures, ancient facial features, 96
Seligman, Charles (1873–1940), 82, 113
Seligsohn, Uri (1937–), 179
Semite. *See* Races of man
Sephardi. *See* Jewish communities
Sequences. *See* DNA
Sex, 24, 42, 60, 66, 79, 83, 128, 144,
  147, 158, 207
Shapira, Zvi Herman (1840–1898), 57
Sheba, Chaim (1908–1971), 2, 145–148, 156,
  163, 168, 198
Sheldon, Wiliam (1898–1977), 96
  ectomorph, mesomorph, endomorph, 96
Sicily, 147, 150, 156
Sickle cell anemia. *See* Genetic diseases
Skills, 58, 60, 130, 134. *See also* Talent
Skorecki, Karl, 184
SNP. *See* Genetic Polymorphism
Sokolov, Nahum (1859–1936), 57
Sombart, Werner (1863–1941), 105
Sommerring, Samuel Thomas von
  (1755–1830), 24
Spain, 1, 7, 24, 44, 80, 82, 85, 94, 99, 113,
  125, 178, 191, 192
Spencer, Herbert (1820–1903), 3, 40, 58
Sperm, 38, 72. *See also* Ovum
Stoecker, Adolf (1835–1909), 32, 44
Survival, 8, 33, 48, 64, 69, 77, 78, 98, 119,
  127, 175–177
Szeinberg, Arieh, 148, 171, 172
Szold, Henrietta (1860–1945), 30

T
Talent, 103, 122, 128
Tay-Sachs. *See* Hereditary diseases
Ten Lost Tribes, 16, 162, 197
Thalassemia. *See* Hereditary diseases
Tiberias, 1
Tndem repeat. *See* DNA
Touroff, Nisan (1877–1952), 130, 131
Treitschke, Heinrich von (1834–1896), 32
  Ein Wort über unser Judenthum, 32
Trellis, 14, 72, 80, 107, 192, 193. *See also*
  Branching pattern

Tuberculosis, 58, 158, 204
Turner syndrome. *See* Hereditary diseases

U
Uganda, 116
United Nations' decision of November 29 1947, 2
Urbanization, 3, 59, 120
Urjude, 75, 108, 163
Usishkin, Menachem (1863–1941), 57

V
Verein für Kultur und Wissenschaft der Juden, 34
Verschuer, Otmar Freiher von (1896–1969), 12
  Forschungen zur Judenfrage, 12
Vertical depiction, 3, 42, 160, 182, 191–193, 195, 198. *See also* Horizontal associations
Vienna-Congress (1814–1815), 23
Virchow, Rudolf (1821–1902), 27, 77
Volk, 4, 5, 35, 52, 56, 59, 61, 63, 69, 99
Voltaire (1694–1778), 20
von Luschan, Felix (1854–1924), 26, 29, 67, 73, 77, 80, 81, 85, 88, 127

W
Wallace, Alfred Russel (1823–1913), 27
Warburg, Otto (1883–1970), 56
Weismann, August (1834–1914), 41
Weissenberg, Samuel (1867–1928), 29, 74, 75, 97, 98, 103, 183
Weizmann, Chaim (1874–1952), 57

Weltsch, Robert (1891–1982), 63
Wild noble, 99
Wissenschaft des Judentums, 34–36, 50, 98, 113
Wolfson, David (1856–1914), 56, 57
World War I, 61, 75, 83, 90, 114, 126, 137, 152

Y
Yavnieli (Warshavsky) Shmuel (1884–1961), 126
Y-chromosome polymorphisms, 185
  Cohen Modal Haplotype (CMH), 160, 185, 186, 191
  non-recombining portion of Y-chromosome (NRY), 158, 187
Yellow Patch, 30, 122
Yemen, 1, 85, 109, 125, 126, 146, 155, 160, 165, 168, 170, 178, 180, 198, xii
Yemenites. *See* Jewish communities

Z
Zangwill, Israel (1864–1926), 71, 82
Zentralverein deutscher Staasbürger jüdischen Glaubes, 30
Zion, 1–3, 5, 50, 51, 53, 54, 57, 90, 141, 160, xi–xiii
Zionist Congress, 2, 49, 51, 55, 57, 61, 71, 74, 75, 110, 119, 124
  First Zionist Congress, 2, 49, 57, 61, 71, 75, 110
Zlenov, Yechiel (1863–1918), 57
Zollschan, Ignatz (1877–1948), 29, 78
Zunz, Leopold (1794–1886), 34, 35